Obstructive Sleep Apnea Syndrome: History, Current Status, Perspectives

Obstructive Sleep Apnea Syndrome: History, Current Status, Perspectives

Editors

David Slouka
Milan Štengl

MDPI

Basel • Beijing • Wuhan • Barcelona • Belgrade • Novi Sad • Cluj • Manchester

Editors

David Slouka
ENT Department
Charles University
Přeštice
Czech Republic

Milan Štengl
Physiology Department
Charles University
Pilsen
Czech Republic

Editorial Office
MDPI AG
Grosspeteranlage 5
4052 Basel, Switzerland

This is a reprint of articles from the Special Issue published online in the open access journal *Life* (ISSN 2075-1729) (available at: https://www.mdpi.com/journal/life/special_issues/Sleep_Syndrome).

For citation purposes, cite each article independently as indicated on the article page online and as indicated below:

Lastname, A.A.; Lastname, B.B. Article Title. *Journal Name* **Year**, *Volume Number*, Page Range.

ISBN 978-3-7258-2064-1 (Hbk)
ISBN 978-3-7258-2063-4 (PDF)
doi.org/10.3390/books978-3-7258-2063-4

Contents

About the Editors

David Slouka

David Slouka is an associate professor at the Medical Faculty in Pilsen, Charles University in Prag. His research interests include otorhinolaryngology, especially in obstructive sleep apnea and ear surgery. David Slouka earned his Ph.D. from Charles University in 2014 and his MBA from the Central European Management Institute in 2016. Since 2017, he has served as the head of the Otorhinolaryngology Department at the Faculty Hospital in Pilsen. Prof. Slouka has published over 50 academic papers in international scientific journals. He has a Hirsch-index of 14, and his scientific work has received more than 700 citations on the Web of Science. Additionally, he has authored three ENT monographs, and he is also the primary author of two textbooks and the co-author of six monographs. He supervises the scientific work of many graduate and postgraduate medical students.

Milan Štengl

Milan Štengl graduated from the Faculty of Medicine in Pilsen, Charles University, in 1993. He completed his Ph.D. studies at the Department of Physiology, Faculty of Medicine in Pilsen, and the Center for Experimental Surgery and Anesthesiology at the University of Leuven in 1999. He has served as an associate professor (Doc., Physiology) since 2006 and as the scientific director of the Biomedical Center at the Faculty of Medicine in Pilsen since 2012. In 2013, he became the head of the Department of Physiology at the Faculty of Medicine in Pilsen. In 2019, he became a professor of physiology and pathological physiology. Prof. Štengl has been involved in international activities, including undergoing training at the University of Leuven for two years and at the Cardiovascular Research Institute of Maastricht for five years; he also spent a year at the University of Utrecht. His research focuses on the physiology and pathophysiology of the cardiovascular system and the remodeling of the heart under pathological conditions. He has published over 60 original articles in IF journals and has more than 1600 citations on the Web of Science. His H-index is 20.

Preface

Sleep apnea syndrome is a significant societal concern, affecting many patients. Despite its long-standing presence, in-depth exploration of this disease only began in the 19th century.

Sleep apnea syndrome is a multidisciplinary issue, encompassing fields such as dentistry, internal medicine, neurology, otorhinolaryngology, psychiatry, pulmonology, and psychology. Treatment options include surgical and conservative approaches. Although the surgical aspect of dealing with sleep apnea holds much promise, the current focus is placed on conservative medicine. Both surgical and nonsurgical approaches face unresolved issues and questions. It is now evident that there is no universal solution to this issue, which remains interdisciplinary, requiring mutual cooperation from both disciplines.

The diagnosis and subsequent care of sleep apnea patients are unique. Its specificities are determined by the anatomy and physiology of the respiratory tract and contemporary medicine's treatment options. Yet, the success of therapy largely depends on the patient's cooperation. Motivating and providing psychological support to patients is not solely the responsibility of medical staff but also of their family members. For treatment to truly bring about change, the patient's personal health habits must be altered. Therefore, patient cooperation and lifestyle changes are crucial for ensuring successful long-term treatment outcomes.

David Slouka and Milan Štengl

Editors

life

MDPI

Article

A Dietary and Lifestyle Intervention Improves Treatment Adherence and Clinical Outcomes in Overweight and Obese Patients with Obstructive Sleep Apnea: A Randomized, Controlled Trial

Izolde Bouloukaki [1,2,*], Eleni Daskalaki [1], Eleni Mavroudi [2], Violeta Moniaki [2], Sophia E. Schiza [2] and Ioanna Tsiligianni [1]

[1] Department of Social Medicine, Faculty of Medicine, University of Crete, 71500 Heraklion, Greece; elenianastasiadaskalaki@gmail.com (E.D.); pdkapa@yahoo.gr (I.T.)

[2] Sleep Disorders Center, Department of Respiratory Medicine, Medical School, University of Crete, 71500 Heraklion, Greece; elenima23@hotmail.com (E.M.); vmoniaki@yahoo.gr (V.M.); schizas@uoc.gr (S.E.S.)

* Correspondence: izolthi@gmail.com; Tel.: +30-2810-394824

Abstract: The study's objective was to assess the impact of Mediterranean diet/lifestyle interventions for weight loss on positive airway pressure (PAP) adherence, body mass index (BMI), sleepiness, and blood pressure measurements (BP) in patients with obstructive sleep apnea (OSA). We designed a randomized, controlled trial, including overweight and obese patients with moderate to severe OSA, randomized to standard care (SCG, n = 37) or a Mediterranean diet group (MDG, n = 37). The SCG received healthy lifestyle advice, while the MDG underwent a 6-month behavioral intervention aiming to enhance weight loss and adherence to a Mediterranean diet. PAP adherence, BMI, Epworth Sleepiness Scale (ESS), and BP measurements were evaluated pre- and post-intervention. Post-intervention PAP use was higher in the MDG compared to the SCG (6.1 vs. 5.4, $p = 0.02$). Diet/lifestyle intervention was one of the most significant predictive factors for PAP adherence (OR = 5.458, 95% CI = 1.144–26.036, $p = 0.03$). The SCG demonstrated a rise in BMI, while the MDG displayed a decline (0.41 vs. −0.75, $p = 0.02$). The MDG also demonstrated a substantial reduction in adjusted SBP (−5.5 vs. 2.8, $p = 0.014$) and DBP (−4.0 vs. 2.5, $p = 0.01$). Ultimately, incorporating a dietary/lifestyle intervention with standard care yields superior PAP adherence, BMI, and BP measurements in contrast to standard care alone, emphasizing the advantages of dedicating more time and support within the MDG.

Keywords: dietary intervention; obstructive sleep apnea; treatment adherence; sleepiness; obesity; Mediterranean diet

Citation: Bouloukaki, I.; Daskalaki, E.; Mavroudi, E.; Moniaki, V.; Schiza, S.E.; Tsiligianni, I. A Dietary and Lifestyle Intervention Improves Treatment Adherence and Clinical Outcomes in Overweight and Obese Patients with Obstructive Sleep Apnea: A Randomized, Controlled Trial. *Life* **2023**, *13*, 1755. https://doi.org/10.3390/life13081755

Academic Editors: David Slouka and Milan Štengl

Received: 9 July 2023
Revised: 1 August 2023
Accepted: 10 August 2023
Published: 16 August 2023

1. Introduction

Obstructive sleep apnea (OSA) is a common and under-recognized public health problem, associated with increased cardiovascular morbidity and mortality and significant increases in health and social costs [1–4]. Obesity has long been acknowledged as one of the most significant risk factors for OSA [5]. Furthermore, the prevalence of OSA increases with adiposity and ranges between 50 and 80% for individuals who are classified as overweight or obese [6].

OSA should be approached as a chronic disease that requires pathophysiological and clinical phenotyping, objective diagnostic testing and individualized treatment plans with positive airway pressure (PAP) as the first-line symptomatic treatment of choice. PAP treatment in adherent patients with OSA yields several benefits, such as improved daytime sleepiness, systemic blood pressure, quality of life, neurobehavioral performance, and a decreased risk of motor vehicle accidents [7–9]. However, PAP acceptance and compliance remain a challenging issue [10]. Therefore, it is crucial to implement strategies for improving and sustaining adherence over time in these patients. These strategies should go beyond the traditional mask adjustment and leak assessment, and further integrate the involvement of multidisciplinary teams [11].

Given the strong association between obesity and OSA, lifestyle interventions have emerged as complementary therapeutic choices. In line with this, the American Heart Association recommends including weight-loss-focused lifestyle interventions alongside conventional OSA treatment [2]. Lifestyle-induced weight loss has been extensively investigated as a treatment approach to reverse OSA pathogenesis and is effective in improvement in both OSA severity and OSA-related symptoms [12–14]. A combination of a 6-month behavioral dietary/lifestyle modification program based on the Mediterranean pattern and PAP therapy was found to effectively reduce weight, improve OSA severity, and result in favorable anti-inflammatory, antioxidant, and cardiometabolic outcomes in OSA patients [15–17]. While weight regain is expected in the long-term, these cardiometabolic benefits appear to be sustainable even after six months following the intervention [18].

Besides improvements in OSA severity and cardiometabolic parameters, no evaluation has been undertaken to determine the effect of diet/lifestyle interventions on objective PAP adherence. Since adherence is one of the major determinants of PAP efficacy, we hypothesized that expanding the scope of diet/lifestyle interventions beyond OSA severity and cardiometabolic parameters could enhance PAP adherence among these patients. Therefore, the aim of our study was to explore the role of a 6-month diet/lifestyle intervention on treatment adherence and clinical outcomes in patients with OSA. Specifically, we evaluated the effects of a combination of PAP and weight-loss Mediterranean diet/lifestyle intervention on improving PAP adherence (hours of device use), body mass index (BMI), and daytime symptoms, mainly sleepiness and arterial blood pressure measurements, over the effect of usual (standard) care alone.

2. Materials and Methods

2.1. Study Patients

We conducted a parallel, randomized, controlled, follow-up study of consecutive patients who were admitted to the Sleep Disorders Center, Department of Respiratory Medicine, University of Crete Medical School, between December 2021 and March 2022. The inclusion criteria were (a) patients aged >18 years with newly diagnosed moderate to severe OSA (apnea-hypopnea index (AHI) $\geq$ 15 events/h) through an attended overnight polysomnography according to standard criteria, (b) overweight and obese (BMI > 25 kg/m^2), (c) eligible for PAP treatment with a follow-up in-laboratory PAP titration with full polysomnography to establish the appropriate PAP settings, (d) with adherence data available in the 6 months after initiation of treatment, and (e) with an above-elementary school education. The exclusion criteria were refusal to participate, patients on PAP treatment, current participation in a weight loss program, central sleep apnea syndromes, obesity hypoventilation syndrome, restrictive ventilator syndromes, severe congestive heart failure, a history of life-threatening arrhythmias, severe cardiomyopathy, long-term oxygen therapy, chronic kidney disease, family or personal history of mental illness, drug or alcohol abuse, severe cognitive impairment, concurrent oncological diseases, pregnancy or lactation, recent hospitalization for acute or chronic respiratory disease, history of narcolepsy, or restless leg syndrome. All subjects provided written informed consent and ethical approval was provided by the University of Crete Research Ethics Committee (REC-UOC) (approval number: 158/29.11.2021). The trial was also registered on ClinicalTrials.gov with Trial registry: ClinicalTrials.gov; No.: NCT05881824.

Individuals were assigned (1:1) following simple randomization procedures (computerized random numbers) to a usual (standard) care group (SCG, n = 39) receiving usual follow-up care or an intervention group—Mediterranean diet group (MDG, n = 37) with follow-up care based on an additional behavioral intervention aiming at weight loss and increasing adherence to the Mediterranean diet (Figure 1). Following randomization, sequentially numbered envelopes that were opaque and sealed were utilized in the allocation concealment process, prepared by an individual not involved in the trial.

Blinding the patients or care team was not feasible since all patients were informed in the consent form that they would be randomized to either the MDG or SCG. Study investigators and the statistician involved in data analysis were blinded to the intervention.

Patients were followed for 6 months.

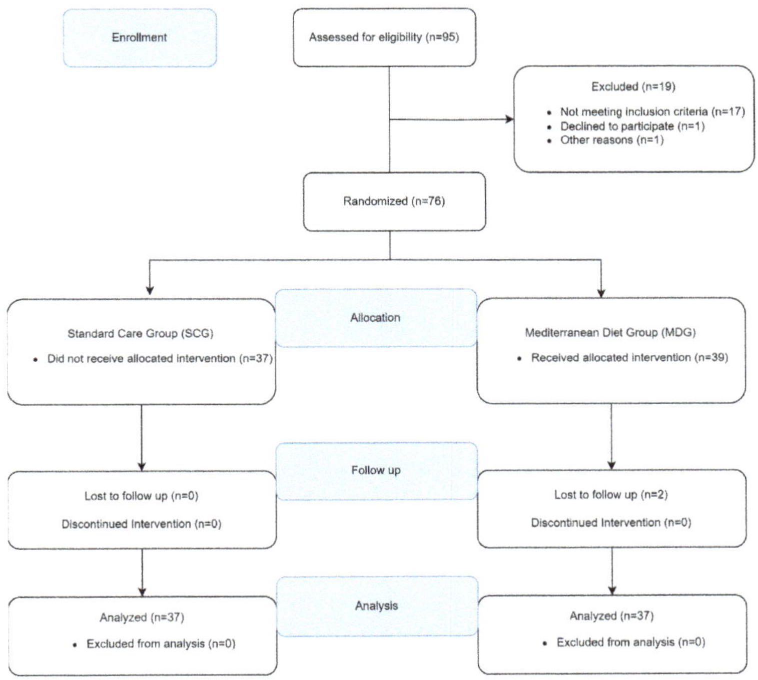

Figure 1. Flowchart of the study.

2.2. *Data Collection*

All patients underwent a detailed evaluation that included anthropometric parameters including BMI, and medical and sleep history, and comorbidities including physician-based diagnosis for depression, smoking history, and alcohol intake. Subjective daytime sleepiness, reflected by the Epworth Sleepiness Scale (ESS), and the patient's level of depression reflected by Beck's depression scale (BDI) were also recorded before PAP initiation. ESS, PAP adherence (hours of device use), BMI, and arterial blood pressure measurements were also evaluated pre- and post-intervention.

2.2.1. Epworth Sleepiness Scale (ESS)

The ESS is currently the most widely used subjective test of daytime sleepiness in clinical practice [19]. A score of 10 or higher represents excessive daytime sleepiness.

2.2.2. Beck Depression Inventory (BDI)

This 21-item questionnaire is a widely used and well-validated self-reported inventory of depressive symptoms [20]. The BDI measures the severity of depressive symptoms over the preceding week. For each item, the respondent chooses one or more options rated from 0 (absence of symptoms) to 3 (most severe level). Total scores range from 0 to 63 and represent the sum of the highest level endorsed on each item. Scores below 10 are considered normal.

2.3. Follow-Up—Usual (Standard) Care Group

All patients attended a PAP clinic before treatment initiation, where they were given specific counseling and education on the proper use and maintenance of PAP and underwent personalized, formal mask fitting by a specialized nurse. The total time for the appointment in the PAP clinic was 20 min/patient. Once PAP was started, patients were reviewed in the outpatient sleep clinic at 1-month and at 3-month intervals during the first year, and every 6 months thereafter. During these appointments, a clinical assessment was made, and patients were further encouraged to use the device. In addition, all patients received oral healthy lifestyle advice and counseling on physical activity and sleep habits and had the opportunity to discuss other health issues related to the condition, such as weight reduction and smoking cessation. At each visit, the compliance data were downloaded from the PAP device and reviewed by the PAP clinic nurse together with the patients. Any concerns or questions, such as pressure sores, persistent air leakage, claustrophobia, nasal congestion, and other side effects resulting from the nasal mask interface that might lead to suboptimal compliance were addressed immediately by the PAP clinic nurse. Changes in the PAP setting, nose/face mask, or circuit were made after consultation with the responsible sleep physician if necessary. In every follow-up visit, residual symptoms, including residual sleepiness, or change in patient's overall health status were recorded by the sleep nurse and sleep physician. This format adhered to a standardized approach according to our PAP clinic's procedures [21].

2.4. Follow-Up—Intervention Group

All patients in the intervention group attended individual weekly 60–90 min sessions led by a dietitian in the first month and twice/month thereafter. In this group, all the features described above for the standard group were included, plus additional visits involving intensive dietitian-led behavioral intervention aiming at weight loss and increasing adherence to the Mediterranean diet [22]. Dietary behavior was assessed through the Food Frequency Questionnaire [23] and Mediterranean Diet Score [24] before PAP initiation. The dietitian conducted a 24 h recall for the participant, aiming to gather data on his customary dietary program, meal timings, preferred food quantities, and quality. At this point, questions were also posed regarding the consumption of water, alcohol, smoking, any allergies, special preferences or aversions, and about physical activity. In addition, a comprehensive approach was adopted by obtaining data on family, cultural, and professional background in order to guide subsequent steps towards the maximum possible outcome.

Subsequently, the dietitian provided every participant in this group with a document containing comprehensive suggestions for a nutritious diet, illustrating all food categories and outlining the specific amounts for each, clarifying which options are healthier for each group (e.g., lean dairy). The dietitian engaged in a dialogue with the participant to gather further details or inquiries about the diet, following which the former devised a personalized diet plan based on the latter's requirements, inclinations, and unique characteristics.

Accordingly, this intervention was adjusted to fulfill the specific needs of these patients. Guidance in physical exercise, optimal sleep length, and sleep hygiene education were also given. Ultimately, a personalized therapeutic diet plan was implemented in this group.

2.4.1. Food Frequency Questionnaire

The FFQ used in this study has previously been demonstrated to be reproducible and relatively valid to assess practically all food groups, as well as macronutrients and energy consumption [23]. It includes 75 items (foods and beverages commonly consumed in Greece and dietary habits). The amounts of food consumed were expressed in grams or milliliters or in other common measures, such as slice, tablespoon or cup, representing the standard serving size. On a 6-point scale, participants were asked to report how frequently they consumed each of the meals and beverages listed in the FFQ on average over the period of one month preceding the study period (never/rarely, 1–3 times/month, 1–2 times/week, 3–6 times/week, 1 time/day, or ≥2 times/day).

2.4.2. Mediterranean Diet Score

The Mediterranean Diet Score (MedDietScore) is a 14-item validated questionnaire produced for each participant to assess their level of adherence to the Mediterranean Diet, taking into consideration their consumption of food items from nine food groups, as well as olive oil and alcoholic beverages [24]. Each of the 14 items is scored one or zero, depending on whether participants adhere to each MedDiet component or not. The Mediterranean Diet Score has a range of 0–55, with higher values indicating greater adherence to the Mediterranean Diet.

2.5. PAP Adherence

PAP usage data included mask type (nasal or full face), number of nights on PAP, average use per night (hours), air leakage, and air pressure delivered. The usage of PAP, effective pressure, and residual AHI was monitored at 1, 6, 12 months after initiation, and patients were contacted by a trained sleep nurse. Patients were also encouraged to contact the telephone helpline during working hours. The humidification was defined based on patients' feedback and was adjusted during the follow-up if necessary. In order to optimize PAP adherence unplanned visits were immediately arranged in case of low adherence to PAP therapy. Regular PAP compliance was defined as using the therapy for an average of 4 h a night for at least 70% of the nights [25]. However, in our study for optimal PAP adherence we used the cut-off point of six hours of PAP use [21,26].

2.6. Statistical Analysis

A pilot study was conducted with 49 individuals to determine the sample size. With the data obtained from the pilot study, the sample size was determined as at least 63 individuals, to obtain at least 80% power to detect a significant difference in the follow-up mean hours of PAP use values between the MDG and the SCG, allowing for a type-I error rate of 0.05.

Results are presented as mean ± standard deviation (SD) for continuous variables if normally distributed and as median (25th–75th percentile) if not. Qualitative variables are presented as absolute number (percentage). For comparisons between groups, a two-tailed t-test for independent samples (for normally distributed data) or a Mann–Whitney U test (for non-normally distributed data) was utilized for continuous variables and the chi-square test for categorical variables. The analysis of covariance was used to test adjusted between-group differences at the end of the 6-month intervention. All models were adjusted for basic confounders, namely age and gender, baseline BMI, smoking status, questionnaires scores, OSA severity indices, and co-morbidities. Analyses were additionally adjusted for weight loss (expressed as BMI difference) to test the weight-loss independent impact of the dietary/lifestyle intervention implemented on PAP adherence. Factors associated with optimal adherence (use of the device ≥6 h) at the end of the 6 months' follow-up were analyzed with bivariate logistic regression after adjustment for various potential basic explanatory confounders. We checked multicollinearity among the predictors using collinearity statistics to ensure that collinearity between predictor variables was in the acceptable range as indicated by the tolerance value variance inflation factor. Age was considered continuously and categorically, as age groups of 18–59 and >60 years, BMI was also considered continuously and categorically, as BMI groups of <30 and ≥30 kg/m^2. For the purpose of this analysis, the term cardiovascular disease (CVD) used, referred to any of the following conditions: coronary disease, stroke, atrial fibrillation, and heart failure. Results were considered significant when p values were <0.05. Data were analyzed using SPSS software (version 25, SPSS Inc., Chicago, IL, USA).

3. Results

Of 96 individuals with suspected OSA screened, 18 were non-eligible (mild OSA, normal weight, presence of other chronic diseases, etc.), one declined participation, and the remaining 76 were enrolled and randomized (SCG: 37, MDG: 39). After enrollment, 2 participants were excluded (lost to follow-up), leaving a final sample of 74 patients for analysis (SCG: 37, MDG: 37).

3.1. Comparison of Baseline Characteristics between the SCG and the MDG

Participants' socio-demographic and health status characteristics are outlined in Table 1. Most of the participants were men (78%), obese (77%), and current or former smokers (64%) with a medium educational level. The most prevalent diseases were hypertension (49%), followed by dyslipidemia (39%), CVD (27%), and COPD (22%). There was a significantly higher proportion of men in the intervention group compared to the control group. Other evaluated features remained relatively insignificant between both groups, such as age, other comorbidities, and smoking status (all $p > 0.05$), except for the presence of cardiovascular disease (CVD; 41 vs. 14%, $p = 0.009$).

Table 1. Baseline anthropometric and clinical characteristics of the study population.

	All OSA Patients ($n = 74$)	OSA SCGroup ($n = 37$)	OSA MDGroup ($n = 37$)	p Value
Demographics				
Gender, males (%)	58 (78%)	36 (97%)	22 (60%)	<0.001
Age (years)	53 ± 11	53 ± 10	53 ± 13	0.79
Age ≥ 60 years	24 (32%)	10 (27%)	14 (38%)	0.32
BMI (kg/m^2)	36 ± 8	35 ± 8	38 ± 9	0.16
BMI ≥ 30	57 (77%)	28 (76%)	29 (78%)	0.78
Neck circumference (cm)	42 ± 4	42 ± 3	43 ± 4	0.54
Waist circumference (cm)	120 ± 16	120 ± 17	121± 15	0.65
Hip circumference (cm)	120 ± 17	118 ± 16	123 ± 18	0.16
Waist/hip circumference ratio	1.0 ± 0.07	1.0 ± 0.06	0.99 ± 0.07	0.07
Educational Level				
Primary level or less	10 (13%)	7 (19%)	3 (8%)	
Secondary level	38 (52%)	18 (49%)	20 (54%)	
Tertiary level or higher	26 (35%)	12 (32%)	14 (38%)	0.75
Smoking status				
Never, n (%)	26 (36%)	11 (30%)	15 (40%)	
Currently smoking, n (%)	21 (28%)	10 (27%)	11 (30%)	
Former, n (%)	26 (36%)	16 (43%)	11 (30%)	0.36
Pack-years	15 (0, 39)	15 (0, 40)	13 (0, 39)	0.49
Co-morbidities				
Hypertension	36 (49%)	19 (51%)	17 (46%)	0.64
Coronary heart disease	5 (7%)	1 (3%)	4 (11%)	0.17
Atrial fibrillation	8 (11%)	2 (5%)	6 (16%)	0.13
Cardiovascular disease	20 (27%)	5 (14%)	15 (41%)	0.009
Diabetes type II	11 (15%)	5 (14%)	6 (16%)	0.74
COPD	16 (22%)	9 (24%)	7 (19%)	0.57
Bronchial asthma	10 (14%)	5 (14%)	5 (14%)	1.00
Hypothyroidism	11 (15%)	4 (11%)	7 (19%)	0.33
Dyslipidemia	29 (39%)	17 (46%)	12 (32%)	0.23
Depression	8 (11%)	2 (5%)	6 (16%)	0.13
BP measurements				
SBP (mmHg)	126 ± 14	127 ± 16	124 ± 13	0.37
DBP (mmHg)	78 ± 10	78 ± 11	77 ± 9	0.60

Data are presented as mean values ± SD or median (25th–75th percentile), unless otherwise indicated. BMI: body mass index, COPD: chronic obstructive pulmonary disease, BP: blood pressure, SBP: systolic blood pressure, DBP: diastolic blood pressure.

Comparison of PSG parameters of the SCG and the MDG population showed no significant differences between groups (Table 2). Although no significant difference was noted between nocturnal and diurnal symptoms (Table 3), frequent awakenings reported were significantly higher in the MDG compared to the SCG (65 vs. 41%, $p = 0.04$).

Table 2. Baseline polysomnography characteristics of the study population.

	All OSA Patients (n = 74)	OSA SCGroup (n = 37)	OSA MDGroup (n = 37)	p Value
Total recording time (min)	406 ± 40	404 ± 47	410 ± 31	0.53
Total sleep time (min)	276 ± 52	274 ± 52	280 ± 52	0.65
Sleep efficiency, %	68 ± 12	68 ± 12	68 ± 12	0.95
Wake after sleep onset time (min)	91 ± 36	95 ± 40	89 ± 33	0.53
Sleep latency	35 (24, 62)	33 (25, 57)	35 (24, 67)	0.75
REM latency	232 ± 79	231 ± 76	232 ± 83	0.96
NREM (%)	91 ± 3	91 ± 3	91 ± 3	0.75
REM (%)	9 ± 3	9 ± 3	9 ± 3	0.75
AHI	47 ± 24	46 ± 22	49 ± 27	0.60
REM AHI	54 ± 26	54 ± 26	55 ± 27	0.89
Arousal index	45 ± 17	46 ± 13	46 ± 20	0.95
Oxygen desaturation index	44 (29, 65)	46 (32, 66)	43 (29, 82)	0.83
Mean SaO_2	92 ± 3	92 ± 2	91 ± 3	0.39
Lowest SaO_2	80 (71, 83)	80 (72, 83)	79 (69, 83)	0.58
TST90 (min)	77 (34, 137)	74 (29, 114)	81 (36, 166)	0.53
Severity of OSA (%)				
$15 \leq$ AHI < 30	18 (24%)	8 (22%)	10 (27%)	
AHI ≥ 30	56 (76%)	29 (78%)	27 (73%)	0.59

OSA: obstructive sleep apnea, AHI: apnea-hypopnea index, TST90: sleep time with oxygen saturation below 90%.

Table 3. Nocturnal and diurnal symptoms of the study population.

	All OSA Patients (n = 74)	OSA SCGroup (n = 37)	OSA MDGroup (n = 37)	p Value
Nocturnal Symptoms				
Snoring	74 (100%)	37 (100%)	37 (100%)	1
Witnessed apneas	73 (99%)	36 (97%)	37 (100%)	0.31
Frequent awakenings	39 (53%)	15 (41%)	24 (65%)	0.04
Nocturia	67 (91%)	33 (89%)	34 (92%)	0.69
Diurnal symptoms				
ESS score	11 ± 5	12 ± 5	11 ± 5	0.43
ESS > 10	43 (58%)	23 (62%)	20 (54%)	0.48
Morning headache	51 (69%)	25 (68%)	26 (70%)	0.80
Driving problems	1 (1%)	0 (0%)	1 (3%)	0.31
BDI score	8 (3, 14)	5 (3, 18)	9 (5, 12)	0.25
BDI ≥ 10	25 (39%)	10 (31%)	15 (47%)	0.20

ESS: Epworth Sleepiness Scale, BDI: Beck Depression Inventory.

Regarding lifestyle habits, the MDG exhibited a moderate level of adherence to the Mediterranean diet as assessed by MedDietScore (29 ± 5).

3.2. PAP Adherence

All patients continued to use their PAP at the end of the follow-up period. Auto-PAP was prescribed to the majority of participants with final levels at the end of the follow-up of PAP pressure of 9.2 for the SCG and 8.4 for the MDG ($p = 0.04$). Post-intervention PAP use was significantly higher in the MDG compared to the SCG (6.1 ± 1.2 vs. 5.4 ± 1.4, $p = 0.02$). Further analysis showed that this difference persisted after adjustments for age, gender, BMI, difference in BMI, ESS, difference in ESS, BDI score, OSA severity, and comorbidities

(5.2 vs. 6.1, $p = 0.03$). Moreover, diet/lifestyle intervention was identified as one of the most significant predictive factors for optimal PAP adherence (OR = 5.458, 95% CI = 1.144–26.036, $p = 0.03$) (Table 4).

Table 4. Factors associated with optimal adherence to treatment at the end of the follow-up period in all patients.

Variable	B	S.E.	*p*-Value	OR (95% CI)
Females vs. males	−0.70	1.024	0.945	0.932 (0.125–26.036)
Age (years)	0.006	0.036	0.870	1.006 (0.938–1.079)
Baseline BMI (kg/m^2)	0.094	0.057	0.101	1.099 (0.982–1.229)
Currently smoking vs. never/formerly smoking	−1.589	0.811	0.050	0.204 (0.042–1.001)
Baseline ESS score > 10	1.047	0.694	0.131	2.850 (0.732–11.098)
Baseline BDI score ≥ 10	−1.649	0.816	0.043	0.192 (0.039–0.951)
Arterial hypertension	1.037	0.778	0.183	2.821 (0.614–12.968)
Cardiovascular disease	−1.103	0.813	0.175	0.332 (0.067–1.634)
Type 2 diabetes	0.372	0.937	0.691	1.451 (0.231–9.109)
COPD	−0.655	0.917	0.475	0.519 (0.086–3.133)
MDG vs. SCG	1.697	0.797	0.033	5.458 (1.144–26.036)
Severe vs. moderate OSA	−0.625	0.863	0.469	0.535 (0.099–2.907)

3.3. *Effect of Diet Intervention on Anthropometric and Daytime Symptoms Parameters*

Regarding BMI, an increase was noted in the SCG, whereas a decrease (improvement) was observed in the MDG, (0.41 ± 1.8 vs. −0.75 ± 1.3, $p = 0.02$) (Figure 2). Specifically, during the 6-month follow-up, patients in the SCG demonstrated an average weight gain of 1.6% of their baseline body weight, while those in the MDG showcased a loss equivalent to 1.5% ($p = 0.04$). BMI difference, although attenuated, persisted after adjustments for age, gender, BMI, ESS, BDI score, OSA severity, PAP adherence, and comorbidities (0.23 vs. −0.55, $p = 0.31$).

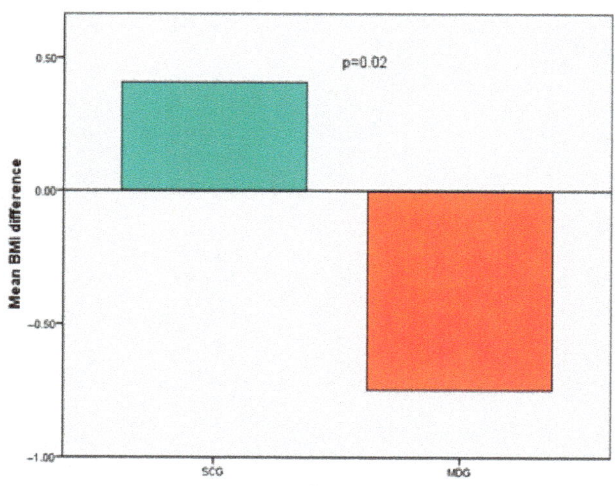

Figure 2. Post-treatment BMI differences between the SCG and the MDG.

In terms of blood pressure measurements a decrease was noticed only in the MDG (SBP −2.5 ± 11, DBP −1.1 ± 7.6) compared to the SCG (SBP 0.14 ± 9.2, DBP 0.40 ± 9.2) after the 6 months of the follow-up period. Nevertheless, the aforementioned changes did not exhibit statistical significance ($p = 0.28$ and $p = 0.45$, respectively) (Figure 3). However, after accounting for confounding variables, a substantial decline in SBP (−5.5 vs. 2.8, $p =$

0.014) and DBP (−4.0 vs. 2.5, $p = 0.01$) was specifically detected in the MDG versus the SCG.

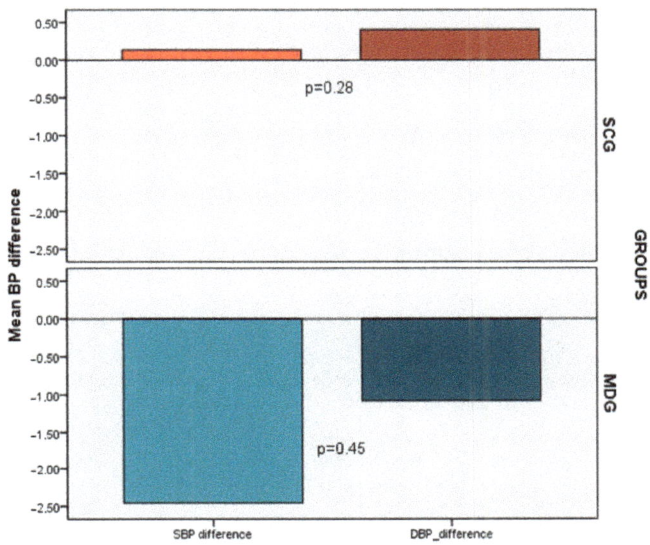

Figure 3. Post-treatment BP differences between the SCG and the MDG.

A significant decrease in ESS was also noted in both groups. Nonetheless, no group manifested a significant predominance in this improvement (−4.5 vs. −3.5, $p = 0.19$), even following adjustment of confounders (Figure 4). Additionally, both groups displayed a noteworthy decline in the proportion of patients experiencing excessive daytime sleepiness (SCG: 19% vs. 62%, $p < 0.001$, MDG: 16% vs. 54%, $p < 0.001$). The proportion of residual sleepiness at the end of the follow-up period was comparable between the groups (16% versus 19%, $p = 0.76$).

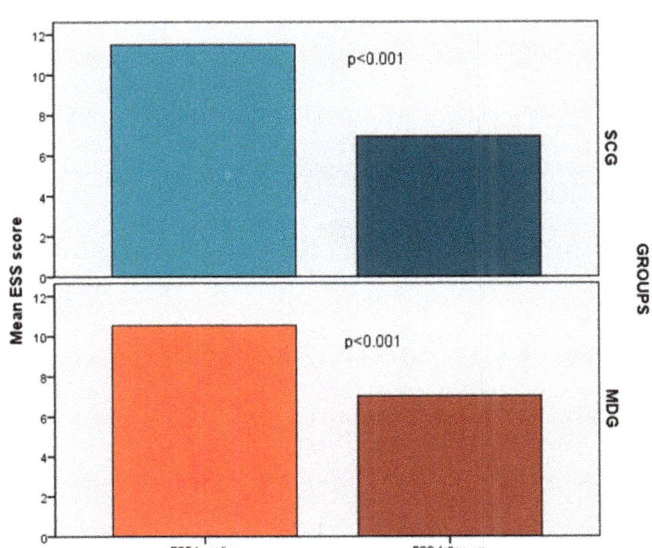

Figure 4. Post-treatment ESS score differences between the SCG and the MDG.

4. Discussion

Our study assessed the impact of dietary/lifestyle intervention along with usual care on PAP adherence, BMI, BP measurements, and sleepiness in moderate to severe OSA patients. The intervention demonstrated significant and clinically meaningful improvements in PAP adherence, BMI, and BP measurements in the intervention group compared to the control group, independent of age, gender, BMI, weight loss, baseline level of sleepiness, depressive symptoms, OSA severity, and comorbidities and even though participants were enrolled in the study for a period of six months.

This is the first study demonstrating the favorable effect of incorporating a dietary/lifestyle intervention in combination with standard care towards objective PAP adherence in Greece. The findings suggest that providing intensive support with additional time allocated to the MDG may prove beneficial in enhancing PAP compliance, irrespective of weight loss. While previous studies with comparable follow-up times have investigated the impact of diet and lifestyle interventions alongside usual care on various OSA severity and cardiometabolic parameters, there has been no objective evaluation of PAP adherence [17,18,27–29]. Only self-reported PAP adherence was reported in a limited number of studies [17,18,28,29]. Compared to our study, Georgoulis et al. found a lower self-reported average PAP use of roughly 4 h/day [17]. Likewise, they observed a greater PAP compliance in the Mediterranean lifestyle group when compared to the Mediterranean diet alone and standard care group six months following the intervention [18]. Although Schiavo et al., in a similar study investigating the effect of a low-calorie ketogenic diet combined with PAP therapy on OSA and cardiometabolic parameters, acknowledged that they monitored adherence to PAP treatment, they did not provide relevant data [30]. Moreover, it is significant to observe that PAP adherence, even in the SCG, was substantially higher than in earlier studies [31,32]. The SCG's adherence of 5.4 h per night is comparable to the one reported in a previous study from our group [21]. This can be attributed to the comprehensive, standardized approach of our PAP clinic. This approach includes personalized counseling and education on the proper use and maintenance of PAP, formal mask fitting, and frequent follow-ups to address residual symptoms, side effects, and the patients' overall health status.

Despite recommendations [33] and studies indicating better clinical outcomes [34] with diet-induced weight loss as part of OSA treatment, interventions targeting weight loss through lifestyle changes are underutilized. Indeed, based on MeditDietScore we found medium adherence to the Mediterranean diet in our patients, suggesting that non-conventional care was not regarded as an essential component of overall patient care. Furthermore, another challenge is the lack of specific educational skills from health care professionals to support diet/lifestyle interventions or promote behavioral changes in these patients. Optimizing OSA treatment is crucial for clinicians, but many fail to recognize the importance of incorporating diet and lifestyle interventions into their treatment plans. Thus, our study has the potential to raise awareness regarding the importance of diet and lifestyle among OSA patients who are overweight or obese. Our results also demonstrated that the Mediterranean diet/lifestyle intervention contributed to improvements in BMI and BP measurements compared with standard care. These benefits were evident even after adjustment for confounders, suggesting that the Mediterranean lifestyle can lead to cardiovascular benefits beyond weight loss in these patients. These findings are noteworthy as they align with prior research indicating more weight loss in patients following diet/lifestyle intervention than standard care [16–18,27,30]. Although the weight loss achieved in our study was below the recommended 5–10% for managing obesity and achieving health benefits, even a slight reduction in weight was seen as beneficial for health, particularly among patients with OSA [17,35,36]. Another study we conducted involved randomly assigning 40 obese individuals with OSA undergoing PAP treatment to either a weight-loss Mediterranean diet or a weight-loss prudent diet for 6 months. The group on the Mediterranean diet showed more significant weight loss, but the results were not statistically significant due to a small number of participants [37]. In the MIMOSA RCT conducted recently, a dietary/lifestyle intervention also based on the Mediterranean pattern was combined with PAP treatment for OSA patients, resulting in a notable improvement in weight reduction, OSA severity, and related symptoms when compared to standard care alone [16,38].

At present, the prescription of PAP remains the first-line treatment for patients with moderate to severe OSA. However, effectiveness is dependent on the patient's usage. Consequently, it is essential to take into account diet/lifestyle intervention as an added strategy in the treatment of OSA. These interventions provide a non-invasive approach that may also help to address PAP adherence issues. Combining diet and exercise has been found to be an effective intervention in mitigating OSA severity [34], therefore, it is probable that lower pressures will be required in the PAP machine [39], making it easier for patients to tolerate high PAP pressures. In support of this, evidence suggests that patients diagnosed with OSA frequently encounter distress and are incapable of tolerating heightened PAP pressures, thus abandoning PAP devices [40]. In addition, our intervention group received personalized and consistent guidance from a specialized dietitian, leading to better PAP adherence. Previously, in a 2-year study, our group found that intensive follow-up support was more effective in improving long-term PAP adherence than standard support [21]. It is plausible that the unequal intervention time spent between the SCG and the MDG in the current study may partially account for our results. The MDG received additional visits and hours of contact compared with the SCG, potentially influencing adherence outcomes. Therefore, the regular implementation of dietary and lifestyle interventions is a valuable strategy for treating OSA and must be consistently applied across all OSA patients.

The main strengths of the current study are the design (RCT) and implementation of a diet and lifestyle intervention readily adaptable to real-world clinical practice. On the other hand, the study has some limitations that deserve comments. Since it was a real-life implementation of a diet/lifestyle program, the results can only be generalized to patients who are motivated to participate in such a program and not to all OSA patients. In addition, the generalizability of our results is limited, due to the fact that the trial was conducted at a single center with a predominantly male patient population. Female under-representation is a common challenge in many studies, with women being referred less for OSA diagnosis and treatment compared to males [41]. Thus, considering the gender bias in the existing literature towards male subjects, it is essential that future studies encompass a more heterogeneous sample that includes both males and females with OSA.

Furthermore, this study was carried out during the second wave of the COVID-19 pandemic, which resulted in a small sample size. Additionally, a 6-month period was insufficient to determine long-term intervention effects and maintenance of benefits. Consequently, large-scale studies may require prolonged follow-up periods to assess the durability and sustainability of the observed improvements in PAP adherence, BMI, and blood pressure.

5. Conclusions

In conclusion, our results provide evidence that overweight/obese patients with moderate to severe OSA can benefit significantly in terms of PAP adherence, BP, and BMI control from behavioral interventions aiming at weight loss through the adoption of appropriate food and lifestyle practices. Therefore, it is essential to consider such type of intervention as an add-on approach to OSA management. However, additional evidence is needed from studies, including larger numbers of patients with longer-term follow-ups to explore the influence of diet/lifestyle interventions on OSA, and especially on the long-term sequelae.

Author Contributions: Conceptualization, I.T. and I.B.; methodology, E.D.; software, I.B.; validation, I.B., S.E.S. and I.T.; formal analysis, I.B.; investigation, E.D.; data curation, E.M. and V.M.; writing—original draft preparation, I.B.; writing—review and editing, I.B.; visualization, I.T.; supervision, S.E.S. All authors have read and agreed to the published version of the manuscript.

Funding: This research received no external funding.

Institutional Review Board Statement: The study was conducted in accordance with the Declaration of Helsinki, and approved by the University of Crete Research Ethics Committee (REC-UOC) (approval number: 158/29.11.2021).

Informed Consent Statement: Informed consent was obtained from all subjects involved in the study.

Data Availability Statement: Data are available upon request.

Conflicts of Interest: The authors declare no conflict of interest.

References

1. Benjafield, A.V.; Ayas, N.T.; Eastwood, P.R.; Heinzer, R.; Ip, M.S.M.; Morrell, M.J.; Nunez, C.M.; Patel, S.R.; Penzel, T.; Pépin, J.L.; et al. Estimation of the global prevalence and burden of obstructive sleep apnoea: A literature-based analysis. *Lancet Resp. Med.* **2019**, *7*, 687–698. [CrossRef] [PubMed]
2. Yeghiazarians, Y.; Jneid, H.; Tietjens, J.R.; Redline, S.; Brown, D.L.; El-Sherif, N.; Mehra, R.; Bozkurt, B.; Ndumele, C.E.; Somers, V.K. Obstructive Sleep Apnea and Cardiovascular Disease: A Scientific Statement From the American Heart Association. *Circulation* **2021**, *144*, e56–e67. [CrossRef] [PubMed]
3. Frost & Sullivan; American Academy of Sleep Medicine. Hidden Health Crisis Costing America billions: Underdiagnosing and Undertreating Obstructive Sleep Apnea Draining Health Care System. Published 8 August 2016. Available online: https://aasm.org/resources/pdf/sleep-apnea-economic-crisis.pdf (accessed on 22 March 2023).
4. Knauert, M.; Naik, S.; Gillespie, M.B.; Kryger, M. Clinical consequences and economic costs of untreated obstructive sleep apnea syndrome. *World J. Otorhinolaryngol. Head Neck Surg.* **2015**, *1*, 17–27. [CrossRef] [PubMed]
5. Patel, S.R. The complex relationship between weight and sleep apnoea. *Thorax* **2015**, *70*, 205–206. [CrossRef]
6. Sareli, A.E.; Cantor, C.R.; Williams, N.N.; Korus, G.; Raper, S.E.; Pien, G.; Hurley, S.; Maislin, G.; Schwab, R.J. Obstructive sleep apnea in patients undergoing bariatric surgery—A tertiary center experience. *Obes. Surg.* **2011**, *21*, 316–327. [CrossRef]
7. Giles, T.L.; Lasserson, T.J.; Smith, B.J.; White, J.; Wright, J.; Cates, C.J. Continuous positive airways pressure for obstructive sleep apnoea in adults. *Cochrane Database Syst. Rev.* **2006**, *1*, CD001106.
8. Pepperell, J.C.; Ramdassingh-Dow, S.; Crosthwaite, N.; Mullins, R.; Jenkinson, C.; Stradling, J.R.; Davies, R.J. Ambulatory blood pressure after therapeutic and subtherapeutic nasal continuous positive airway pressure for obstructive sleep apnoea: A randomised parallel trial. *Lancet* **2002**, *359*, 204–210. [CrossRef]
9. Walia, H.K.; Thompson, N.R.; Pascoe, M.; Faisal, M.; Moul, D.E.; Katzan, I.; Mehra, R.; Foldvary-Schaefer, N. Effect of Positive Airway Pressure Therapy on Drowsy Driving in a Large Clinic-Based Obstructive Sleep Apnea Cohort. *J. Clin. Sleep Med.* **2019**, *15*, 1613–1620. [CrossRef]
10. Weaver, T.E.; Grunstein, R.R. Adherence to continuous positive airway pressure therapy: The challenge to effective treatment. *Proc. Am. Thorac. Soc.* **2008**, *5*, 173–178. [CrossRef]
11. Pepin, J.L.; Baillieul, S.; Tamisier, R. Reshaping sleep apnea care: Time for value-based strategies. *Ann. Am. Thorac. Soc.* **2019**, *16*, 1501–1503. [CrossRef]
12. Johansson, K.; Neovius, M.; Lagerros, Y.T.; Harlid, R.; Rössner, S.; Granath, F.; Hemmingsson, E. Effect of a very low energy diet on moderate and severe obstructive sleep apnoea in obese men: A randomised controlled trial. *BMJ* **2009**, *339*, b4609. [CrossRef] [PubMed]
13. Tuomilehto, H.P.; Seppä, J.M.; Partinen, M.M.; Peltonen, M.; Gylling, H.; Tuomilehto, J.O.; Vanninen, E.J.; Kokkarinen, J.; Sahlman, J.K.; Martikainen, T.; et al. Lifestyle intervention with weight reduction: First-line treatment in mild obstructive sleep apnea. *Am. J. Respir. Crit. Care Med.* **2009**, *179*, 320–327. [CrossRef] [PubMed]
14. Stelmach-Mardas, M.; Brajer-Luftmann, B.; Kuśnierczak, M.; Batura-Gabryel, H.; Piorunek, T.; Mardas, M. Body Mass Index Reduction and Selected Cardiometabolic Risk Factors in Obstructive Sleep Apnea: Meta-Analysis. *J. Clin. Med.* **2021**, *10*, 1485. [CrossRef]
15. Georgoulis, M.; Yiannakouris, N.; Kechribari, I.; Lamprou, K.; Perraki, E.; Vagiakis, E.; Kontogianni, M.D. Cardiometabolic Benefits of a Weight-Loss Mediterranean Diet/Lifestyle Intervention in Patients with Obstructive Sleep Apnea: The "MIMOSA" Randomized Clinical Trial. *Nutrients* **2020**, *12*, 1570. [CrossRef]
16. Georgoulis, M.; Yiannakouris, N.; Tenta, R.; Fragopoulou, E.; Kechribari, I.; Lamprou, K.; Perraki, E.; Vagiakis, E.; Kontogianni, M.D. A weight-loss Mediterranean diet/lifestyle intervention ameliorates inflammation and oxidative stress in patients with obstructive sleep apnea: Results of the "MIMOSA" randomized clinical trial. *Eur. J. Nutr.* **2021**, *60*, 3799–3810. [CrossRef] [PubMed]
17. Georgoulis, M.; Yiannakouris, N.; Kechribari, I.; Lamprou, K.; Perraki, E.; Vagiakis, E.; Kontogianni, M.D. Dose-response relationship between weight loss and improvements in obstructive sleep apnea severity after a diet/lifestyle interventions: Secondary analyses of the "MIMOSA" randomized clinical trial. *J. Clin. Sleep Med.* **2022**, *18*, 1251–1261. [CrossRef]
18. Georgoulis, M.; Yiannakouris, N.; Kechribari, I.; Lamprou, K.; Perraki, E.; Vagiakis, E.; Kontogianni, M.D. Sustained improvements in the cardiometabolic profile of patients with obstructive sleep apnea after a weight-loss Mediterranean diet/lifestyle intervention: 12-month follow-up (6 months post-intervention) of the "MIMOSA" randomized clinical trial. *Nutr. Metab. Cardiovasc. Dis.* **2023**, *33*, 1019–1028. [CrossRef]

19. Johns, M.W. A new method for measuring daytime sleepiness:the Epworth sleepiness scale. *Sleep* **1991**, *14*, 540–545. [CrossRef]
20. Beck, A.T.; Steer, R.A.; Garbin, M.G. Psychometric properties of the Beck depression inventory: Twenty-five years of evaluation. *Clin. Psychol. Rev.* **1988**, *8*, 77–100. [CrossRef]
21. Bouloukaki, I.; Giannadaki, K.; Mermigkis, C.; Tzanakis, N.; Mauroudi, E.; Moniaki, V.; Michelakis, S.; Siafakas, N.M.; Schiza, S.E. Intensive versus standard follow-up to improve continuous positive airway pressure compliance. *Eur. Respir. J.* **2014**, *44*, 1262–1274. [CrossRef]
22. Bayram, S.Ş.; Aktaş, N. Mediterranean diet and frequently used indexes for measuring Mediterranean diet quality. In Proceedings of the International Eurasian Congress on "Natural Nutrition and Healthy Life", Ankara, Turkey, 12–15 July 2018; Karaman, M.R., Artık, N., Şanlıer, N., Eds.; Ankara University Institute of Food Safety "Pelin Ofset": Ankara, Turkey, 2018.
23. Bountziouka, V.; Bathrellou, E.; Giotopoulou, A.; Katsagoni, C.; Bonou, M.; Vallianou, N.; Barbetseas, J.; Avgerinos, P.C.; Panagiotakos, D.B. Development, repeatability and validity regarding energy and macronutrient intake of a semi-quantitative food frequency questionnaire: Methodological considerations. *Nutr. Metab. Cardiovasc. Dis.* **2012**, *22*, 659–667. [CrossRef] [PubMed]
24. Panagiotakos, D.B.; Pitsavos, C.; Arvaniti, F.; Stefanadis, C. Adherence to the Mediterranean food pattern predicts the prevalence of hypertension, hypercholesterolemia, diabetes and obesity, among healthy adults; the accuracy of the MedDietScore. *Prev. Med.* **2007**, *44*, 335–340. [CrossRef] [PubMed]
25. Kribbs, N.B.; Pack, A.I.; Kline, L.R.; Smith, P.L.; Schwartz, A.R.; Schubert, N.M.; Redline, S.; Henry, J.N.; Getsy, J.E.; Dinges, D.F. Objective measurement of patterns of nasal CPAP use by patients with obstructive sleep apnea. *Am. J. Respir. Crit. Care Med.* **1993**, *147*, 887–895. [CrossRef]
26. Weaver, T.E.; Maislin, G.; Dinges, D.F.; Bloxham, T.; George, C.F.; Greenberg, H.; Kader, G.; Mahowald, M.; Younger, J.; Pack, A.I. Relationship between hours of CPAP use and achieving normal levels of sleepiness and daily functioning. *Sleep* **2007**, *30*, 711–719. [CrossRef]
27. Carneiro-Barrera, A.; Amaro-Gahete, F.J.; Guillén-Riquelme, A.; Jurado-Fasoli, L.; Sáez-Roca, G.; Martín-Carrasco, C.; Buela-Casal, G.; Ruiz, J.R. Effect of an Interdisciplinary Weight Loss and Lifestyle Intervention on Obstructive Sleep Apnea Severity: The INTERAPNEA Randomized Clinical Trial. *JAMA Netw. Open* **2022**, *5*, e228212. [CrossRef] [PubMed]
28. Igelström, H.; Åsenlöf, P.; Emtner, M.; Lindberg, E. Improvement in obstructive sleep apnea after a tailored behavioural sleep medicine intervention targeting healthy eating and physical activity: A randomised controlled trial. *Sleep Breath.* **2018**, *22*, 653–661. [CrossRef]
29. Spörndly-Nees, S.; Åsenlöf, P.; Lindberg, E.; Emtner, M.; Igelström, H. Effects on obstructive sleep apnea severity following a tailored behavioral sleep medicine intervention aimed at increased physical activity and sound eating: An 18-month follow-up of a randomized controlled trial. *J. Clin. Sleep Med.* **2020**, *16*, 705–713. [CrossRef]
30. Schiavo, L.; Pierro, R.; Asteria, C.; Calabrese, P.; Di Biasio, A.; Coluzzi, I.; Severino, L.; Giovanelli, A.; Pilone, V.; Silecchia, G. Low-Calorie Ketogenic Diet with Continuous Positive Airway Pressure to Alleviate Severe Obstructive Sleep Apnea Syndrome in Patients with Obesity Scheduled for Bariatric/Metabolic Surgery: A Pilot, Prospective, Randomized Multicenter Comparative Study. *Obes. Surg.* **2022**, *32*, 634–642. [CrossRef]
31. McEvoy, R.D.; Antic, N.A.; Heeley, E.; Luo, Y.; Ou, Q.; Zhang, X.; Mediano, O.; Chen, R.; Drager, L.F.; Liu, Z.; et al. SAVE Investigators and Coordinators CPAP for Prevention of Cardiovascular Events in Obstructive Sleep Apnea. *N. Engl. J. Med.* **2016**, *375*, 919–931. [CrossRef]
32. Batool-Anwar, S.; Goodwin, J.L.; Kushida, C.A.; Walsh, J.A.; Simon, R.D.; Nichols, D.A.; Quan, S.F. Impact of continuous positive airway pressure (CPAP) on quality of life in patients with obstructive sleep apnea (OSA). *J. Sleep Res.* **2016**, *25*, 731–738. [CrossRef]
33. Hudgel, D.W.; Patel, S.R.; Ahasic, A.M.; Bartlett, S.J.; Bessesen, D.H.; Coaker, M.A.; Fiander, P.M.; Grunstein, R.R.; Gurubhagavatula, I.; Kapur, V.K.; et al. The role of weight management in the treatment of adult obstructive sleep apnea. An Official American Thoracic Society Clinical Practice Guideline. *Am. J. Respir. Crit. Care Med.* **2018**, *198*, e70–e87. [CrossRef]
34. Rokou, A.; Eleftheriou, A.; Tsigalou, C.; Apessos, I.; Nena, E.; Dalamaga, M.; Voulgaris, A.; Steiropoulos, P. Effect of the Implementation of a Structured Diet Management Plan on the Severity of Obstructive Sleep Apnea: A Systematic Review. *Curr. Nutr. Rep.* **2023**, *12*, 26–38. [CrossRef]
35. Jensen, M.D.; Ryan, D.H.; Apovian, C.M.; Ard, J.D.; Comuzzie, A.G.; Donato, K.A.; Hu, F.B.; Hubbard, V.S.; Jakicic, J.M.; Kushner, R.F.; et al. Obesity Society. 2013 AHA/ACC/TOS guideline for the management of overweight and obesity in adults: A report of the American College of Cardiology/American Heart Association Task Force on Practice Guidelines and The Obesity Society. *Circulation* **2014**, *129*, S102–S138. [CrossRef]
36. Yumuk, V.; Tsigos, C.; Fried, M.; Schindler, K.; Busetto, L.; Micic, D.; Toplak, H. Obesity Management Task Force of the European Association for the Study of Obesity. European Guidelines for obesity management in adults. *Obes Facts* **2015**, *8*, 402–424. [CrossRef] [PubMed]
37. Papandreou, C.; Schiza, S.E.; Bouloukaki, I.; Hatzis, C.M.; Kafatos, A.G.; Siafakas, N.M.; Tzanakis, N.E. Effect of Mediterranean diet versus prudent diet combined with physical activity on: A randomised trial. *Eur. Respir. J.* **2012**, *39*, 1398–1404. [CrossRef]
38. Georgoulis, M.; Yiannakouris, N.; Kechribari, I.; Lamprou, K.; Perraki, E.; Vagiakis, E.; Kontogianni, M.D. The effectiveness of a weight-loss Mediterranean diet/lifestyle intervention in the management of obstructive sleep apnea: Results of the "MIMOSA" randomized clinical trial. *Clin. Nutr.* **2021**, *40*, 850–859. [CrossRef] [PubMed]

39. Lankford, D.A.; Proctor, C.D.; Richard, R. Continuous positive airway pressure (CPAP) changes in bariatric surgery patients undergoing rapid weight loss. *Obes. Surg.* **2005**, *15*, 336–341. [CrossRef] [PubMed]
40. Ghadiri, M.; Grunstein, R.R. Clinical side effects of continuous positive airway pressure in patients with obstructive sleep apnoea. *Respirology* **2020**, *25*, 593–602. [CrossRef] [PubMed]
41. Lindberg, E.; Benediktsdottir, B.; Franklin, K.A.; Holm, M.; Johannessen, A.; Jögi, R.; Gislason, T.; Real, F.G.; Schlünssen, V.; Janson, C. Women with symptoms of sleep-disordered breathing are less likely to be diagnosed and treated for sleep apnea than men. *Sleep Med.* **2017**, *35*, 17–22. [CrossRef] [PubMed]

life

MDPI

Brief Report

Serum Cytokines as Biomarkers in Heart Failure with Preserved Ejection Fraction and Sleep Apnea: A Prospective Cohort Study

Alexey Yakovlev [1,2], Alexander Teplyakov [3], Elena Grakova [4], Sergey Shilov [5], Natalia Yakovleva [6], Kristina Kopeva [4], Valery Shirinsky [7] and Ivan Shirinsky [7,*]

1 Department of Therapy, Hematology and Transfusiology, Novosibirsk State Medical University, Krasny Prospect, 52, 630091 Novosibirsk, Russia
2 Cardiology Department, Non-State Healthcare Institution "Road Clinical Hospital in Novosibirsk", Russian Railways Joint Stock Company, Vladimirovsky Spusk, 630003 Novosibirsk, Russia
3 Cardiology Research Institute, Tomsk National Research Medical Center (NRMC), Kievskaya Str., 111a, 634012 Tomsk, Russia
4 Department of Myocardial Pathology, Cardiology Research Institute, Tomsk National Research Medical Center (NRMC), Kievskaya Str., 111a, 634012 Tomsk, Russia
5 Department of Pathological Physiology and Clinical Pathophysiology, Novosibirsk State Medical University, Krasny Prospect, 52, 630091 Novosibirsk, Russia
6 Department of Policlinic Therapy and General Medical Practice, Novosibirsk State Medical University, Krasny Prospect, 52, 630091 Novosibirsk, Russia
7 Laboratory of Clinical Immunopharmacology, Federal State Budgetary Scientific Institution, Research Institute of Fundamental and Clinical Immunology, 6 Zalesskogo Str., 630099 Novosibirsk, Russia
* Correspondence: ivan.shirinsky@niikim.ru; Tel.: +73-832-282-547

Abstract: Heart failure with preserved ejection fraction (HFpEF) and obstructive sleep apnea (OSA) frequently co-occur and this comorbidity represents a separate phenotype of HFpEF. While many research attempts are focused on biomarkers of HFpEF, currently, there is a lack of validated biomarkers of HFpEF and OSA. In this study, we aimed to evaluate prognostic significance of several serum cytokines in patients with HFpEF and OSA. The patients with HFpEF and OSA were recruited from the Sleep Apnea Center of Novosibirsk, Russian Federation and followed up for 12 months. The main analyzed outcomes were five-point major adverse cardiovascular events (MACE) and the 6-min walk test (6MWT). The analyzed cytokines were circulating IL-6, IL-10, and VEGF measured at baseline. We recruited 77 male patients with HFpEF and OSA, the data of 71 patients were available for analyses. Patients who developed MACE had four-fold elevated concentrations of serum IL-10. There was no association between baseline cytokine levels and longitudinal changes in 6MWT. Circulating IL-10 levels are positively associated with MACE in men with HFpEF and OSA and thus may be a potential prognostic biomarker in this subgroup of patients. These results should be confirmed in larger studies encompassing both males and females.

Keywords: heart failure; cytokines; prognosis; sleep apnea; obstructive; biomarkers; comorbidity; phenotype; OSA; HFpEF; HF

Citation: Yakovlev, A.; Teplyakov, A.; Grakova, E.; Shilov, S.; Yakovleva, N.; Kopeva, K.; Shirinsky, V.; Shirinsky, I. Serum Cytokines as Biomarkers in Heart Failure with Preserved Ejection Fraction and Sleep Apnea: A Prospective Cohort Study. *Life* **2023**, *13*, 628. https://doi.org/10.3390/life13030628

Academic Editors: David Slouka and Milan Štengl

Received: 1 February 2023
Revised: 17 February 2023
Accepted: 20 February 2023
Published: 23 February 2023

1. Introduction

Heart failure (HF) is a significant health problem with prevalence reaching 1–2% of the adult population in the Western world [1]. There are two major phenotypes of HF based on presence or absence of reduced left ventricular ejection fraction (LVEF)—HF with reduced ejection fraction (HFrEF) and HF with preserved ejection fraction (HFpEF). LVEF is a surrogate measurement of systolic function of the left ventricle of the heart and is frequently assessed using echocardiography [2]. HFpEF is defined as a "clinical syndrome in patients with current or prior symptoms of HF with a left ventricular ejection fraction (LVEF) $\geq$ 50 percent and evidence of cardiac dysfunction as a cause of symptoms". Previously, HFpEF was termed "diastolic HF", in contrast to HFrEF which was termed "systolic HF" [3]. HFrEF and HFpEF are pathogenetically distinct conditions and are associated with different biomarkers [4]. Currently, HFpEF is of particular concern forglobal healthcare [5] as its prevalence is rising [6] and there are still no effective treatment options for this type of HF [7].

HF is frequently accompanied with different comorbidities which can further impair prognosis of this condition [8]. One of the most important comorbidities in HF is sleep disordered breathing (SDB), which in turn can be central (central sleep apnea, CSA) or obstructive (obstructive sleep apnea, OSA) [9]. Both HFrEF and HFpEF can be linked to SDB. The predominant type of SDB in HFpEF is OSA, which occurs in up to 69–81% of these patients [10,11]. The relationships between HFpEF and OSA seem to be bidirectional. The hallmark of OSA is intermittent hopoxia. In spite of the short duration of hypoxia episodes, over the years, the cumulative burden of hypoxia-related changes becomes high. Hypoxia-induced tissue injury and lipid peroxidation cause systemic inflammation, triggering endothelial expression of adhesion molecules, which attracts monocytes, lowers endothelial production of nitric oxide, and raises endothelial production of reactive oxygen species. This in turn leads to local myocardial proinflammatory/fibrogenic signaling and finally, to myocardial fibrosis, which is the central mechanism of HFpEF [12]. On the other hand, HFpEF is associated with fluid retention [13]. Increased fluid accumulation in the neck results in narrowing of the pharynx and increasing its propensity to collapse during sleep. This represents a possible mechanism by which HFpEF can lead to increased risk of OSA [14].

There is a need for novel biomarkers for both HF [15] and OSA [16]. Due to high heterogeneity of HF and involvement of multiple factors in its pathogenesis, it can be assumed that each HF phenotype may be characterized by its own set of biomarkers. While a recent study for the first time evaluated serum biomarkers in HFrEF and CSA [17], there is a lack of research assessing biomarkers of HFpEF and OSA.

Given the role of cytokines in inflammation and angiogenesis in both HF [18] and SDB [19], we hypothesized that serum cytokines could be potential biomarkers in HFpEF and OSA. We chose three cytokines to evaluate as possible biomarkers. First, we used IL-6 as it is one of the major pro-inflammatory cytokines [20]. To characterize ongoing anti-inflammatory processes, we assessed serum IL-10 as a prototypical anti-inflammatory cytokine having a protective role against atherosclerosis [21]. For the assessment of angiogenesis, we chose vascular endothelial growth factor (VEGF), which is considered a pivotal regulator of angiogenesis [22]. We sought to evaluate an angiogenesis biomarker as recent research shows that coronary microvascular rarefaction (reduced myocardial capillary density) is a major contributor to diastolic dysfunction in HFpEF [23].

2. Materials and Methods

The study design was a prospective cohort study. The study protocol was approved by the Local Ethics Committee attached to the Clinical Hospital of Rossiyskie Zheleznye Dorogi, approval number 76. All patients provided written informed consent.

2.1. Patient Population

The patients were recruited from a population of male railroad workers from Novosibirsk Oblast attending an annual required medical checkup in the period from 2017 to 2019. People having three risk factors for OSA (BMI > 30, hypertension, snoring) were then referred to the Sleep Apnea Center for further evaluation. In order to improve efficiency of the screening, we did not use common screening tools like STOP-BANG, as in some studies, it lacked good performance in younger patients [24]. Thus, we used two objective risk factors of OSA (high BMI, verified hypertension) with one subjective factor, snoring, which was shown to be more associated with moderate/severe OSA than other risk factors [25].

The people attending the Sleep Apnea Center were invited to participate in the study. The patients were included in this study if, on baseline visit, they had confirmed HFpEF and OSA and fulfilled the following criteria:

2.2. Inclusion Criteria

(1) Symptoms of HF, New York Heart Association (NYHA) Functional Classification class I–II;
(2) Moderate to severe OSA (Apnea–Hypopnea Index (AHA)) > 15 in hour;
(3) Arterial hypertension;
(4) Abdominal obesity (waist circumference ≥ 92 cm, BMI ≥ 30;
(5) N-terminal (NT)-pro hormone BNP (NT-proBNP) > 125 pg/mL.

2.3. Exclusion Criteria

1. Reduced (≥50%) left ventricular ejection fraction;
2. Primary pulmonary hypertension;
3. History of pulmonary embolism with pulmonary hypertension ≥ 45 mm Hg;
4. Severe asthma or COPD;
5. Significant valvular abnormality;
6. Hypertrophic or dilated cardiomyopathy;
7. Coronary artery disease;
8. Persistent atrial fibrillation;
9. Thyroid disease, renal failure with creatinine clearance < 30 mL/m^2;
10. Significant CSA (≥15 episodes of CSA in hour).

To diagnose OSA, all patients underwent polysomnography using the Somnolab2PSG diagnostic system (Weinemann, Germany). We used the Apnea Hypopnea Index (AHI) to assess the severity of OSA. To evaluate serum cytokines levels, we used the enzyme-linked immunosorbent assay (ELISA). Echocardiography was performed in all patients using standard protocol on the EPIQ device (Philips Ultrasound, Inc., Bothell, WA, USA).

2.4. Serum Cytokines

The ELISA analyses were performed using commercial ELISA kits (IL-6, IL-10, and VEGF ELISA-Best, Vector-Best, Novosibirsk, Russia). Based on the manufacturer's instructions, the detection limits for IL-6, IL-10, and VEGF were 0.5 pg/mL, 2.5 pg/mL, and 10 pg/mL, respectively. Concentrations below these thresholds were considered non-detects.

The assessments were made on the baseline visit and on 12 months follow-up. On the baseline visit, we evaluated clinical and demographic parameters, polysomnography, echocardiography, 6-min walk test, and serum cytokine levels. On the follow-up visit, we re-performed all assessments except serum cytokines.

2.5. Outcomes

In this study, we assessed the following outcomes: five-point Major Adverse Cardiac Events (MACE) (primary endpoint) and six-minute walk test (6MWT) (secondary endpoint). The reason for choosing MACE as an outcome is that it is one of the most commonly used composite endpoints (CE) in both epidemiological studies and clinical trials on HF [26]. CE is defined as a "single measure of effect based on a combination of a variety of clinically relevant individual end points" [27]. The are many benefits of using CE as an outcome, including their clinical relevance, ease of use by all patients, capability of unbiased assessment, sensitivity, and low cost [27]. The choice of 6MWT was used because it is available, well-tolerated, and a highly reproducible test of functional capacity in HF patients [28].

2.6. Five-Point MACE

The criteria of five-point MACE were as follows [26]:

- Total death;
- Myocardial infarction;
- Stroke;
- Hospitalization because of HF;

- Revascularization, including percutaneous coronary intervention, and coronary artery bypass graft.

2.7. 6MWT

The 6MWT was performed according to the previous guidelines [29]. We advised patients to not engage in physical activities for 24 h and to not smoke or ingest alcohol for at least 3 h before the test. All 6MWT were performed outdoors along an 18-m corridor.

We instructed all patients to walk as fast as they could along the corridor. They were also informed to slow down their walk or to interrupt it if necessary. During the tests, we monitored the patients verbally by the Borg modified scale every 2 min. By the sixth minute, we requested each subject to stop and the number of runs and the remaining distance of the last run were summed. Before the test and during the first and sixth minute after finishing the test in the sitting position, we measured the blood pressure and H. Two tests were performed with a 30-min interval, and the average of the two 6MWD values was used to best approximate a representative true value.

2.8. Statistical Analysis

Serum cytokine values usually have a substantial proportion of non-detects. For a particular cytokine, we considered the concentrations below the reporting threshold as non-detectable and further treated them as left-censored. We performed Tobit regression, a recommended statistical approach for left-censored data [30] for each cytokine separately. This method allows adjustment for the effects of potential confounders such as age and BMI. Summary statistics and regression equations for the left-censored data were computed using maximum likelihood estimation (MLE) [31]. Using MLE allows data to be analyzed with up to 80% of censored values [32]. Using R package censReg version 0.5-26 [33], both the Student t test and censored regressions with and without potential confounders were performed to test the differences in cytokine levels between the patients with dichotomous outcomes (worsening of heart failure, hospital admission). In the case of continuous outcome (6MWT), we used censored regressions adjusted and non-adjusted for age and BMI. Due to the exploratory nature of this study, we did not perform sample size calculations.

3. Results

Figure 1 shows a flow chart of the participants. We enrolled 77 patients, 71 (92.2%) of them completed the follow-up, six (7.8%) were lost to follow-up, and one (1.4%) died. The patients who completed the study or died during the follow-up were included in the analyses.

The baseline characteristics of included patients are found in Table 1. All patients were middle aged men. Every fourth patient had concomitant COPD, every third patient was a smoker. All included patients had abdominal obesity and medication-controlled hypertension. Most of the patients had general (BMI > 30 kg/m^2) and abdominal (waist circumference > 92 cm) obesity. All of the patients were diagnosed with NYHA grades I-II HF.

During the follow-up, 14 (19.17%) patients developed MACE. The incidence of individual components of MACE are presented in Table 2. As shown in the table, the hospitalization due to HF accounted for all cases of MACEs that occurred during the follow-up. In addition, one patient died due to worsening of HF, and one patient had a stroke.

Table 3 shows a comparison of baseline serum cytokine levels between patients with and without five-point MACE during the follow-up. In patients who developed MACE, the baseline concentrations of IL10 were four-fold higher; this difference was significant using the Student's t test, and remained significant after adjustment for left censoring with Tobit regression modeling and after concomitant adjustment for age and BMI.

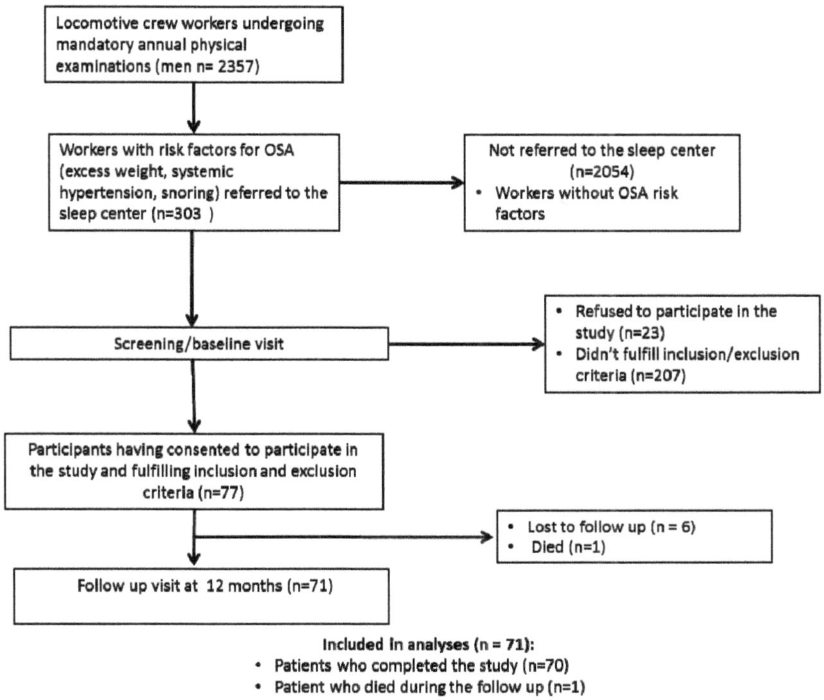

Figure 1. Study flow chart.

Table 1. Baseline characteristics of patients.

n	71
Age, years	46.5 (10.7)
Male	71 (100)
COPD	18 (25.35)
Smoking	27 (38.03)
Dyslipidemia	32 (43.8)
Diabetes mellitus type 2	12 (16.4)
6MWT, m	494.21 (100.8)
BMI, kg/m^2	34.26 (5.72)
NYHA class:	
I	33 (46.48)
II	28 (38.36)
III	0
IV	0
Hypertension	71 (100)
Paroxysmal atrial fibrillation	13 (17.8)

Table 1. *Cont.*

Cardiovascular medication profile	
ACE inhibitors	43 (58.9)
Angiotensin 2 receptor antagonists	35 (47.95)
Beta blockers	45 (61.64)
Diuretics	35 (47.95)
Calcium channel blockers	29 (39.73)

Data are presented as the mean (SD) or *n* (%). HF: heart failure, NYHA: New York Heart Association (NYHA) Functional Classification, VPB: ventricular premature beats, BP: blood pressure, ACE: Angiotensin-converting enzyme, SD: standard deviation.

Table 2. Components of major adverse cardiovascular events during the follow-up (n = 14).

Cardiovascular Event	*n*, %
Total death	1 (7.1)
Myocardial infarction	0 (0)
Stroke	1 (7.1)
Hospitalization because of HF	14 (100)
Revascularization, including percutaneous coronary intervention, and coronary artery bypass graft.	0 (0)

HF: heart failure.

Table 3. Association between baseline serum cytokines and five-point major adverse cardiac events during the follow-up.

Cytokine	Patients without MACE (n = 57)	Patients with MACE (n = 14)	p Value [a]	p Value [b]	p Value [c]
IL6, pg/mL	5.29 (11.70)	6.23 (12.22)	0.790	0.493	0.674
VEGF, pg/mL	339.34 (274.51)	405.02 (178.59)	0.398	0.371	0.5742
IL10, pg/mL	6.85 (8.80)	26.01 (55.88)	0.014	0.0102	0.00574

Cytokine concentrations are presented as mean (SD). [a] Student's *t* test, [b] Tobit regression taking censoring into account, [c] Tobit regression taking censoring into account and adjusting for age and BMI. MACE: Major adverse cardiac events, IL: interleukin, VEGF: Vascular endothelial growth factor, SD: standard deviation.

As shown in Table 4, there were no associations between baseline cytokine levels and changes in 6MWT over time.

Table 4. Associations between baseline serum cytokines and changes in 6MWT after 1 year of follow-up.

	Linear Regression		Non-Adjusted Tobit Regression		Tobit Regression Adjusted for Age and BMI	
	β-Coefficient (95 CI)	p Value	β-Coefficient (95 CI)	p Value	β-Coefficient (95 CI)	p Value
IL-6	−0.01 (−0.1–0)	0.6	−0.02 (−0.1–0)	0.34	0.01 (−0.1–0)	0.63
VEGF	−0.01 (−1.1–1.0)	0.86	0 (−1.2–1.0)	0.9	0.01 (−1.1–1.1)	0.19
IL-10	−0.02 (−0.1–0.1)	0.67	−0.02 (−0.1–0.1)	0.65	−0.13 (−0.1–0.1)	0.59

IL: interleukin, VEGF: Vascular endothelial growth factor, CI: confidence interval.

4. Discussion

In this study, we found increased circulating IL-10 levels in patients with HFpEF and OSA who later developed MACE during the follow-up. There were no links between studied serum cytokines and physical function as measured by 6MWT.

To our knowledge, this was the first study evaluating serum cytokine biomarkers in a subset of patients with HFpEF and comorbid CSA.

Circulating cytokine levels have been assessed in many cross-sectional studies on OSA or HF. A recent meta-analysis showed increased circulating IL-6 levels in patients with OSA [34]. Reduced systemic levels of IL-10 were associated with the severity of OSA and insulin resistance [35] while VEGF levels were found to be frequently elevated in OSA [36]. In a large cohort of HF patients, elevated IL-6 levels were detected in more than 50% of patients [37]. In a recent study, patients with HFrEF exhibited a significant decrease in circulating VEGF [38]. IL-10 levels were also shown to be increased in HF patients [39].

Although many studies cross-sectionally assessed serum cytokines in HF and OS, there has been a lack of research prospectively evaluating prognostic value of IL-6, IL-10, and VEGF.

Our results conflict with the findings of the BIOSTAT-CHF study, showing IL-6 to be a predictor of worse CV outcomes [37]. The differences may be explained by the different study populations. Thus, the BIOSTAT-CHF study encompassed a more heterogeneous group of HF patients while our study focused on a subset of patients with HFpEF and comorbid CSA. It may therefore be hypothesized that different cytokines are involved in particular subtypes of HF.

Our findings are in line with the studies showing impaired CV outcomes in patients with elevated serum IL-10 [40–42]. These data have been difficult to interpret given the biology of IL-10, which is a prototypical anti-inflammatory cytokine [43] and theoretically should promote better cardiovascular outcomes [44]. The possible explanations for the positive associations between impaired CV outcomes and higher IL-10 levels are that the latter may exert unknown harmful action that could overcome any of its favorable anti-inflammatory effects. Alternatively, increased levels of this anti-inflammatory marker may represent a compensatory or counterregulatory mechanism. Any inflammation is accompanied by IL-10 production. The purpose of IL-10 in the setting of inflammation is to diminish excessive inflammation and to prevent unnecessary tissue damage [45]. Thus, higher IL-10 levels found in our study may represent a secondary increase of IL-10 production in response to higher inflammation [41].

From the first glance, the absence of a link between baseline cytokines and longitudinal changes in 6MWT contradicts our findings on elevated IL-10 in patients who had developed MACE. This discrepancy can be explained by suboptimal performance of 6MWT as an outcome in HF patients. Thus, a meta-analysis showed 6MWT improvement in only nine of 47 randomized controlled trials of pharmacological therapy in HF [46].

Our study has several important limitations: it was a single-center study, performed on a rather small sample of middle-aged men. These might restrict the generalizability of our findings.

The strengths of our study are that it had a longitudinal design and that we performed a statistical analyses accounting for left censoring inherent to immunological data.

This study represents the first screening step in evaluating IL-10 as a candidate predictor biomarker of adverse CV outcomes in patients with HFpEF with OSA. The future studies on larger populations will need to confirm these results. If IL-10 is confirmed to be significantly elevated in patients with increased incidence of negative CV outcomes, further research will have to evaluate the effectiveness of serum IL-10 as a predictive biomarker. Determining clinical cutoffs using receiver operating characteristics analysis will be necessary to allow using serum cytokines as biomarkers in clinical practice.

An "ideal" predictive biomarker should be noninvasive, inexpensive, and effective [47]. The first two requirements are probably met for serum cytokines as their measurement requires a draw of just a small amount of venous blood and quantification using ELISA is relatively inexpensive. The effectiveness of the measurement of serum cytokines for the prediction of clinical outcomes is more questionable. Using serum cytokines as a biomarker in clinical settings is challenging due to several reasons. Serum cytokine concentrations might be affected by comorbid diseases. Blood cytokines have a short half-life, their blood levels are relatively low [48]. In addition, there is marked variability in serum cytokine concentrations, including between-day [49] and diurnal [50] variability.

In spite of these challenges, there is still a hope that some cytokines can be used as effective biomarkers for some diseases in the future. The reason for this hope is that there are at least several probable ways to overcome the said difficulties: using a combination of several cytokines to build more accurate predictive models, establishing valid reference cytokine concentrations on large populations of healthy people, and a rigorous standardization of the measurement techniques [51,52].

In conclusion, higher serum concentrations of IL10 in men with HFpEF and CSA are associated with MACE during the follow-up. These findings need to be replicated on a more general population with a larger sample size.

Author Contributions: Conceptualization, A.Y., A.T. and I.S.; methodology, A.Y., A.T., I.S. and V.S.; formal analysis, A.Y. and I.S.; investigation, A.Y., N.Y. and K.K.; data curation, A.Y., N.Y., E.G. and S.S.; A.Y. and I.S., writing—original draft preparation; writing, A.Y., I.S. and V.S.; review and editing, A.Y., A.T., E.G., S.S., N.Y., K.K., V.S. and I.S.; project administration, A.Y., S.S. and K.K. All authors have read and agreed to the published version of the manuscript.

Funding: This research received no external funding.

Institutional Review Board Statement: The study was conducted in accordance with the Declaration of Helsinki, and approved by the Ethics Committee of Clinical Hospital of Rossiyskie Zheleznye Dorogi (protocol #76, 26 February 2019).

Informed Consent Statement: Informed consent was obtained from all subjects involved in the study.

Data Availability Statement: The data presented in this study are available on request from the corresponding author. The data are not publicly available due to privacy and ethical restrictions.

Conflicts of Interest: The authors declare no conflict of interest.

References

1. Mosterd, A.; Hoes, A.W. Clinical epidemiology of heart failure. *Heart* **2007**, *93*, 1137–1146. [CrossRef]
2. Solomon, S.D.; Anavekar, N.; Skali, H.; McMurray, J.J.; Swedberg, K.; Yusuf, S.; Granger, C.B.; Michelson, E.L.; Wang, D.; Pocock, S.; et al. Influence of ejection fraction on cardiovascular outcomes in a broad spectrum of heart failure patients. *Circulation* **2005**, *112*, 3738–3744. [CrossRef]
3. Borlaug, B. Heart failure with preserved ejection fraction: Clinical manifestations and diagnosis. *UpToDate* **2022**.
4. Tromp, J.; Khan, M.A.; Klip, I.T.; Meyer, S.; de Boer, R.A.; Jaarsma, T.; Hillege, H.; van Veldhuisen, D.J.; van der Meer, P.; Voors, A.A. Biomarker Profiles in Heart Failure Patients With Preserved and Reduced Ejection Fraction. *J. Am. Heart Assoc.* **2017**, *6*, e003989. [CrossRef]
5. Naing, P.; Forrester, D.; Kangaharan, N.; Muthumala, A.; Mon Myint, S.; Playford, D. Heart failure with preserved ejection fraction: A growing global epidemic. *Aust. J. Gen. Pract.* **2019**, *48*, 465–471. [CrossRef] [PubMed]
6. Owan, T.E.; Hodge, D.O.; Herges, R.M.; Jacobsen, S.J.; Roger, V.L.; Redfield, M.M. Trends in prevalence and outcome of heart failure with preserved ejection fraction. *N. Engl. J. Med.* **2006**, *355*, 251–259. [CrossRef] [PubMed]
7. Roh, J.; Houstis, N.; Rosenzweig, A. Why Don't We Have Proven Treatments for HFpEF? *Circ. Res.* **2017**, *120*, 1243–1245. [CrossRef]
8. Kheirbek, R.E.; Alemi, F.; Fletcher, R. Heart failure prognosis: Comorbidities matter. *J. Palliat. Med.* **2015**, *18*, 447–452. [CrossRef]
9. Cowie, M.R.; Gallagher, A.M. Sleep Disordered Breathing and Heart Failure: What Does the Future Hold? *JACC Heart Fail.* **2017**, *5*, 715–723. [CrossRef]
10. Bitter, T.; Faber, L.; Hering, D.; Langer, C.; Horstkotte, D.; Oldenburg, O. Sleep-disordered breathing in heart failure with normal left ventricular ejection fraction. *Eur. J. Heart Fail.* **2009**, *11*, 602–608. [CrossRef]
11. Herrscher, T.E.; Akre, H.; Øverland, B.; Sandvik, L.; Westheim, A.S. High prevalence of sleep apnea in heart failure outpatients: Even in patients with preserved systolic function. *J. Card. Fail.* **2011**, *17*, 420–425. [CrossRef] [PubMed]
12. Paulus, W.J.; Zile, M.R. From Systemic Inflammation to Myocardial Fibrosis: The Heart Failure With Preserved Ejection Fraction Paradigm Revisited. *Circ. Res.* **2021**, *128*, 1451–1467. [CrossRef] [PubMed]

13. Fudim, M.; Ashur, N.; Jones, A.D.; Ambrosy, A.P.; Bart, B.A.; Butler, J.; Chen, H.H.; Greene, S.J.; Reddy, Y.; Redfield, M.M.; et al. Implications of peripheral oedema in heart failure with preserved ejection fraction: A heart failure network analysis. *ESC Heart Fail.* **2021**, *8*, 662–669. [CrossRef] [PubMed]
14. Levy, P.; Naughton, M.T.; Tamisier, R.; Cowie, M.R.; Bradley, T.D. Sleep apnoea and heart failure. *Eur. Respir. J.* **2022**, *59*. [CrossRef]
15. Ibrahim, N.E.; Januzzi, J.L., Jr. Established and Emerging Roles of Biomarkers in Heart Failure. *Circ. Res.* **2018**, *123*, 614–629. [CrossRef]
16. Lorenzi-Fillho, G.; Drager, L.F. Sleep apnea: Why should we look for cardiac biomarkers? *J. Clin. Sleep Med.* **2010**, *6*, 420–421. [CrossRef]
17. Ferreira, J.P.; Duarte, K.; Woehrle, H.; Cowie, M.R.; Wegscheider, K.; Angermann, C.; d'Ortho, M.P.; Erdmann, E.; Levy, P.; Simonds, A.K.; et al. Biomarkers in patients with heart failure and central sleep apnoea: Findings from the SERVE-HF trial. *ESC Heart Fail.* **2020**, *7*, 503–511. [CrossRef]
18. Suthahar, N.; Meijers, W.C.; Sillje, H.H.W.; de Boer, R.A. From Inflammation to Fibrosis-Molecular and Cellular Mechanisms of Myocardial Tissue Remodelling and Perspectives on Differential Treatment Opportunities. *Curr. Heart Fail. Rep.* **2017**, *14*, 235–250. [CrossRef]
19. Kheirandish-Gozal, L.; Gozal, D. Obstructive Sleep Apnea and Inflammation: Proof of Concept Based on Two Illustrative Cytokines. *Int. J. Mol. Sci.* **2019**, *20*, 30459. [CrossRef]
20. Fernando, M.R.; Reyes, J.L.; Iannuzzi, J.; Leung, G.; McKay, D.M. The pro-inflammatory cytokine, interleukin-6, enhances the polarization of alternatively activated macrophages. *PLoS ONE* **2014**, *9*, e94188. [CrossRef]
21. Han, X.; Boisvert, W.A. Interleukin-10 protects against atherosclerosis by modulating multiple atherogenic macrophage function. *Thromb. Haemost.* **2015**, *113*, 505–512. [CrossRef] [PubMed]
22. Dabravolski, S.A.; Khotina, V.A.; Omelchenko, A.V.; Kalmykov, V.A.; Orekhov, A.N. The Role of the VEGF Family in Atherosclerosis Development and Its Potential as Treatment Targets. *Int. J. Mol. Sci.* **2022**, *23*, 20931. [CrossRef] [PubMed]
23. Zeng, H.; Chen, J.X. Microvascular Rarefaction and Heart Failure With Preserved Ejection Fraction. *Front. Cardiovasc. Med.* **2019**, *6*, 15. [CrossRef] [PubMed]
24. Lyons, R.; Barbir, L.A.; Owens, R.; Colvonen, P.J. STOP-BANG screener vs objective obstructive sleep apnea testing among younger veterans with PTSD and insomnia: STOP-BANG does not sufficiently detect risk. *J. Clin. Sleep Med.* **2022**, *18*, 67–73. [CrossRef] [PubMed]
25. Borsini, E.; Ernst, G.; Salvado, A.; Bosio, M.; Chertcoff, J.; Nogueira, F.; Nigro, C. Utility of the STOP-BANG components to identify sleep apnea using home respiratory polygraphy. *Sleep. Breath* **2015**, *19*, 1327–1333. [CrossRef]
26. Bosco, E.; Hsueh, L.; McConeghy, K.W.; Gravenstein, S.; Saade, E. Major adverse cardiovascular event definitions used in observational analysis of administrative databases: A systematic review. *BMC Med. Res. Methodol.* **2021**, *21*, 241. [CrossRef]
27. Gomez, G.; Gomez-Mateu, M.; Dafni, U. Informed choice of composite end points in cardiovascular trials. *Circ. Cardiovasc. Qual. Outcomes* **2014**, *7*, 170–178. [CrossRef]
28. Giannitsi, S.; Bougiakli, M.; Bechlioulis, A.; Kotsia, A.; Michalis, L.K.; Naka, K.K. 6-minute walking test: A useful tool in the management of heart failure patients. *Ther. Adv. Cardiovasc. Dis.* **2019**, *13*, 1753944719870084. [CrossRef]
29. ATS statement: Guidelines for the six-minute walk test. *Am. J. Respir. Crit. Care Med.* **2002**, *166*, 111–117. [CrossRef]
30. Ballenberger, N.; Lluis, A.; von Mutius, E.; Illi, S.; Schaub, B. Novel statistical approaches for non-normal censored immunological data: Analysis of cytokine and gene expression data. *PLoS ONE* **2012**, *7*, e46423. [CrossRef]
31. Helsel, D.R. *Nondetects and Data Analysis: Statistics for Censored Environmental Data*; Wiley-Interscience: Hoboken, NJ, USA, 2005; p. xv. 250p.
32. Huynh, T.; Ramachandran, G.; Banerjee, S.; Monteiro, J.; Stenzel, M.; Sandler, D.P.; Engel, L.S.; Kwok, R.K.; Blair, A.; Stewart, P.A. Comparison of methods for analyzing left-censored occupational exposure data. *Ann. Occup. Hyg.* **2014**, *58*, 1126–1142. [CrossRef] [PubMed]
33. Henningsen, A. censReg: Censored Regression (Tobit) Models. Available online: https://CRAN.R-project.org/package=censReg (accessed on 11 December 2019).
34. Imani, M.M.; Sadeghi, M.; Khazaie, H.; Emami, M.; Sadeghi Bahmani, D.; Brand, S. Evaluation of Serum and Plasma Interleukin-6 Levels in Obstructive Sleep Apnea Syndrome: A Meta-Analysis and Meta-Regression. *Front. Immunol.* **2020**, *11*, 1343. [CrossRef]
35. Leon-Cabrera, S.; Arana-Lechuga, Y.; Esqueda-Leon, E.; Teran-Perez, G.; Gonzalez-Chavez, A.; Escobedo, G.; Velazquez Moctezuma, J. Reduced systemic levels of IL-10 are associated with the severity of obstructive sleep apnea and insulin resistance in morbidly obese humans. *Mediat. Inflamm.* **2015**, *2015*, 493409. [CrossRef] [PubMed]
36. Gozal, D.; Lipton, A.J.; Jones, K.L. Circulating vascular endothelial growth factor levels in patients with obstructive sleep apnea. *Sleep* **2002**, *25*, 59–65. [CrossRef] [PubMed]
37. Markousis-Mavrogenis, G.; Tromp, J.; Ouwerkerk, W.; Devalaraja, M.; Anker, S.D.; Cleland, J.G.; Dickstein, K.; Filippatos, G.S.; van der Harst, P.; Lang, C.C.; et al. The clinical significance of interleukin-6 in heart failure: Results from the BIOSTAT-CHF study. *Eur. J. Heart Fail.* **2019**, *21*, 965–973. [CrossRef] [PubMed]
38. Chaar, D.; Dumont, B.; Vulesevic, B.; Neagoe, P.E.; Rakel, A.; Sirois, M.G.; White, M. Neutrophils pro-inflammatory and anti-inflammatory cytokine release in patients with heart failure and reduced ejection fraction. *ESC Heart Fail.* **2021**, *8*, 3855–3864. [CrossRef] [PubMed]

39. Loppnow, H.; Werdan, K.; Werner, C. The enhanced plasma levels of soluble tumor necrosis factor receptors (sTNF-R1; sTNF-R2) and interleukin-10 (IL-10) in patients suffering from chronic heart failure are reversed in patients treated with beta-adrenoceptor antagonist. *Auton. Autacoid. Pharmacol.* **2002**, *22*, 83–92. [CrossRef]
40. Yilmaz, M.I.; Solak, Y.; Saglam, M.; Cayci, T.; Acikel, C.; Unal, H.U.; Eyileten, T.; Oguz, Y.; Sari, S.; Carrero, J.J.; et al. The relationship between IL-10 levels and cardiovascular events in patients with CKD. *Clin. J. Am. Soc. Nephrol.* **2014**, *9*, 1207–1216. [CrossRef]
41. Cavusoglu, E.; Marmur, J.D.; Hojjati, M.R.; Chopra, V.; Butala, M.; Subnani, R.; Huda, M.S.; Yanamadala, S.; Ruwende, C.; Eng, C.; et al. Plasma interleukin-10 levels and adverse outcomes in acute coronary syndrome. *Am. J. Med.* **2011**, *124*, 724–730. [CrossRef]
42. Amir, O.; Rogowski, O.; David, M.; Lahat, N.; Wolff, R.; Lewis, B.S. Circulating interleukin-10: Association with higher mortality in systolic heart failure patients with elevated tumor necrosis factor-alpha. *Isr. Med. Assoc. J.* **2010**, *12*, 158–162.
43. Moore, K.W.; de Waal Malefyt, R.; Coffman, R.L.; O'Garra, A. Interleukin-10 and the interleukin-10 receptor. *Annu. Rev. Immunol.* **2001**, *19*, 683–765. [CrossRef] [PubMed]
44. Tedgui, A.; Mallat, Z. Cytokines in atherosclerosis: Pathogenic and regulatory pathways. *Physiol. Rev.* **2006**, *86*, 515–581. [CrossRef] [PubMed]
45. Couper, K.N.; Blount, D.G.; Riley, E.M. IL-10: The master regulator of immunity to infection. *J. Immunol.* **2008**, *180*, 5771–5777. [CrossRef] [PubMed]
46. Olsson, L.G.; Swedberg, K.; Clark, A.L.; Witte, K.K.; Cleland, J.G. Six minute corridor walk test as an outcome measure for the assessment of treatment in randomized, blinded intervention trials of chronic heart failure: A systematic review. *Eur. Heart J.* **2005**, *26*, 778–793. [CrossRef] [PubMed]
47. Yang, C.; Luo, G.; Cheng, H.; Lu, Y.; Jin, K.; Wang, Z.; Liu, C.; Yu, X. Potential biomarkers to evaluate therapeutic response in advanced pancreatic cancer. *Transl. Cancer Res.* **2018**, *7*, 208–218. [CrossRef]
48. Radonjic-Hoesli, S.; Pavlov, N.; Simon, H.U.; Simon, D. Are blood cytokines reliable biomarkers of allergic disease diagnosis and treatment responses? *J. Allergy Clin. Immunol.* **2022**, *150*, 251–258. [CrossRef] [PubMed]
49. Rose, G.L.; Farley, M.J.; Flemming, N.B.; Skinner, T.L.; Schaumberg, M.A. Between-day reliability of cytokines and adipokines for application in research and practice. *Front. Physiol* **2022**, *13*, 967169. [CrossRef]
50. Altara, R.; Manca, M.; Hermans, K.C.; Daskalopoulos, E.P.; Brunner-La Rocca, H.P.; Hermans, R.J.; Struijker-Boudier, H.A.; Blankesteijn, M.W. Diurnal rhythms of serum and plasma cytokine profiles in healthy elderly individuals assessed using membrane based multiplexed immunoassay. *J. Transl. Med.* **2015**, *13*, 129. [CrossRef]
51. Monastero, R.N.; Pentyala, S. Cytokines as Biomarkers and Their Respective Clinical Cutoff Levels. *Int. J. Inflamm.* **2017**, *2017*, 4309485. [CrossRef]
52. Liu, C.; Chu, D.; Kalantar-Zadeh, K.; George, J.; Young, H.A.; Liu, G. Cytokines: From Clinical Significance to Quantification. *Adv. Sci.* **2021**, *8*, e2004433. [CrossRef]

life

MDPI

Communication

No Difference in Sleep Desaturations Severity between Patients with Wake-Up and Non-Wake-Up Stroke: A PRESS Study Results

Katarína Klobučníková [1], Branislav Kollár [1,*], Matúš Jurík [1], Katarína Valovičová [1], Miroslava Hardoňová [1], Michal Poddaný [2], Miroslav Tedla [3,4], Michal Riant [5], Pavel Klail [5], Peter Turčáni [1] and Pavel Šiarnik [1]

1 1st Department of Neurology, Faculty of Medicine, Comenius University, 814 69 Bratislava, Slovakia
2 Department of Neurology, General Hospital, 031 23 Liptovsky Mikulas, Slovakia
3 Department of ENT and HNS, Faculty of Medicine, University Hospital Bratislava, Comenius University, 814 69 Bratislava, Slovakia
4 Institute of Cancer and Genomic Sciences, University of Birmingham, Birmingham B15 2T8, UK
5 Department of Otorhinolaryngology, University Hospital, Faculty of Medicine in Pilsen, Charles University, 100 00 Prague, Czech Republic
* Correspondence: b.kollar.md@gmail.com; Tel.: +421-2572-90147

Citation: Klobučníková, K.; Kollár, B.; Jurík, M.; Valovičová, K.; Hardoňová, M.; Poddaný, M.; Tedla, M.; Riant, M.; Klail, P.; Turčáni, P.; et al. No Difference in Sleep Desaturations Severity between Patients with Wake-Up and Non-Wake-Up Stroke: A PRESS Study Results. *Life* **2023**, *13*, 517. https://doi.org/10.3390/life13020517

Academic Editor: Larry D. Sanford

Received: 19 December 2022
Revised: 8 February 2023
Accepted: 11 February 2023
Published: 13 February 2023

Abstract: Background: Wake-up stroke (WUS) is a certain type of ischemic stroke in which a patient wakes up with a new neurological deficit due to cerebral ischemia. Sleep-disordered breathing is an independent risk factor for stroke, but the role of nocturnal oxygen desaturation in the pathophysiology of WUS is still insufficiently explored. According to several studies, patients with WUS have a significantly more severe sleep apnea syndrome and lower mean blood oxygen saturation. This study aimed to assess the severity of nocturnal desaturations in acute WUS and non-WUS patients using nocturnal pulse oximetry. Material and Methods: The cohort of 225 consecutive patients with neuroimaging-verified acute cerebral ischemia was prospectively enrolled. For further analyses, 213 subjects with known WUS/non-WUS status were selected (111 males and 102 females, average age 70.4 ±12.9, median baseline NIHSS = 5, median baseline mRS = 3). Patients were divided into the WUS group (n = 45) and the non-WUS group (n = 168). Overnight pulse oximetry was performed within 7 days of the stroke onset and data of both of the studied groups were compared. Results: We found oxygen desaturation index (ODI) in the WUS group was 14.5 vs. 16.6 (p = 0.728) in the non-WUS group, basal O2 saturation was 92.2% vs. 92.5% (p = 0.475), average low O2 saturation was 90.3% vs. 89.6% (p = 0.375), minimal O2 saturation was 79.5% vs. 80.6% (p = 0.563), and time with O2 saturation <90% (T90) was 4.4% vs. 4.7% (p = 0.729). Conclusions: In the studied sample, monitored respiratory parameters including ODI, basal O2 saturation, average low O2 saturation, minimal O2 saturation, and T90 did not significantly differ between groups of WUS and non-WUS patients.

Keywords: ischemic stroke; wake-up stroke; pulse oximetry; sleep-disordered breathing

1. Introduction

Wake-up stroke (WUS) is a certain type of ischemic stroke in which a patient wakes up with a new neurological deficit due to cerebral ischemia. Up to 14–25% of patients with acute stroke suffer WUS [1,2]. As the exact time of stroke onset is unknown, these patients are at risk of losing the potential benefit of revascularization therapy [3]. Sleep-disordered breathing (SDB) is an independent risk factor for stroke, but the role of nocturnal oxygen desaturation in the pathophysiology of WUS is still insufficiently explored [4]. According to several studies, patients with WUS have significantly higher severity of sleep apnea and lower mean blood oxygen saturation [5–7]. A recent meta-analysis of 13 studies' results showed that sleep apnea syndrome (SAS) is significantly higher in WUS patients, with significantly higher apnea-hypopnea index (AHI; 95% confidence interval: 1.38–14.11; p = 0.017) and oxygen desaturation index (ODI; 95% confidence interval: 0.261–7.438; p = 0.035) [8]. Increased risk of stroke in these patients may be mediated by hemodynamic changes due to sleep apnea, oxidative stress, endothelial dysfunction, sympathetic overactivity, metabolic dysregulation, accelerated hypertension, paradoxical embolism, or arrhythmogenesis [9,10]. Polysomnography (PSG) is a gold standard diagnostic tool for the complex estimation of SDB. However, this method is suitable only for cooperating patients. The clear advantage of PSG is the information on the relationship between the occurrence of apneic episodes and the stages of sleep, including the hypnogram, but limits of PSG include that it requires educated medical staff and monitoring devices, and causes patients discomfort. Several studies proved overnight pulse oximetry is a simple method to evaluate nocturnal desaturations in patients with stroke [11–13]. This method can be used as a sensitive screening tool in this population [14–16]. This study aimed to explore and compare the severity of nocturnal desaturations in a "real world" population of patients with acute WUS and non-WUS.

2. Materials and Methods

2.1. Materials

The study subjects were enrolled from the population of stroke subjects who were recruited within the PRESS study (pulse oximetric routine examination of sleep apnea in acute stroke), as previously described [14] (see flowchart in Figure 1).

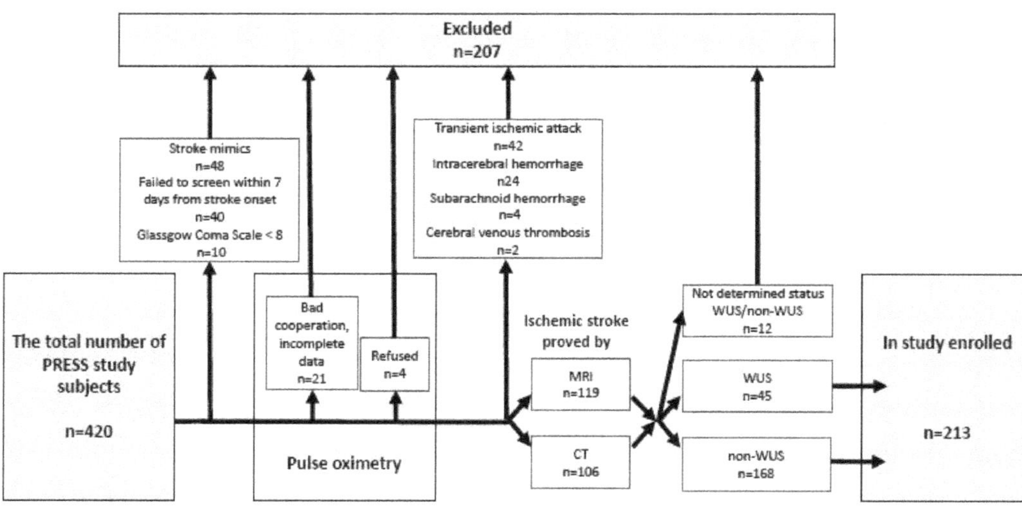

Figure 1. Flowchart.

We prospectively enrolled patients hospitalized in the 1st Department of Neurology, Comenius University Bratislava. The diagnosis of stroke was confirmed clinically, and computed tomography (CT) or magnetic resonance imaging (MRI) was used to confirm the site of the ischemic lesion. Only the subjects with neuroimaging-verified acute brain infarction were enrolled. WUS was defined as the occurrence of new stroke symptoms detected upon waking up from sleep and non-WUS status was set when symptoms were detected during the awake state; for demographic details see Table 1. Baseline stroke severity was measured by the National Institutes of Health Stroke Scale (NIHSS) and modified Rankin Scale (mRS) [17,18].

Inclusion criteria: patients from PRESS study, clinical diagnosis of acute stroke confirmed by CT or MRI, age over 18.

Table 1. Demographic data of the studied sample.

Group	Female				Male				Total				
	n	%	Age		n	%	Age		n	%	Age		
			Mean	Median			Mean	Median			Mean	Median	*p*-Level
WUS	22	21.6	71.8	73.5	23	20.7	67.3	70.0	45	21.1	69.5	72.0	0.611
non-WUS	80	78.4	73.9	76.5	88	79.3	67.6	67.5	168	78.9	70.6	72.0	
Total	102	100	73.4	75.5	111	100	67.6	68.0	213	100	70.4	72	

WUS: Wake-up stroke.

Exclusion criteria: not determined WUS/non-WUS status, diagnosis of acute stroke not-confirmed by CT or MRI, other than ischemic etiology of stroke, stroke mimics, stroke onset > 7 days before pulse oximetry, impairment of consciousness with the Glasgow coma scale < 8 on admission, bad cooperation (including severe disability, agitated confusion, acute cardiac/respiratory comorbidity, acute exacerbation of chronic cardiac/respiratory comorbidity), incomplete clinical data (impossibility to record at least 4 h of continuous pulse oximetric monitoring), refusal to participate.

2.2. Methods

Single-night pulse oximetry monitoring was performed as previously described [14]. Briefly, using the WristOx2 device (model 3150, Nonin Medical, Plymouth, MA, USA) nocturnal monitoring was performed from 10 p.m. to 6 a.m.

The following variables were recorded and analyzed:

Desaturation was defined as an event with a drop of oxygen level >3% and with a duration >10 s.

Oxygen desaturation index (ODI) was defined as the average number of desaturations during 1 h of recording.

Basal O2 saturation (steady-state O2 saturation) was defined as an average of the O2 saturation readings that were not included in any desaturation event.

Average low O2 saturation was defined as the average of the lowest O2 saturations seen over all respiratory events.

Average event duration was defined as the average duration of desaturation and expressed as an average time in seconds per event.

Minimal O2 saturation was defined as the lowest single O2 saturation seen in the recording.

Time with O2 saturation <90% (T90) was defined as the proportion of time period with O2 saturation <90% from the total duration of the recording.

2.3. Statistics

Continuous variables were expressed as means ± standard deviation or median, interquartile range (IQR), minimal and maximal values. Categorical variables were expressed as numbers and proportions (%). To compare variables in the WUS vs. non-WUS population, the Student t test, Mann–Whitney test, and $\chi 2$ test were used for particular variables. SPSS version 21 (SPSS Inc., Chicago, USA) was used for statistical analyses. p values < 0.05 were considered statistically significant.

2.4. Ethics

The study was approved by the Ethics Committee of the Old Town Hospital, University Hospital Bratislava on 15th April 2019. All the patients or their next of kin signed the informed consent before being recruited.

3. Results

In our study, out of 225 patients with neuroimaging verified acute cerebral ischemia, 213 subjects had known WUS/non-WUS status (111 males and 102 females, average age 70.4 ±12.9, median baseline NIHSS = 5, median baseline mRS = 3). This sample was selected for further analyses. We compared the data of the 45 patients WUS group and the 168 patients in the non-WUS group. Between WUS and non-WUS groups, no significant gender and age differences were found (p = 0.443 and p = 0.611).

Stroke severity was measured by the National Institutes of Health Stroke Scale and modified Rankin Scale. Both methods showed insignificant differences in stroke severity in compared groups (p = 0.743, p = 0.763, consecutively).

Oximetry-related indices commonly used in SAS examination were also without disparity. Oxygen desaturation index was 14.5 vs. 16.6 (p = 0.728), basal O2 saturation 92.2% vs. 92.5% (p = 0.475), average low O2 saturation 90.3% vs. 89.6 % (p = 0.375), minimal O2 saturation 79.5% vs. 80.6% (p = 0.563) and T90 was 4.4% vs. 4.7% (p = 0.729). See Table 2 for more details.

Table 2. Comparison of monitored variables in WUS vs. non-WUS stroke groups.

Group		WUS					Non-WUS					p
		N = 45					N = 168					—
Gender	Male (n)	23		51.1%			88		52.4%			0.443
	Female (n)	22		48.9%			80		47.6%			
		Mean	SD	Median	IQR	Range	Mean	SD	Median	IQR	Range	
Age (years)		69.5	12.6	72.0	14.5	28.0–94.0	70.6	13.0	72.0	18.0	23.0–97.0	0.611
NIHSS on admission		6.8	5.3	5.0	6.0	1.0–20.0	6.8	5.7	5.0	8.0	1.0–33.0	0.743
mRS on admission		3.1	1.5	3	2.5	1.0–5.0	3.1	1.5	3.0	3.0	1.0–5.0	0.763
ODI (n/h)		21.6	17.8	14.5	19.6	2.8–69.4	23.2	19.1	16.6	24.4	1.1–74.9	0.728
	ODI ≥ 5 (n)	41		91.1%			148		88.1%			0.570
	ODI ≥ 15 (n)	25		55.5%			77		45.8%			0.246
	ODI ≥ 30 (n)	10		22.2%			46		27.4%			0.485
Average event duration (s)		38.9	15.0	35.6	16.9	21.1–87.1	39.0	12.4	37.0	15.5	20.6–98.4	0.985
Basal O2 saturation (%)		92.2	2.2	92.6	2.3	83.3–95.6	92.5	2.0	92.9	2.5	86.4–97.3	0.475
Average low O2 saturation (%)		90.3	2.3	90.7	2.7	81.6–93.8	89.6	4.9	90.5	3.1	34.9–95.1	0.375
Minimal O2 saturation (%)		79.5	9.6	82.0	10.5	50.0–90.0	80.6	9.2	83.0	9.0	39.0–93.0	0.563
T 90 (%)		13.4	19.4	4.4	21.6	0–95.8	15.7	22.6	4.7	21.6	0–95.8	0.729

IQR: interquartile range, mRS: modified Rankin Scale, NIHSS: National Institutes of Health Stroke Scale, ODI: oxygen desaturation index, SD: standard deviation, T90: time with O2 saturation < 90%, WUS: Wake-up stroke.

4. Discussion

From the total number of 420 patients from PRESS study [14], only 213 subjects fulfilled the inclusion criteria (102 females, 111 males). The studied sample included in the work represented the "real world" population (female age median 75.5; male 68.0 years) and the respondents were divided into two well comparable groups: WUS group with 22 females (48.9%), age median 73.5 years and 23 males (51.1%), age median 70.0 years; versus NWUS-group with 80 females (47.6%), age median 76.5 years and 88 males (52.4%), age median 67.5 years.

A "gold standard" diagnostic tool for sleep apnea, polysomnography, is used in the majority of the studies [19–21], although other simpler sleep monitoring tools of home polygraphy (PG) are also used [22,23]. The use of questionnaires (Berlin questionnaire, STOP-BANG, Epworth Sleepiness Scale) in sleep apnea assessment in stroke subjects is limited by their low sensitivity and specificity [24,25]. PSG, like the most complex investigation, provides information on all important markers of SAS including a hypnogram but it is connected with patient hospitalization-associated stress and is equipment-demanding (educated staff as well as expensive devices). PG gives less information, but it is performed on patients in their home environment. For these reasons, research into less demanding options for sleep monitoring examinations is ongoing [26,27]. Except for the fact that PSG is a technically demanding and costly examination with limited availability, the feasibility of PSG in acute stroke patients represents one of the most challenging issues even in departments with close cooperation between the stroke unit and sleep laboratory [16,19]. All these issues limit routine polysomnographic assessment in a "real world" population of patients with acute stroke. Nocturnal pulse oximetry in stroke subjects represents an alternative diagnostic method with excellent adherence and good sensitivity (90.5%) [14,28,29]. However, lower specificity (75%) may limit the interpretation of the findings described [14].

The high prevalence of sleep apnea in stroke subjects is well-known, and was described in 2005 by Bassetti and Artzt et al. [4,30]. According to a meta-analysis from 2019 by Seiler and his team, it is present in 71% [31]. A large recent work from 2022 by Lu and Liu proved the prevalence of sleep apnea might be even higher in the WUS population. Their meta-analysis included 13 studies and showed that SAS-prevalence is significantly higher in WUS patients [8], with a significantly higher severity of sleep apnea according to AHI and ODI ($p = 0.017$, and $p = 0.035$). This topic was researched by Barreto et al., Haula et al., and Schütz et al. [32–34] and the results of cited studies suggested that preexisting sleep apnea is associated with the occurrence of WUS or even with worse short-term outcomes measured in mRS [33]. Kim et al. in 2018, using the Berlin Questionnaire [10], found that preexisting witnessed or self-recognized sleep apnea was significantly more frequent in WUS when compared to the non-WUS population (28.3% vs. 16.6%, $p = 0.036$). In the same study, the presence of preexisting witnessed or self-recognized sleep apnea was the only risk factor for WUS (adjusted odds ratio = 2.055, 95% confidence interval = 1.035–4.083, $p = 0.040$) [10]. Their conclusion is not in agreement with our study. We did not prove the higher severity of sleep apnea in WUS. In a comparison of WUS and non-WUS groups, we found median ODI 14.5 vs. 16.6 ($p = 0.728$) and T90 4.4% vs. 4.7% ($p = 0.729$). For details, see Table 2. Similarly to our study, also other studies using portable devices failed to find an association between sleep apnea and WUS [32,35]. In our case, we suppose that the population enrolled in the current study may better reflect the "real world" acute stroke population than populations in polysomnographic studies.

Early sleep apnea assessment (nocturnal pulse oximetry evaluation within seven days from the stroke onset) was performed in the current study. It is necessary to admit that Slonková et al. previously described spontaneous improvement of sleep apnea in stroke patients [36], and these findings were supported also by the results of a recent study by Šiarnik et al. [16]. It remains unknown if there is any significant difference in spontaneous OSA improvement between WUS and non-WUS subjects and future research in this field is warranted. However, later sleep apnea assessment could influence the results of our current study. The optimal timing of sleep apnea assessment post-stroke also needs to be elucidated by future prospective studies, because it can influence further therapeutic approaches and treatment adherence in this high-risk population [16].

We are aware of the fact that the low specificity of the pulse oximetry examination in sleep apnea assessment represents one of the main limitations of the current study. PG seems to be a promising alternative diagnostic tool in stroke subjects [22,23]. We have to admit that the PRESS study [14] was not initially designed to search for predictors of nocturnal desaturations. Multiple variables, including stroke comorbidities and therapy, could play a more important role in nocturnal oxygen desaturation than WUS/non-WUS status itself. In the retrospective search of the patients' records, we found none of the 213 analyzed subjects was treated with positive airway pressure therapy pre-stroke. Similarly none of these patients was treated by oxygen therapy during the diagnostic night. However, multiple other sleep apnea-related conditions beyond these should be considered in future prospective studies, including respiratory tract disorders and cardiac disorders.

5. Conclusions

In our studied sample, monitored respiratory parameters including ODI, basal O2 saturation, average low O2 saturation, minimal O2 saturation, and T90 did not significantly differ between groups of WUS and non-WUS patients.

Author Contributions: Conceptualization, P.Š., B.K. and P.T.; methodology, K.K., K.V.; investigation, M.J., K.K.; resources, P.Š., M.P., M.H.; writing—original draft preparation, K.K., P.Š; writing—review and editing, P.T., M.T., M.R., P.K.; supervision, B.K.; project administration, P.Š. All authors have read and agreed to the published version of the manuscript.

Funding: The ITMS: 26240120023 grant supported this work.

Institutional Review Board Statement: This study was approved by the Ethics Committee of the Old Town Hospital, University Hospital Bratislava (Approval Code: 04/2019, Approval Date: 15 April 2019). The study was conducted according to the guidelines of the Declaration of Helsinki.

Informed Consent Statement: Informed consent was obtained from all subjects involved in the study.

Data Availability Statement: The data presented in this study are available on request from the corresponding author.

Conflicts of Interest: The authors declare no conflict of interest.

Abbreviations

AHI	apnea-hypopnea index
CT	computed tomography
MRI	magnetic resonance imaging
mRS	modified Rankin Scale
NIHSS	National Institutes of Health Stroke Scale
ODI	oxygen desaturation index
OSA	obstructive sleep apnea
PG	polygraphy
PSG	polysomnography
SAS	sleep apnea syndrome
SDB	sleep-disordered breathing
T90	time with O_2 saturation < 90%
WUS	wake-up stroke

References

1. Mackey, J.; Kleindorfer, D.; Sucharew, H.; Moomaw, C.J.; Kissela, B.M.; Alwell, K.; Flaherty, M.L.; Woo, D.; Khatri, P.; Adeoye, O.; et al. Population-based study of wake-up strokes. *Neurology* **2011**, *76*, 1662–1667. [CrossRef] [PubMed]
2. Sun, T.; Xu, Z.; Diao, S.S.; Zhang, L.L.; Fang, Q.; Cai, X.Y.; Kong, Y. Safety and cost-effectiveness thrombolysis by diffusion-weighted imaging and fluid attenuated inversion recovery mismatch for wake-up stroke. *Clin. Neurol. Neurosurg.* **2018**, *170*, 47–52. [CrossRef] [PubMed]
3. Tsai, J.P.; Albers, G.W. Wake-Up Stroke: Current Understanding. *Top. Magn. Reson. Imaging TMRI* **2017**, *26*, 97–102. [CrossRef] [PubMed]
4. Arzt, M.; Young, T.; Finn, L.; Skatrud, J.B.; Bradley, T.D. Association of sleep-disordered breathing and the occurrence of stroke. *Am. J. Respir. Crit. Care Med.* **2005**, *172*, 1447–1451. [CrossRef]
5. Šiarnik, P.; Kollár, B.; Čarnická, Z.; Šurda, P.; Klobučníková, K.; Sýkora, M.; Turčáni, P. Association of Sleep Disordered Breathing with Wake-Up Acute Ischemic Stroke: A Full Polysomnographic Study. *J. Clin. Sleep Med. JCSM: Off. Publ. Am. Acad. Sleep Med.* **2016**, *12*, 549–554. [CrossRef]
6. Hsieh, S.W.; Lai, C.L.; Liu, C.K.; Hsieh, C.F.; Hsu, C.Y. Obstructive sleep apnea linked to wake-up strokes. *J. Neurol.* **2012**, *259*, 1433–1439. [CrossRef]
7. Kim, T.J.; Ko, S.B.; Jeong, H.G.; Lee, J.S.; Kim, C.K.; Kim, Y.; Nam, K.; Mo, H.; Joon An, S.; Choi, H.A.; et al. Nocturnal Desaturation in the Stroke Unit Is Associated With Wake-Up Ischemic Stroke. *Stroke* **2016**, *47*, 1748–1753. [CrossRef]

8. Lu, X.; Liu, W.; Wang, H. Investigating the Association between Wake-Up Stroke and Obstructive Sleep Apnea: A Meta-Analysis. *Eur. Neurol.* **2022**, *85*, 14–23. [CrossRef]
9. Ott, S.R.; Korostovtseva, L.; Schmidt, M.; Horvath, T.; Brill, A.K.; Bassetti, C.L. Sleep-disordered breathing: Clinical features, pathophysiology and diagnosis. *Swiss Med. Wkly.* **2017**, *147*, w14436. [CrossRef]
10. Kim, J.S.; Kim, S.; Lee, S.H.; Lee, H.Y.; Lee, S.Y.; Im, K.B. Increased Risk of Ischemic Stroke during Sleep in Apneic Patients. *J. Clin. Neurol.* **2018**, *14*, 174–178. [CrossRef]
11. Lin, S.H.; Branson, C.; Leung, J.; Park, L.; Doshi, N.; Auerbach, S.H. Oximetry as an Accurate Tool for Identifying Moderate to Severe Sleep Apnea in Patients With Acute Stroke. *J. Clin. Sleep Med.* **2018**, *14*, 2065–2073. [CrossRef]
12. Chung, F.; Liao, P.; Elsaid, H.; Islam, S.; Shapiro, C.M.; Sun, Y. Oxygen desaturation index from nocturnal oximetry: A sensitive and specific tool to detect sleep-disordered breathing in surgical patients. *Anesth Analg.* **2012**, *114*, 993–1000. [CrossRef]
13. Linz, D.; Kadhim, K.; Brooks, A.G.; Elliott, A.D.; Hendriks, J.M.L.; Lau, D.H.; Mahajan, R.; Gupta, A.K.; Middeldorp, M.E.; Hohl, M.; et al. Diagnostic accuracy of overnight oximetry for the diagnosis of sleep-disordered breathing in atrial fibrillation patients. *Int. J. Cardiol.* **2018**, *272*, 155–161. [CrossRef]
14. Šiarnik, P.; Jurík, M.; Veverka, J.; Klobučníková, K.; Kollár, B.; Turčáni, P.; Sýkora, M. Pulse oximetric routine examination of sleep apnea in acute stroke (PRESS). *Sleep Med.* **2020**, *73*, 208–212. [CrossRef]
15. Ward, N.R.; Cowie, M.R.; Rosen, S.D.; Roldao, V.; De Villa, M.; McDonagh, T.A.; Simonds, A.; Morrell, M.J. Utility of overnight pulse oximetry and heart rate variability analysis to screen for sleep-disordered breathing in chronic heart failure. *Thorax* **2012**, *67*, 1000–1005. [CrossRef]
16. Šiarnik, P.; Jurík, M.; Valovičová, K.; Klobučníková, K.; Kollár, B.; Poddaný, M.; Rovňák, M.; Turčáni, P.; Sýkora, M. Adherence to positive airway pressure therapy initiated in the early subacute phase of ischemic stroke: A PRESS study follow-up. *Sleep Breath* **2022**, *5*, 1–8. [CrossRef]
17. Brott, T.; Adams, H.P., Jr.; Olinger, C.P.; Marler, J.R.; Barsan, W.G.; Biller, J.; Spilker, J.; Holleran, R.; Eberle, R.; Hertzberg, V.; et al. Measurements of acute cerebral infarction: A clinical examination scale. *Stroke* **1989**, *20*, 864–870. [CrossRef]
18. Sulter, G.; Steen, C.; De Keyser, J. Use of the Barthel index and modified Rankin scale in acute stroke trials. *Stroke* **1999**, *30*, 1538–1541. [CrossRef]
19. Šiarnik, P.; Kollár, B.; Čarnická, Z.; Šutovský, S.; Klobučníková, K.; Turčáni, P. Characteristics of Sleep-Disordered Breathing in Etiologic Subtypes of Minor-to-Moderate Acute Ischemic Stroke. *J. Stroke Cerebrovasc. Dis. Off. J. Natl. Stroke Assoc.* **2015**, *24*, 1087–1093. [CrossRef]
20. Rundo, J.V.; Downey, R., 3rd. Polysomnography. *Handb. Clin. Neurol.* **2019**, *160*, 381–392. [CrossRef]
21. Lee, J.J.; Sundar, K.M. Evaluation and Management of Adults with Obstructive Sleep Apnea Syndrome. *Lung* **2021**, *199*, 87–101. [CrossRef] [PubMed]
22. Slouka, D.; Honnerova, M.; Hosek, P.; Gal, B.; Trcka, O.; Kostlivy, T.; Landsmanova, J.; Havel, D.; Baneckova, M.; Kucera, R. Improved prediction of CPAP failure using T90, age and gender. *J. Appl. Biomed.* **2019**, *17*, 81. [CrossRef] [PubMed]
23. Sánchez-de-la-Torre, M.; Sánchez-de-la-Torre, A.; Bertran, S.; Abad, J.; Duran-Cantolla, J.; Cabriada, V.; Mediano, O.; Masdeu, M.J.; Alonso, M.L.; Masa, J.F.; et al. Effect of obstructive sleep apnoea and its treatment with continuous positive airway pressure on the prevalence of cardiovascular events in patients with acute coronary syndrome (ISAACC study): A randomised controlled trial. *Lancet Respir. Med.* **2020**, *8*, 359–367. [CrossRef] [PubMed]
24. Solecka, Š.; Matler, K.; Kostlivý, T.; Kubec, V.; Tomášková, H.; Betka, J. A Comparison of the Reliability of Five Sleep Questionnaires for the Detection of Obstructive Sleep Apnea. *Life* **2022**, *12*, 1416. [CrossRef]
25. Reuter, H.; Herkenrath, S.; Treml, M.; Halbach, M.; Steven, D.; Frank, K.; Castrogiovanni, A.; Kietzmann, I.; Baldus, S.; Randerath, W.J. Sleep-disordered breathing in patients with cardiovascular diseases cannot be detected by ESS, STOP-BANG, and Berlin questionnaires. *Clin. Res. Cardiol. Off. J. Ger. Card. Soc.* **2018**, *107*, 1071–1078. [CrossRef]
26. Slouka, D.; Windrichova, J.; Rezackova, H.; Houfkova, K.; Kucera, R.; Cerna, V.; Kostlivy, T.; Topolcan, O.; Pesta, M. The potential of miR-499 plasmatic level as a biomarker of obstructive sleep apnea syndrome. *Biomark. Med.* **2021**, *15*, 1011–1019. [CrossRef]
27. Slouka, D.; Kucera, R.; Gal, B.; Betka, J.; Skalova, A. Biomarkers—A possibility for monitoring of obstructive sleep apnea syndrome. *Neuro Endocrinol. Lett.* **2019**, *40*, 85–92.
28. Hornero, R.; Alvarez, D.; Abásolo, D.; del Campo, F.; Zamarrón, C. Utility of approximate entropy from overnight pulse oximetry data in the diagnosis of the obstructive sleep apnea syndrome. *IEEE Trans. Bio-Med. Eng.* **2007**, *54*, 107–113. [CrossRef]
29. Yatsu, S.; Naito, R.; Kasai, T.; Matsumoto, H.; Shitara, J.; Shimizu, M.; Murata, A.; Kato, T.; Suda, S.; Hiki, M.; et al. Influence of sleep-disordered breathing assessed by pulse oximetry on long-term clinical outcomes in patients who underwent percutaneous coronary intervention. *Clin. Res. Cardiol. Off. J. Ger. Card. Soc.* **2018**, *107*, 711–718. [CrossRef]
30. Bassetti, C.L. Sleep and stroke. *Semin. Neurol.* **2005**, *25*, 19–32. [CrossRef]
31. Seiler, A.; Camilo, M.; Korostovtseva, L.; Haynes, A.G.; Brill, A.K.; Horvath, T.; Egger, M.; Bassetti, C.L. Prevalence of sleep-disordered breathing after stroke and TIA: A meta-analysis. *Neurology* **2019**, *92*, e648–e654. [CrossRef]
32. Schütz, S.G.; Lisabeth, L.D.; Gibbs, R.; Shi, X.; Case, E.; Chervin, R.D.; Brown, D.L. Wake-up stroke is not associated with obstructive sleep apnea. *Sleep Med.* **2021**, *81*, 158–162. [CrossRef]
33. Haula, T.M.; Puustinen, J.; Takala, M.; Holm, A. Wake-up strokes are linked to obstructive sleep apnea and worse early functional outcome. *Brain Behav.* **2021**, *11*, e2284. [CrossRef]

34. Barreto, P.R.; Diniz, D.L.O.; Lopes, J.P.; Barroso, M.C.; Daniele, T.; de Bruin, P.F.C.; de Bruin, V.M.S. Obstructive Sleep Apnea and Wake-up Stroke—A 12 Months Prospective Longitudinal Study. *J. Stroke Cerebrovasc. Dis. Off. J. Natl. Stroke Assoc.* **2020**, *29*, 104564. [CrossRef]
35. Brown, D.L.; Li, C.; Chervin, R.D.; Case, E.; Garcia, N.M.; Tower, S.D.; Lisabeth, L.D. Wake-up stroke is not associated with sleep-disordered breathing in women. *Neurology. Clin. Pract.* **2018**, *8*, 8–14. [CrossRef]
36. Slonkova, J.; Bar, M.; Nilius, P.; Berankova, D.; Salounova, D.; Sonka, K. Spontaneous improvement in both obstructive sleep apnea and cognitive impairment after stroke. *Sleep Med.* **2017**, *32*, 137–142. [CrossRef]

life

MDPI

Communication

Lipoprotein Subfractions Associated with Endothelial Function in Previously Healthy Subjects with Newly Diagnosed Sleep Apnea—A Pilot Study

Alzbeta Hluchanova [1], Branislav Kollar [2,*], Katarina Klobucnikova [2], Miroslava Hardonova [2], Michal Poddany [3], Ingrid Zitnanova [4], Monika Dvorakova [4], Katarina Konarikova [4], Miroslav Tedla [5,6], Milan Urik [7,8], Pavel Klail [9], Petr Skopek [9], Peter Turcani [2] and Pavel Siarnik [2]

1 Department of Neurology, University Hospital Bratislava, 85107 Bratislava, Slovakia
2 1st Department of Neurology, Faculty of Medicine, Comenius University, University Hospital, 81499 Bratislava, Slovakia
3 Department of Neurology, General Hospital, 03123 Liptovsky Mikulas, Slovakia
4 Institute of Medical Chemistry, Biochemistry and Clinical Biochemistry, Faculty of Medicine, Comenius University, 81499 Bratislava, Slovakia
5 Department of ENT and HNS, Faculty of Medicine, Comenius University, University Hospital, 81499 Bratislava, Slovakia
6 Institute of Cancer and Genomic Sciences, University of Birmingham, Birmingham B15 2SQ, UK
7 Department of Pediatric Otorhinolaryngology, University Hospital Brno, 61300 Brno, Czech Republic
8 Faculty of Medicine, Masaryk University, 62500 Brno, Czech Republic
9 Department of Otorhinolaryngology, University Hospital in Pilsen, Faculty of Medicine, Charles University, 11000 Pilsen, Czech Republic
* Correspondence: b.kollar.md@gmail.com; Tel.: +421-257290147

Abstract: Background: Obstructive sleep apnea (OSA) activates several pathophysiological mechanisms which can lead to the development of vascular diseases. Endothelial dysfunction (ED) is an initial step in the development of atherosclerosis. The association between ED and OSA has been described in several studies, even in previously healthy subjects. High-density lipoproteins (HDL) were generally considered to be atheroprotective, and low-density lipoprotein (LDL) to be an atherogenic component of lipoproteins. However, recent findings suggest a pro-atherogenic role of small HDL subfractions (8–10) and LDL subfractions (3–7). This study aimed to evaluate the relationship between endothelial function and lipid subfractions in previously healthy OSA subjects. Material and Methods: We prospectively enrolled 205 subjects with sleep monitoring. Plasma levels of triacylglycerols, total cholesterol, LDL, HDL, and their subfractions were assessed. Endothelial function was determined using peripheral arterial tonometry, and reperfusion hyperemia index (RHI) was assessed. Results: Plasma levels of small and intermediate HDL subfractions have statistically significant pro-atherogenic correlations with endothelial function ($p = 0.015$ and $p = 0.019$). In other lipoprotein levels, no other significant correlation was found with RHI. In stepwise multiple linear regression analysis, small HDL (beta = -0.507, $p = 0.032$) was the only significant contributor in the model predicting RHI. Conclusions: In our studied sample, a pro-atherogenic role of small HDL subfractions in previously healthy subjects with moderate-to-severe OSA was proven.

Keywords: lipoprotein subfractions; obstructive sleep apnea; polysomnography

Citation: Hluchanova, A.; Kollar, B.; Klobucnikova, K.; Hardonova, M.; Poddany, M.; Zitnanova, I.; Dvorakova, M.; Konarikova, K.; Tedla, M.; Urik, M.; et al. Lipoprotein Subfractions Associated with Endothelial Function in Previously Healthy Subjects with Newly Diagnosed Sleep Apnea—A Pilot Study. *Life* **2023**, *13*, 441. https://doi.org/10.3390/life13020441

Academic Editors: Larry D. Sanford, David Slouka and Milan Štengl

Received: 28 December 2022
Revised: 27 January 2023
Accepted: 2 February 2023
Published: 4 February 2023

1. Introduction

Obstructive sleep apnea (OSA) is a common disorder that occurs in approximately one-quarter of adults. The prevalence of OSA (apnea/hypopnea index (AHI) over 5) is about 24% in men and 9% in women [1]. Unrefreshing sleep with excessive sleepiness is the most common presenting symptom of OSA. Patients experience snoring as well as awakenings accompanied by gasping or choking. Effective therapeutic approaches include weight loss, positive airway pressure, oral appliances, surgical modification of the pharyngeal soft tissues or facial skeleton, and hypoglossal nerve stimulation in selected cases [2]. OSA is characterized by repeated partial or complete obstructions in the upper airways, leading to large changes in intrathoracic pressure, consequent hemodynamic changes, chronic intermittent hypoxia, and sleep fragmentation [3]. Intermittent hypoxemia with concomitant hypercapnia activates the sympathetic nervous system and contributes to elevation of blood pressure. Repetitive respiratory events increase reactive oxygen species, which may also contribute to vascular disease [3]. OSA activates several other pathophysiological mechanisms which can lead to the development of vascular diseases including endothelial dysfunction, systemic inflammatory response, and impaired glucose and lipid metabolism [4,5]. Endothelial dysfunction is an important underlying pathophysiological mechanism of atherogenesis that occurs at the early stages of vascular diseases and is associated with the occurrence of vascular events in the future [6,7]. Multiple mechanisms underlying OSA, including oxidative stress and systemic inflammation, are also important mechanisms in the pathogenesis of decreased nitric oxide bioavailability and subsequent endothelial dysfunction [8]. Endothelial dysfunction can be assessed by various methods, including peripheral arterial tonometry (PAT). The reperfusion hyperemia index (RHI) measured by PAT is a validated marker of endothelial function and a predictor of future vascular events [9–11]. Endothelial dysfunction is defined by RHI as lower than 1.67. Endothelial dysfunction is present also in previously healthy subjects with newly diagnosed OSA, and OSA is associated with the severity-dependent impairment of endothelial function assessed by RHI in both adults and children [12–14]. Dyslipidemia, defined as an excessive increase in total cholesterol (TC) or triacylglycerols (TAG), with or without a concomitant decrease in high-density lipoproteins (HDL), leads to the acceleration of the atherosclerotic process in predisposed individuals and is also one of the most important risk factors for vascular disease [15]. Along with increased sympathetic nervous system activity, oxidative stress, systemic inflammation, subsequent hypertension and glucose metabolism impairment [16], dyslipidemia is one of the possible mechanisms linking OSA with increased vascular morbidity [17]. The mechanisms of dyslipidemia induced by OSA include the up-regulation of lipoprotein biosynthesis, increased lipolysis, impaired clearance of lipoproteins, lipoprotein peroxidation, and HDL dysfunction [18]; this may represent a pathway by which the increase in cardiovascular risk is mediated [19]. There is also increasing evidence that small dense LDL and HDL subfractions have pro-atherogenic properties [20,21]. This study aimed to evaluate the relationship between endothelial function and lipid subfractions in previously healthy OSA subjects.

2. Materials and Methods

2.1. Study Population

A total of 205 patients were included in this prospective monocentric study. The cohort includes patients who were suspected of suffering from OSA and were hospitalized in the sleep laboratory of the 1st Department of Neurology, Comenius University, and University Hospital Bratislava. The study was approved by the Ethics Committee of the Old Town Hospital, University Hospital Bratislava (with reference number 26/2021). All participants signed informed consent before enrollment.

Inclusion criteria were set up as follows: suspicion of OSA, sleep monitoring by polysomnography, and aged over 18. A detailed search for premorbid diseases was performed and exclusion criteria included: cardiovascular disease, cerebrovascular disease, diabetes mellitus, other endocrinopathies, cancer, or any other chronic diseases. Sleep monitoring by limited polygraphy (PG), bad cooperation, incomplete data, use of any current medication, or smoking belonged to the additional exclusion criteria. For details, see flowchart Figure 1.

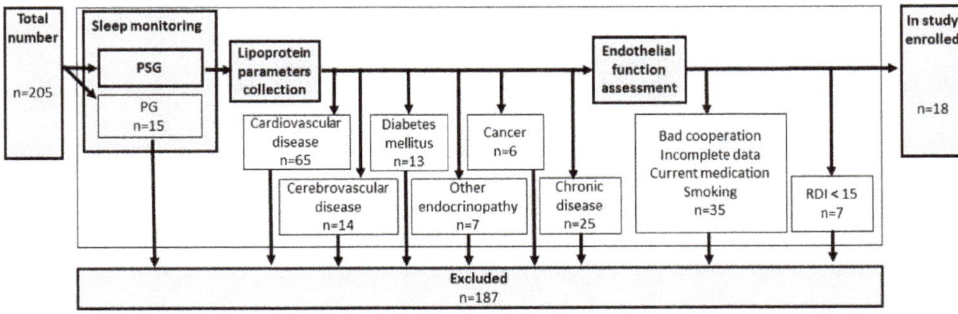

Figure 1. Flowchart of the study.

Finally, apparently healthy 18 male subjects, the "real-world" population with no previous history of sleep apnea, age 47.9 ± 11.5 (Figure 2), body mass index (BMI) 32.3 ± 3.8 kg/m^2, investigated in this tertiary hospital from March 2021 to March 2022 were prospectively enrolled.

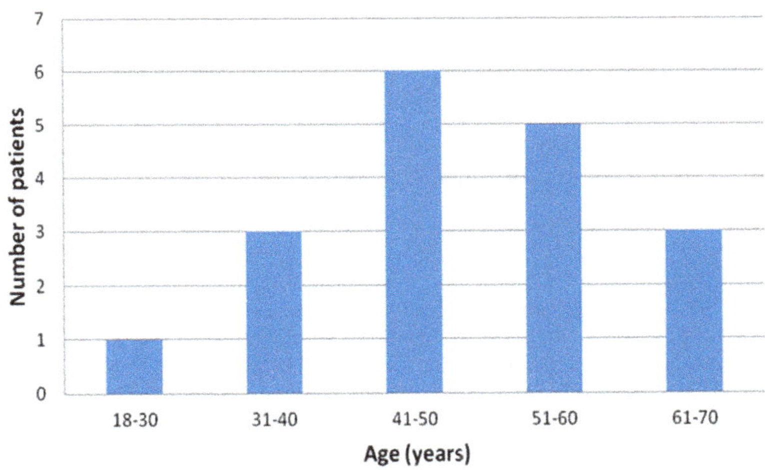

Figure 2. Age distribution of the final enrolled patients' sample.

2.2. Methods

Sleep monitoring

All subjects underwent overnight sleep monitoring. In the study, only patients with polysomnography (Alice 6 device, Philips-Respironics, Murrysville, PA, USA) and recorded respiratory disturbance index (RDI) were enrolled. Only subjects with RDI ≥ 15 were included for further assessment. Other recorded indices included oxygen desaturation index (ODI), arousal index, average nocturnal O_2 saturation, and minimal nocturnal O_2 saturation. Standardized criteria were used for the scoring of sleep characteristics and respiratory events [22].

Monitored variables:

BMI (body mass index)—defined as body weight divided by the square of height; a measure of the degree of obesity.
Apnea—defined as the reduction in airflow ≥90% (or the airflow cessation) lasting >10 s.
Hypopnea—defined as a reduction in airflow ≥30% lasting >10 s with oxygen desaturation ≥3% or arousal.

Respiratory disturbance index (RDI)—defined as an average number of apneas, and hypopneas per 1 h of sleep.
ODI (oxygen desaturation index)—defined as a number of desaturations ≥3% with a duration of >10 s per hour of sleep.
Arousal index—defined as the total number of arousals per hour of sleep.
Average nocturnal O_2 saturation—defined as the mean O_2 saturation during sleep.
Minimal nocturnal O_2 saturation—defined as the lowest single O_2 saturation seen in the recording.
Time with O_2 saturation <90% (T90)—defined by the percentage of sleep time below 90% O_2 saturation.

Lipoprotein parameters

Blood plasma samples were obtained in the morning after polysomnography and after overnight fasting. Blood samples with ethylenediaminetetraacetic acid (EDTA) were collected. Immediately after the collection of plasma samples, levels of TAG, TC, LDL, and HDL were determined in a local certified hospital laboratory with an enzymatic method (Roche Diagnostics, Mannheim, Germany). The quantitative analysis of lipoprotein families and lipoprotein subfractions including very low-density lipoprotein (VLDL), intermediate-density lipoprotein (IDL), and plasma lipoprotein subfractions were analyzed by the Lipoprotein system (Quantimetrix Corp., Redondo Beach, CA, USA) using a polyacrylamide gel electrophoresis [23]. The following subfractions were evaluated: large LDL subfractions 1–2 (which are considered atheroprotective), small dense LDL subfractions 3–7 (which are considered atherogenic), large HDL subfractions 1–3 (which are considered atheroprotective), small dense HDL subfractions 8–10 (which are considered atherogenic), and intermediate HDL subfractions 4–7 (their atherogenic/atheroprotective role remains controversial) [20,21].

Assessment of endothelial function

Endothelial function was assessed by PAT (EndoPAT 2000 device, Itamar Medical Ltd., Caesarea, Israel) as previously described [14]. RHI was calculated as the ratio of the average amplitude of the PAT signal post-to-pre occlusion of the tested arm, normalized to the concurrent signal from the contralateral finger. Calculations were performed using the computer algorithm (software 3.1.2) supplied with the device. RHI value < 1.67 indicated endothelial dysfunction [14,24].

Statistical analysis

Statistical analyses were performed by SPSS ver. 18 (SPSS Inc., Chicago, IL, USA). The results of normally distributed data are expressed as a mean ± standard deviation, and the results of not normally distributed data are expressed as median, interquartile range, minimal and maximal values. Pearson or Spearman correlation coefficients were used to determine the relationships between RHI and the baseline characteristics of the study population. We used stepwise multiple linear regression to create the prediction model and identify the most important contributors to this model. A model with the highest number of significant predictors was chosen. The dependent variable in the model was RHI, independent variables in the model were anthropometric characteristics (age, gender, BMI), sleep characteristics (T90, RDI, ODI, arousal index, average, and minimal nocturnal O_2 saturation), and lipoprotein levels (TAG, TC, LDL, HDL, VLDL, IDL, large LDL, small LDL, large HDL, intermediate HDL, and small HDL). Each model was assessed for the presence of multicollinearity of included variables. The variance inflation factor (VIF) ≥5 was indicative of multicollinearity. The p value < 0.05 was considered statistically significant.

3. Results

The average age of participants was 47.9 years, the average BMI value was 32.31 kg/m^2. All participants were in the category of moderate-to-severe sleep apnea with a mean RDI of

45.7. The average nocturnal O_2 saturation was 87.8, the average minimal nocturnal O_2 saturation was 76.2. None of these monitored variables were statistically significantly correlated with RHI. The results of normally distributed data are expressed as a mean ± standard deviation, and the results of not normally distributed data are expressed as median, interquartile range, minimal and maximal values. For details, see Table 1.

Table 1. Sleep monitoring indices and their correlations with reperfusion hyperemia index.

N = 18	Mean	Median	Standard Deviation	Interquartile Range	Minimal Value	Maximal Value	Correlation		Regression	
							r	p	Beta	p
Age (years)	47.9	49.0	11.5	18	26	66	0.087	0.732	0.139	0.539
BMI (kg/m^2)	32.3	31.9	3.8	6.5	26.6	39.6	−0.278	0.264	−0.048	0.842
RDI (n/h)	45.7	43.6	20.5	35.6	15.8	84.6	−0.088	0.729	−0.092	0.685
ODI (n/h)	39.9	36.8	19.4	29.7	12.8	85.7	−0.061	0.810	−0.069	0.760
Arousal index (n/h)	25.9	24.8	14.0	21.8	7.9	53.7	0.034	0.893	−0.071	0.752
Average nocturnal O_2 sat. (%)	87.8	88.0	3.3	6	81	92	0.283	0.256	0.298	0.176
Minimal nocturnal O_2 sat. (%)	76.2	77.5	8.7	10	55	87	0.284	0.254	0.336	0.124
T90 (%)	11.3	7.0	14.1	11.0	0.2	47.4	−0.425	0.079	−0.192	0.393

BMI: body mass index, RDI: respiratory disturbance index, ODI: oxygen desaturation index, sat.: saturation, T90: Time with O_2 saturation <90%.

In our sample, the average RHI was 1.9. Except for the significant inverse correlation of small HDL and intermediate HDL with RHI ($r = -0.561$, $p = 0.015$ and $r = -0.548$, $p = 0.019$, consecutively), no other significant correlation was found between RHI and other lipoprotein levels (see Table 2). In stepwise multiple linear regression analysis, small HDL (beta = −0.507, $p = 0.032$) was the only significant contributor in the model predicting RHI. The VIF of all variables assessed in this model was <5. For details, see Tables 1 and 2, Figures 3 and 4.

Table 2. Baseline laboratory characteristics and correlations of the reperfusion hyperemia index with variables.

N = 18	Mean	Median	Standard Deviation	Interquartile Range	Minimal Value	Maximal Value	Correlation		Regression	
							r	p	Beta	p
TC (mmol/L)	5.3	5.5	0.8	1.3	3.7	6.4	−0.116	0.646	−0.016	0.947
LDL (mmol/L)	4.0	4.1	0.7	1.0	2.6	5.2	−0.126	0.618	−0.080	0.724
HDL (mmol/L)	1.2	1.1	0.3	0.4	0.7	1.8	−0.086	0.735	−0.031	0.893
TAG (mmol/L)	2.1	1.8	0.9	0.9	1.0	4.4	0.218	0.385	0.279	0.276
VLDL (mg/dL)	56.6	60.5	14.0	23.0	32.0	85.0	0.146	0.564	0.152	0.502
IDL (mg/dL)	45.9	45.0	14.3	27.0	26.0	72.0	−0.281	0.259	−0.106	0.643
Large LDL (mg/dL)	48.3	48.5	12.5	13.8	21.0	70.0	−0.090	0.722	−0.141	0.535
Small LDL (mg/dL)	10.2	6.0	10.1	17.3	0	29.0	0.278	0.264	0.237	0.286
Large HDL (mg/dL)	27.8	22.5	23.8	35.25	1.0	81.0	−0.415	0.087	−0.249	0.309

Table 2. *Cont.*

N = 18	Mean	Median	Standard Deviation	Interquartile Range	Minimal Value	Maximal Value	Correlation		Regression	
							r	*p*	Beta	*p*
Intermediate HDL (mg/dL)	73.0	90.5	50.1	93.75	13.0	141.0	−0.548	***0.019***	−0.224	0.602
Small HDL (mg/dL)	30.2	24.0	21.7	31.5	5.0	74.0	−0.561	***0.015***	***−0.507***	***0.032***
RHI	1.9	1.8	0.4	0.6	1.2	2.7	-	-	-	-

TC: cholesterol, TAG: triacylglycerols, HDL: high-density lipoprotein cholesterol, LDL: low-density lipoprotein cholesterol, VLDL: very low-density lipoprotein, IDL: intermediate-density lipoprotein, RHI: reperfusion hyperemia index. Significant associations in bold.

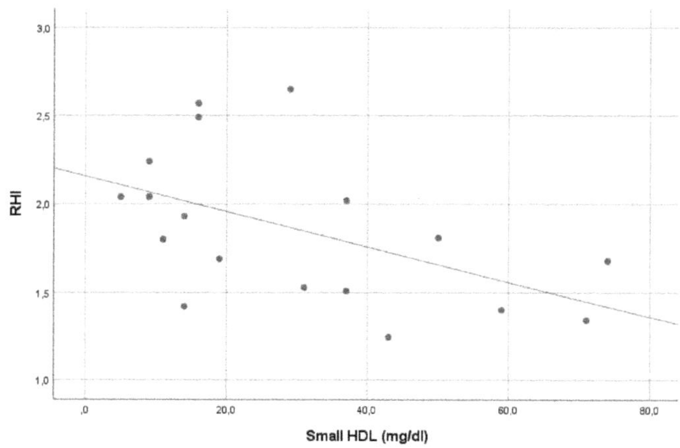

Figure 3. Correlation of small high-density lipoprotein levels with reperfusion hyperemia index (r = −0.561, p = 0.015).

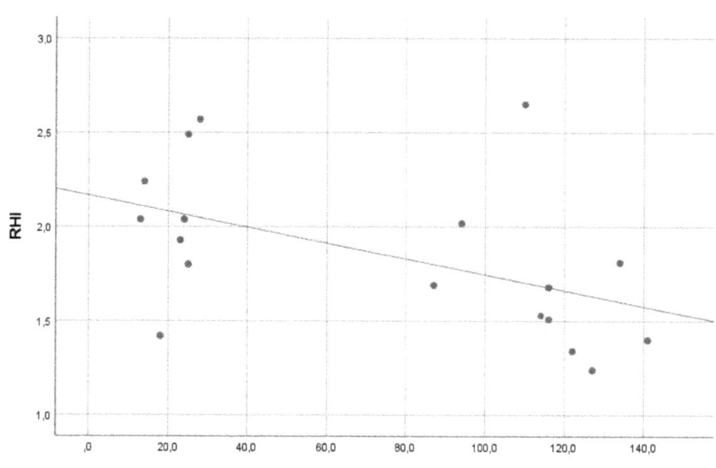

Figure 4. Correlation of intermediate high-density lipoprotein levels with reperfusion hyperemia index (r = −0.548, p = 0.019).

4. Discussion

OSA activates several pathophysiological mechanisms that lead to the development of vascular diseases, as discussed in recent studies by Ott et al. in 2017 [25], Kollar et al. in 2021 [21], and many other authors [26–28]. The potential underlying pathomechanism linking OSA with the development of vascular diseases include endothelial dysfunction [29,30], activation of the sympathetic nervous system [3], oxidative stress [31], metabolic dysregulation [32], activation of inflammatory processes [29,33], and alteration of the coagulation cascade [29]. Levy et al. in 2009 [34] already described, in agreement with Seiler et al. in 2019 [35], that acceleration of atherogenesis could be one of the most important mechanisms involved in the development of vascular diseases in OSA patients. Endothelial dysfunction as the initial step and key process of atherogenesis was proven, for example by Bonetti et al. in 2003 or Gimbrone et al. in 2016 [36,37]. The association between endothelial dysfunction and OSA has been described in several studies [12–14] and was found also in patients with OSA who are not treated for any other diseases, like the findings presented in the work of Ip et al. in 2004 or Siarnik et al. in 2014 [14,38].

Our study was based on a strict exclusion process, and the final sample of apparently healthy subjects of our "real-world" population with no previous history of sleep apnea had only 18 responders; at the time, this was a pilot project on the topic of the possible association of lipoprotein subfractions with endothelial function in previously healthy individuals (with newly diagnosed sleep apnea). Juházs in 2014 [17] suggested dyslipidemia as one of the possible mechanisms linking OSA with increased vascular morbidity. The same opinion was presented by Helkin et al. in 2016 [17,39]. LDL is generally considered to be atherogenic and HDL to be atheroprotective. However, there are increasing data that small LDL (3–7) and small HDL (8–10) subfractions have atherogenic properties [20,21,40]. Our results suggest a pro-atherogenic role of small HDL subfractions in previously healthy subjects with moderate-to-severe OSA. Small HDL (beta = -0.507, $p = 0.032$) was the only significant contributor in the model predicting RHI—a measure of endothelial function, in stepwise multiple linear regression analysis (see Results). The pro-atherogenic lipoprotein phenotype characterized by increased levels of atherogenic lipoprotein subfractions and reduced levels of atheroprotective subfractions was found in individuals with OSA. In this population, significantly lower levels of atheroprotective LDL1 and large HDL subfractions were detected as well as significantly higher levels of atherogenic small dense LDL 3–7 subfractions [21]. Our results are consistent with the findings of previously mentioned studies [17,41–45]. Among our previously healthy patients with newly diagnosed moderate-to-severe OSA, small HDL was the only significant predictor of RHI, suggesting a pro-atherogenic role of small HDL subfractions in this population. We are not aware of any similar study so far.

Although only previously healthy subjects were enrolled and the use of any current medication or smoking belonged to additional exclusion criteria, the co-administration of other supplements as antioxidants and anti-inflammatory substances with beneficial effect on vascular status were not taken in consideration. This fact limits the findings of the current study. For example, the use of micronized purified flavonoid fraction of Rutaceae aurantiae in type 2 diabetic patients proved to reduce the risk of cardiovascular disease [46]. In another study, omega-3 proved its beneficial effect on serum lipid profile and oxidative stress [47]. For endothelial dysfunction, it was found that ramipril [48] as well as febuxostat have a direct ameliorating effect on inflammation and oxidative stress in patients with endothelial dysfunction, which is an important risk factor for cardiovascular diseases [49,50]. Similarly, the oral cholecalciferol effect on vascular calcification and 25(OH)D levels was investigated, which significantly increased serum levels of 25(OH)D and fetuin-A [51].

The enrollment of previously healthy subjects with newly diagnosed sleep apnea is the strength of the current study as it limits the effect of other possible pro-atherogenic confounders. However, a strict exclusion process leads to a small sample of respondents. This pilot study shows that in future large multicenter prospective studies with detailed blood pressure assessment, glycemia testing, a search for anthropometric parameters and physical activity measures, as well as a search for the effect of CPAP on lipoprotein subfractions and endothelial function measures should be beneficial. The effect of lifestyle interventions on LDL and HDL subfractions is known from the results of previous studies [52–55].

5. Conclusions

In our studied sample of previously healthy subjects with moderate-to-severe OSA, the plasma levels of small and intermediate HDL subfractions have statistically significant pro-atherogenic correlations with endothelial function ($p = 0.015$ and $p = 0.019$), but after stepwise multiple linear regression analysis, we conclude that a pro-atherogenic role was proven only for small HDL. Small HDL was the only significant contributor in the model predicting RHI (beta = -0.507, $p = 0.032$). We are not aware of any similar findings so far. No other significant lipoprotein level correlation was found.

Author Contributions: Conceptualization, P.S. (Pavel Siarnik), B.K., and P.T.; methodology, I.Z., K.K. (Katarina Klobucnikova), A.H.; investigation, A.H., I.Z., M.H., M.D., K.K. (Katarina Konarikova); resources, P.S. (Pavel Siarnik), M.D., M.P.; writing—original draft preparation, A.H., P.K., P.S. (Petr Skopek); writing—review and editing, P.T., M.T., M.U.; supervision, B.K.; project administration, I.Z., P.S. (Pavel Siarnik). All authors have read and agreed to the published version of the manuscript.

Funding: VEGA Grant 1/0022/23 supported this work.

Institutional Review Board Statement: This study was approved by the Ethics Committee of the Old Town Hospital, University Hospital Bratislava on 15 February 2021, with reference number 26/2021. The study was conducted according to the guidelines of the Declaration of Helsinki.

Informed Consent Statement: Informed consent was obtained from all subjects involved in the study.

Data Availability Statement: The data presented in this study are available on request from the corresponding author.

Conflicts of Interest: The authors declare no conflict of interest.

Abbreviations

BMI	body mass index
CPAP	continuous positive airway pressure
EDTA	ethylenediaminetetraacetic acid
HDL	high-density lipoprotein cholesterol
IDL	intermediate-density lipoprotein
LDL	low-density lipoprotein cholesterol
ODI	oxygen desaturation index
OSA	obstructive sleep apnea
PAT	peripheral arterial tonometry
PG	polygraphy
PSG	polysomnography
RDI	respiratory disturbance index
RERAs	respiratory effort-related arousal
RHI	reactive hyperemia index
VLDL	very low-density lipoprotein

References

1. Hankey, G.J.; Wardlaw, J.M.; Gorelick, P.B.; Testai, F.D. Hankey's Clinical Neurology preface. In *Hankey's Clinical Neurology*, 2nd ed.; CRC Press: Boca Raton, FL, USA, 2014; p. 9.
2. Gottlieb, D.J.; Punjabi, N.M. Diagnosis and Management of Obstructive Sleep Apnea: A Review. *JAMA* **2020**, *323*, 1389–1400. [CrossRef]
3. Dempsey, J.A.; Veasey, S.C.; Morgan, B.J.; O'Donnell, C.P. Pathophysiology of sleep apnea. *Physiol. Rev.* **2010**, *90*, 47–112. [CrossRef] [PubMed]

4. Drager, L.F.; Polotsky, V.Y.; O'Donnell, C.P.; Cravo, S.L.; Lorenzi-Filho, G.; Machado, B.H. Translational approaches to understanding metabolic dysfunction and cardiovascular consequences of obstructive sleep apnea. *Am. J. Physiol. Heart Circ. Physiol.* **2015**, *309*, H1101–H1111. [CrossRef] [PubMed]
5. Liu, A.; Cardell, J.; Ariel, D.; Lamendola, C.; Abbasi, F.; Kim, S.H.; Holmes, T.H.; Tomasso, V.; Mojaddidi, H.; Grove, K.; et al. Abnormalities of lipoprotein concentrations in obstructive sleep apnea are related to insulin resistance. *Sleep* **2015**, *38*, 793–799. [CrossRef] [PubMed]
6. Hill, J.M.; Zalos, G.; Halcox, J.P.; Schenke, W.H.; Waclawiw, M.A.; Quyyumi, A.A.; Finkel, T. Circulating endothelial progenitor cells, vascular function, and cardiovascular risk. *N. Engl. J. Med.* **2003**, *348*, 593–600. [CrossRef]
7. Perticone, F.; Ceravolo, R.; Pujia, A.; Ventura, G.; Iacopino, S.; Scozzafava, A.; Ferraro, A.; Chello, M.; Mastroroberto, P.; Verdecchia, P.; et al. Prognostic significance of endothelial dysfunction in hypertensive patients. *Circulation* **2001**, *104*, 191–196. [CrossRef]
8. Victor, V.M.; Rocha, M.; Sola, E.; Banuls, C.; Garcia-Malpartida, K.; Hernandez-Mijares, A. Oxidative stress, endothelial dysfunction and atherosclerosis. *Curr. Pharm. Des.* **2009**, *15*, 2988–3002. [CrossRef]
9. Flammer, A.J.; Anderson, T.; Celermajer, D.S.; Creager, M.A.; Deanfield, J.; Ganz, P.; Hamburg, N.M.; Lüscher, T.F.; Shechter, M.; Taddei, S.; et al. The assessment of endothelial function: From research into clinical practice. *Circulation* **2012**, *126*, 753–767. [CrossRef]
10. Rubinshtein, R.; Kuvin, J.T.; Soffler, M.; Lennon, R.J.; Lavi, S.; Nelson, R.E.; Pumper, G.M.; Lerman, L.O.; Lerman, A. Assessment of endothelial function by non-invasive peripheral arterial tonometry predicts late cardiovascular adverse events. *Eur. Heart J.* **2010**, *31*, 1142–1148. [CrossRef]
11. Matsuzawa, Y.; Sugiyama, S.; Sugamura, K.; Nozaki, T.; Ohba, K.; Konishi, M.; Matsubara, J.; Sumida, H.; Kaikita, K.; Kojima, S.; et al. Digital assessment of endothelial function and ischemic heart disease in women. *J. Am. Coll. Cardiol.* **2010**, *55*, 1688–1696. [CrossRef]
12. Azuma, M.; Chihara, Y.; Yoshimura, C.; Murase, K.; Hamada, S.; Tachikawa, R.; Matsumoto, T.; Inouchi, M.; Tanizawa, K.; Handa, T.; et al. Association between endothelial function (assessed on reactive hyperemia peripheral arterial tonometry) and obstructive sleep apnea, visceral fat accumulation, and serum adiponectin. *Circ. J. Off. J. Jpn. Circ. Soc.* **2015**, *79*, 1381–1389. [CrossRef] [PubMed]
13. Kheirandish-Gozal, L.; Etzioni, T.; Bhattacharjee, R.; Tan, H.L.; Samiei, A.; Molero Ramirez, H.; Abu Eta, B.; Pillar, G. Obstructive sleep apnea in children is associated with severity-dependent deterioration in overnight endothelial function. *Sleep Med.* **2013**, *14*, 526–531. [CrossRef]
14. Siarnik, P.; Carnicka, Z.; Krizova, L.; Wagnerova, H.; Sutovsky, S.; Klobucnikova, K.; Kollar, B.; Turcani, P.; Sykora, M. Predictors of impaired endothelial function in obstructive sleep apnea syndrome. *Neuro Endocrinol. Lett.* **2014**, *35*, 142–148. [PubMed]
15. Third Report of the National Cholesterol Education Program (NCEP) Expert Panel on Detection, Evaluation, and Treatment of High Blood Cholesterol in Adults (Adult Treatment Panel III) final report. *Circulation* **2002**, *106*, 3143–3421. [CrossRef]
16. Gottlieb, D.J. Sleep Apnea and Cardiovascular Disease. *Curr. Diabetes Rep.* **2021**, *21*, 64. [CrossRef]
17. Juhász, J. Dyslipidemia: Another brick in the wall. A feasible link in the OSA-cardiovascular disease axis. *Sleep Breath. Schlaf. Atm.* **2014**, *18*, 5–6. [CrossRef]
18. Šiarnik, P.; Klobučníková, K.; Mucska, I.; Černá, K.; Kollár, B.; Turčáni, P. Obstructive sleep apnea and dyslipidemia. *Vnitr. Lek.* **2018**, *64*, 934–938. [CrossRef]
19. Drager, L.F.; Jun, J.; Polotsky, V.Y. Obstructive sleep apnea and dyslipidemia: Implications for atherosclerosis. *Curr. Opin. Endocrinol. Diabetes Obes.* **2010**, *17*, 161–165. [CrossRef] [PubMed]
20. Oravec, S.; Dostal, E.; Dukát, A.; Gavorník, P.; Kucera, M.; Gruber, K. HDL subfractions analysis: A new laboratory diagnostic assay for patients with cardiovascular diseases and dyslipoproteinemia. *Neuro Endocrinol. Lett.* **2011**, *32*, 502–509. [PubMed]
21. Kollar, B.; Siarnik, P.; Hluchanova, A.; Klobucnikova, K.; Mucska, I.; Turcani, P.; Paduchova, Z.; Katrencikova, B.; Janubova, M.; Konarikova, K.; et al. The impact of sleep apnea syndrome on the altered lipid metabolism and the redox balance. *Lipids Health Dis.* **2021**, *20*, 175. [CrossRef]
22. Berry, R.B.; Brooks, R.; Gamaldo, C.E.; Harding, S.M.; Marcus, C.; Vaughn, B.V. The AASM manual for the scoring of sleep and associated events. *Rules Terminol. Tech. Specif. Darien Ill. Am. Acad. Sleep Med.* **2012**, *176*, 2012.
23. Hoefner, D.M.; Hodel, S.D.; O'Brien, J.F.; Branum, E.L.; Sun, D.; Meissner, I.; McConnell, J.P. Development of a rapid, quantitative method for LDL subfractionation with use of the Quantimetrix Lipoprint LDL System. *Clin. Chem.* **2001**, *47*, 266–274. [CrossRef]
24. Moerland, M.; Kales, A.J.; Schrier, L.; van Dongen, M.G.; Bradnock, D.; Burggraaf, J. Evaluation of the EndoPAT as a Tool to Assess Endothelial Function. *Int. J. Vasc. Med.* **2012**, *2012*, 904141. [CrossRef] [PubMed]
25. Ott, S.R.; Korostovtseva, L.; Schmidt, M.; Horvath, T.; Brill, A.K.; Bassetti, C.L. Sleep-disordered breathing: Clinical features, pathophysiology and diagnosis. *Swiss. Med. Wkly.* **2020**, *147*, w14436 . [CrossRef]
26. Slouka, D.; Windrichova, J.; Rezackova, H.; Houfkova, K.; Kucera, R.; Cerna, V.; Kostlivy, T.; Topolcan, O.; Pesta, M. The potential of miR-499 plasmatic level as a biomarker of obstructive sleep apnea syndrome. *Biomark. Med.* **2021**, *15*, 1011–1019. [CrossRef] [PubMed]
27. Javaheri, S.; Barbe, F.; Campos-Rodriguez, F.; Dempsey, J.A.; Khayat, R.; Malhotra, A.; Martinez-Garcia, M.A.; Mehra, R.; Pack, A.I.; Polotsky, V.Y.; et al. Sleep Apnea: Types, Mechanisms, and Clinical Cardiovascular Consequences. *J. Am. Coll. Cardiol.* **2017**, *69*, 841–858. [CrossRef]

28. Tuleta, I.; França, C.N.; Wenzel, D.; Fleischmann, B.; Nickenig, G.; Werner, N.; Skowasch, D. Intermittent Hypoxia Impairs Endothelial Function in Early Preatherosclerosis. *Adv. Exp. Med. Biol.* **2015**, *858*, 1–7 . [CrossRef] [PubMed]
29. Shamsuzzaman, A.S.; Gersh, B.J.; Somers, V.K. Obstructive sleep apnea: Implications for cardiac and vascular disease. *JAMA* **2003**, *290*, 1906–1914. [CrossRef]
30. Orrù, G.; Storari, M.; Scano, A.; Piras, V.; Taibi, R.; Viscuso, D. Obstructive Sleep Apnea, oxidative stress, inflammation and endothelial dysfunction-An overview of predictive laboratory biomarkers. *Eur. Rev. Med. Pharm. Sci.* **2020**, *24*, 6939–6948.
31. Stanek, A.; Brożyna-Tkaczyk, K.; Myśliński, W. Oxidative Stress Markers among Obstructive Sleep Apnea Patients. *Oxid. Med. Cell Longev.* **2021**, *2021*, 9681595 . [CrossRef]
32. Trakada, G.; Lombardi, C.; Knechtle, B. Editorial: The Complex Interaction Between Biological, Metabolic and Neurologic Dysregulation in Obstructive Sleep Apnea. *Front. Psychiatry.* **2021**, *12*, 770930 . [CrossRef]
33. Bhatt, S.P.; Guleria, R.; Kabra, S.K. Metabolic alterations and systemic inflammation in overweight/obese children with obstructive sleep apnea. *PLoS ONE* **2021**, *16*, e0252353. [CrossRef] [PubMed]
34. Lévy, P.; Pépin, J.-L.; Arnaud, C.; Baguet, J.-P.; Dematteis, M.; Mach, F. Obstructive sleep apnea and atherosclerosis. *Prog. Cardiovasc. Dis.* **2009**, *51*, 400–410. [CrossRef] [PubMed]
35. Seiler, A.; Camilo, M.; Korostovtseva, L.; Haynes, A.G.; Brill, A.K.; Horvath, T.; Egger, M.; Bassetti, C.L. Prevalence of sleep-disordered breathing after stroke and TIA: A meta-analysis. *Neurology* **2019**, *92*, e648–e654. [CrossRef]
36. Bonetti, P.O.; Lerman, L.O.; Lerman, A. Endothelial dysfunction: A marker of atherosclerotic risk. *Arterioscler. Thromb. Vasc. Biol.* **2003**, *23*, 168–175. [CrossRef]
37. Gimbrone, M.A., Jr.; García-Cardeña, G. Endothelial Cell Dysfunction and the Pathobiology of Atherosclerosis. *Circ. Res.* **2016**, *118*, 620–636. [CrossRef]
38. Ip, M.S.; Tse, H.-F.; Lam, B.; Tsang, K.W.; Lam, W.-K. Endothelial function in obstructive sleep apnea and response to treatment. *Am. J. Respir. Crit. Care Med.* **2004**, *169*, 348–353. [CrossRef] [PubMed]
39. Helkin, A.; Stein, J.J.; Lin, S.; Siddiqui, S.; Maier, K.G.; Gahtan, V. Dyslipidemia Part 1–Review of Lipid Metabolism and Vascular Cell Physiology. *Vasc Endovasc. Surg.* **2016**, *50*, 107–118. [CrossRef]
40. Oravec, S.; Dukat, A.; Gavornik, P.; Caprnda, M.; Kucera, M.; Ocadlik, I. Contribution of the atherogenic lipoprotein profile to the development of arterial hypertension. *Bratisl. Lek. Listy* **2011**, *112*, 4–7.
41. Oravec, S.; Krivosikova, Z.; Krivosik, M.; Gruber, K.; Gruber, M.; Dukát, A.; Gavorník, P. Lipoprotein profile in patients who survive a stroke. *Neuro Endocrinol. Lett.* **2011**, *32*, 496–501.
42. Oravec, S.; Gruber, K.; Dostal, E.; Mikl, J. Hyper-betalipoproteinemia LDL 1,2: A newly identified nonatherogenic hypercholesterolemia in a group of hypercholesterolemic subjects. *Neuro Endocrinol. Lett.* **2011**, *32*, 322–327. [PubMed]
43. Šiarnik, P.; Čarnická, Z.; Krivošíková, Z.; Klobučníková, K.; Žitňanová, I.; Kollár, B.; Sýkora, M.; Turčáni, P. Association of lipoprotein subfractions with endothelial function and arterial stiffness in acute ischemic stroke. *Scand. J. Clin. Lab. Invest.* **2017**, *77*, 36–39. [CrossRef]
44. Sopkova, Z.; Berneis, K.; Rizzo, M.; Spinas, G.A.; Dorkova, Z.; Tisko, R.; Tkacova, R. Size and subclasses of low-density lipoproteins in patients with obstructive sleep apnea. *Angiology* **2012**, *63*, 617–621. [CrossRef] [PubMed]
45. Meszaros, M.; Bikov, A. Obstructive Sleep Apnoea and Lipid Metabolism: The Summary of Evidence and Future Perspectives in the Pathophysiology of OSA-Associated Dyslipidaemia. *Biomedicines* **2022**, *10*, 2754. [CrossRef] [PubMed]
46. Rizk, S.M.; Sabri, N.A. Evaluation of clinical activity and safety of Daflon 500 mg in type 2 diabetic female patients. *Saudi Pharm. J. SPJ Off. Publ. Saudi Pharm. Soc.* **2009**, *17*, 199–207. [CrossRef]
47. Ateya, A.M.; Sabri, N.A.; El Hakim, I.; Shaheen, S.M. Effect of Omega-3 Fatty Acids on Serum Lipid Profile and Oxidative Stress in Pediatric Patients on Regular Hemodialysis: A Randomized Placebo-Controlled Study. *J. Ren. Nutr. Off. J. Counc. Ren. Nutr. Natl. Kidney Found.* **2017**, *27*, 169–174. [CrossRef]
48. Ateya, A.M.; El Hakim, I.; Shahin, S.M.; El Borolossy, R.; Kreutz, R.; Sabri, N.A. Effects of Ramipril on Biomarkers of Endothelial Dysfunction and Inflammation in Hypertensive Children on Maintenance Hemodialysis: The SEARCH Randomized Placebo-Controlled Trial. *Hypertension* **2022**, *79*, 1856–1865. [CrossRef]
49. Alshahawey, M.; Shaheen, S.M.; Elsaid, T.; Sabri, N.A. Effect of febuxostat on oxidative stress in hemodialysis patients with endothelial dysfunction: A randomized, placebo-controlled, double-blinded study. *Int. Urol. Nephrol.* **2019**, *51*, 1649–1657. [CrossRef]
50. Alshahawey, M.; Shahin, S.M.; Elsaid, T.W.; Sabri, N.A. Effect of Febuxostat on the Endothelial Dysfunction in Hemodialysis Patients: A Randomized, Placebo-Controlled, Double-Blinded Study. *Am. J. Nephrol.* **2017**, *45*, 452–459. [CrossRef]
51. Alshahawey, M.; El Borolossy, R.; El Wakeel, L.; Elsaid, T.; Sabri, N.A. The impact of cholecalciferol on markers of vascular calcification in hemodialysis patients: A randomized placebo controlled study. *Nutr. Metab. Cardiovasc. Dis. NMCD* **2021**, *31*, 626–633. [CrossRef]
52. Dutheil, F.; Walther, G.; Chapier, R.; Mnatzaganian, G.; Lesourd, B.; Naughton, G.; Verney, J.; Fogli, A.; Sapin, V.; Duclos, M.; et al. Atherogenic subfractions of lipoproteins in the treatment of metabolic syndrome by physical activity and diet-the RESOLVE trial. *Lipids Health Dis.* **2014**, *13*, 112. [CrossRef] [PubMed]
53. Bajer, B.; Rádiková, Ž.; Havranová, A.; Žitňanová, I.; Vlček, M.; Imrich, R.; Sabaka, P.; Bendžala, M.; Penesová, A. Effect of 8-weeks intensive lifestyle intervention on LDL and HDL subfractions. *Obes. Res. Clin. Pract.* **2019**, *13*, 586–593. [CrossRef] [PubMed]

54. Habanova, M.; Holovicova, M.; Scepankova, H.; Lorkova, M.; Gazo, J.; Gazarova, M.; Pinto, C.A.; Saraiva, J.A.; Estevinho, L.M. Modulation of Lipid Profile and Lipoprotein Subfractions in Overweight/Obese Women at Risk of Cardiovascular Diseases through the Consumption of Apple/Berry Juice. *Antioxidants* **2022**, *11*, 2239. [CrossRef] [PubMed]
55. Xu, H.; Yi, H.; Guan, J.; Yin, S. Effect of continuous positive airway pressure on lipid profile in patients with obstructive sleep apnea syndrome: A meta-analysis of randomized controlled trials. *Atherosclerosis* **2014**, *234*, 446–453. [CrossRef] [PubMed]

life

MDPI

Review

A Scoping Review of Sleep Apnea: Where Do We Stand?

Rahim Hirani [1,2] and Abbas Smiley [1,2,*]

1 School of Medicine, New York Medical College, Valhalla, NY 10595, USA
2 Department of Surgery, Westchester Medical Center, New York, NY 10595, USA
* Correspondence: abbaset4@gmail.com

Abstract: Obstructive sleep apnea (OSA), a condition in which there is a recurrent collapse of the upper airway while sleeping, is a widespread disease affecting 5% to 10% people worldwide. Despite several advances in the treatment modalities for OSA, morbidity and mortality remain a concern. Common symptoms include loud snoring, gasping for air during sleep, morning headache, insomnia, hypersomnia, attention deficits, and irritability. Obese individuals, male gender, older age (65+), family history, smoking, and alcohol consumption are well recognized risk factors of OSA. This condition holds the ability to increase inflammatory cytokines, cause metabolic dysfunction, and increase the sympathetic output, all of which exacerbate OSA due to their effect on the cardiovascular system. In this review, we discuss its brief history, risk factors, complications, treatment modalities, and the role of clinicians in curbing its risk.

Keywords: sleep apnea; morbidity; upper airway; risk factors; pathogenesis; CPAP

Citation: Hirani, R.; Smiley, A. A Scoping Review of Sleep Apnea: Where Do We Stand? *Life* **2023**, *13*, 387. https://doi.org/10.3390/life13020387

Academic Editors: David Slouka and Milan Štengl

Received: 8 December 2022
Revised: 11 January 2023
Accepted: 28 January 2023
Published: 31 January 2023

1. Introduction

Obstructive sleep apnea (OSA) is an illness of the upper airway that causes intermittent cessation in ventilation, causing hypoxia and hypercapnia due to the periodic collapse of the trachea [1,2]. It is estimated that around 936 million adults aged 30–69 years, both men and women, suffer from mild to severe OSA, and approximately 425 million adults aged 30–69 years have moderate to severe OSA globally [3], making it a worldwide concern for healthcare providers and health delivery systems. Along with a relatively higher incidence, it is particularly concerning given that a high number of patients remain undiagnosed, resulting in an unpredictable epidemiology of this disease [4].

Although various diagnostic and therapeutic advances have been made over the decades to better manage this illness, morbidity and mortality rates remain high; OSA is reportedly associated with a 1.9-times-higher risk in all-cause mortality and 2.65-times-higher risk of mortality related to cardiovascular issues [5,6]. Along with cardiovascular issues, OSA has been shown to be independently associated with stroke, cancer, and many other illnesses [6,7]. Therefore, further studies are required to better understand the associated risk factors and devise novel treatment modalities.

The economic burden of undiagnosed sleep apnea in the US is approximately USD 150 billion, and experts suggest an annual saving of around USD 100 billion if we are able to diagnose and treat every patient in the US who has OSA [8]. Therefore, along with the advances in the management of this illness, it is also important to investigate the causes and solutions in order to increase the diagnosis rate and risk factors associated with OSA.

In this article, we review the etiology and various established risk factors associated with OSA. Additionally, we review the evidence and discuss various treatment modalities used to better manage patients with OSA. Lastly, we review some complications and the role of healthcare providers to reduce associated mortality and morbidity rates.

2. Sleep Apnea: A Brief History and Risk Factors

Although the official naming and discovery of sleep apnea reportedly occurred in the 1960s, it is by no means a new disorder [9]. While it is true that it is only receiving relatively more attention due to further advances in diagnosing and managing the disease, the symptoms first appeared approximately 2000 years ago and were lumped together using the term "Pickwickian syndrome" in the 19th century [10,11]. For years, the initial focus was on the process of understanding the intermittent closure of the upper airway; however, the late 1960s brought a fresh perspective to observe the various symptoms and risk factors at the same time, though limited by methodological difficulties at the time [9,12]. Research studies were mainly conducted via observing dogs and treating the condition with tracheotomies [13]. Though an earlier concept of continuous positive airway pressure (CPAP) using a customized mask in the 1970 and 1980s further advanced modern management [14]. To date, polysomnography, including electrocardiogram, sleep staging, electromyogram, and electroencephalogram, is the gold standard for diagnosing OSA, while home sleep apnea testing (HSAT) is an alternative method with some limitations [15,16].

There are four reported endotypes of OSA: loop gain, upper-airway collapsibility, arousal threshold, and upper-airway dilator muscle response, which is also knows as compensation [17]. Loop gain is basically the ventilatory response-to-disturbance ratio estimated by ventilation characteristics during obstructed breathing episodes [18]. It is usually noted as a drop in CPAP, suggesting a decrease in ventilation as compared to the holding pressure. This drop in ventilation leads to the accumulation of CO2, and thus, an increase in ventilatory drive. This increase can be estimated by measuring the ventilatory overshoot from the holding pressure of CPAP, providing a ratio for response and disturbance in ventilation [17,19,20]. Upper-airway collapsibility measures the propensity for collapse as reported in patients among OSA. Although there are several techniques to measure this mechanistic variable, negative pressure pulses seem to provide a reliable estimate as it is rapid and thus less likely to be influenced by external behaviors [21]. Arousal threshold is essentially the compensatory drive of ventilation that produces arousal [18]. This variable is a measure of the propensity to wake up from sleep given the changes in negative intrathoracic pressure [18,20,22]. Lastly, patients with ineffective upper-airway dilator muscle endotypes have a decreased tone of dilator muscles, particularly genioglossus, the largest extrinsic muscle of the tongue [23]. The absence or presence of these endotypes can manifest differently in individuals that can subsequently have an impact on the severity of the disease. These individual endotypes can also be targeted individually, or in combination, using various techniques that are briefly discussed later in this review article.

Over the years, there have been several advances, including the identification of symptoms (Figure 1) and various risk factors, to better diagnose and manage individual conditions. Loud snoring, gasping for air during sleep, xerostomia, insomnia, hypersomnia, nocturnal choking, and attention deficits are some of the many symptoms that can be observed in patients with OSA [24–27]. Some of the most important risk factors of sleep apnea are discussed in this review (Figure 2).

2.1. *Obesity*

Obesity has been identified as one of the main components contributing to OSA [28,29]. Many correlations have been established between weight, BMI, waist-to-hip ratio, neck circumference, and severity of OSA. The sleep heart healthy study was one of the landmark studies that established the same ideas; the study showed an increase in the apnea-hypopnea index (AHI) by approximately five-fold in men and two-fold in women over the course of their study [30]. Peppard et al. in their population-based prospective cohort study of 690 randomly selected Wisconsin residents also demonstrated a six-fold increase in the odds of developing OSA with a mere 10% increase in weight, while weight loss resulted in decreasing severity among patients [31], suggesting a reciprocal relationship between these two variables.

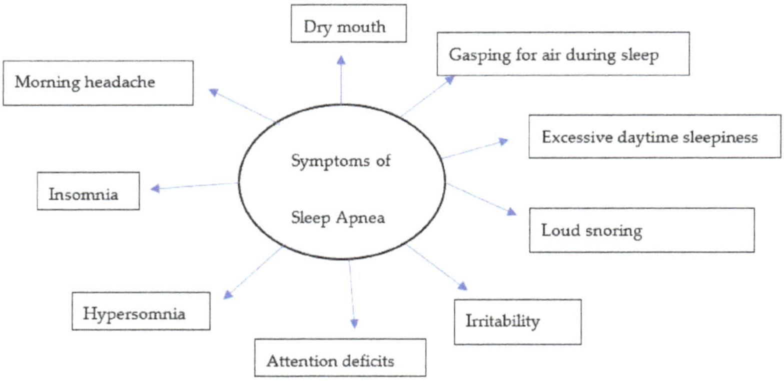

Figure 1. A model for common symptoms of sleep apnea.

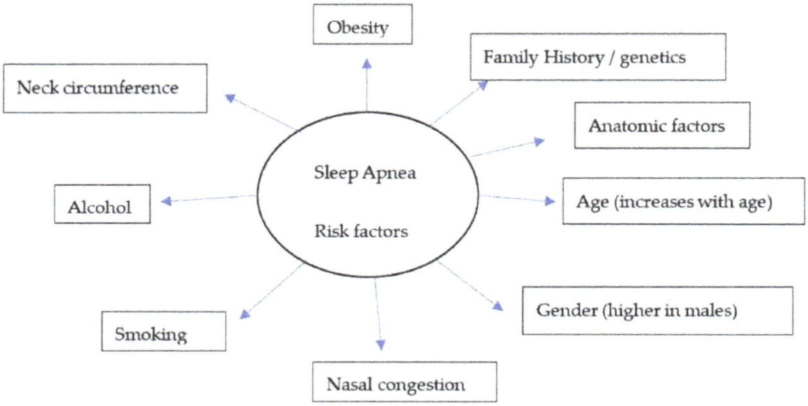

Figure 2. A model for widely recognized risk factors of sleep apnea.

2.2. *Family History/Genetics*

Several studies have reported some underlying causes of OSA to have a genetic component, suggesting its hereditary nature. However, these results should be carefully evaluated as OSA as a disorder is a complex interplay between genetics and environmental factors. The Cleveland Study was another landmark study in investigating OSA and its link to genetics. This genetic–epidemiologic study concluded that OSA is more prevalent in relatives of index probands of OSA as compared to their control counterparts [32]. Additionally, Ferini-Strambi et al. in their comparative study showed a higher prevalence of snoring among monozygotic twins, which is one of the primary symptoms of OSA [33]. Other biomarkers via genome-wide linkage studies have also been investigated to establish an association between OSA and various genes. One study showed such an association between a polymorphism in the angiopoietin-2 gene (ANGPT2) and mean nocturnal oxygen saturation, which is a commonly used marker to determine severity in OSA [34]. Furthermore, similar studies have established polymorphisms in tumor necrosis factor-a (TNF-a), prostaglandin E2 receptor EP3 subtype (PTGER3), and Lysophosphatidic acid receptor 1 (LPAR1) to be a risk factor associated with OSA [35,36]. Therefore, further studies and the biological significance of these polymorphisms in conjunction with OSA are warranted.

2.3. Age and Gender

Generally, OSA has been shown to be more prevalent in men with a two-fold-greater likelihood in people older than 65 years as compared to middle-aged adults aged 30–50 years [37]. Additionally, the prevalence was shown to be around 5% in middle-aged females and 12% in their male counterparts [38]. These estimates are around 12–32% in patients aged 65 years or older [37,39]. However, the severity of the disease varies among elderly individuals and could even be milder than the severity observed in adults [40]. In the elderly, the disease is said to be manifested differently, resembling behavioral and cognitive impairments mimicking dementia [41,42].

According to the current literature, there is a higher prevalence of OSA among men compared to women [43–46]. In a recent study of 1208 people between 20 and 81 years of age with 46% of the cohort being female, an estimated prevalence of OSA was 33% among women when AHI was more than equal to 5%, while it was 59% among men [44]. Even with an AHI of more than or equal to 15%, the prevalence was higher in men when compared to women; 30% vs. 13%, respectively [44]. This effect was not only restricted to prevalence; OSA has been reported to be more severe in men with more specific symptoms suggestive of OSA when compared to women. While men frequently report snoring, gasping, attention deficits, insomnia, snorting, and apnea, women are reportedly presented with more non-specific symptoms, such as headache, fatigue, depression, and anxiety [43]. The presence of only non-specific symptoms then could make it challenging for a physician to perform a correct diagnosis. This also helps explain the results that women are diagnosed at advanced ages and with a higher BMI as compared to men. The difference in gender, generally speaking, could also be due to a differing body-fat distribution in males vs. females, with males having more adipose tissue in the neck region, resulting in a higher susceptibility to airway collapse [47,48]. Although the pharyngeal cross-sectional area is reported to be similar among men and women, men are noted to exhibit greater upper-airway collapsibility. This could be accounted for by the presence of a longer airway length and larger volume of soft tissues on the lateral pharyngeal walls in men [43,49]. Hormonal differences are yet another factor that plays a role in the differing prevalence of OSA among men and women. Previous studies have shown how ventilatory response is affected and AHI is increased in hypogonadal men with an acute administration of testosterone [43,50]. In one study of testosterone replacement therapy among hypogonadal men, Matsumoto et al. reported not only a significant decrease in the ventilatory drive in patients receiving testosterone, but also noted the new induction of OSA, and an exacerbation of symptoms in patients previously diagnosed with OSA [50]. Years later, after this study, it was deciphered that an acute administration of testosterone enhances the ventilatory instability and the loop-gain of the ventilatory system as a consequence of an increase in the ventilatory response to hypoxia [19], increasing the predisposition to OSA in men.

OSA among pregnant and post-menopausal women is one area where there is a lack of research. While it has been reported that sleep-disordered breathing is more severe in postmenopausal when compared to premenopausal women [51], it is unclear whether a decreased production of female hormones plays a role in this exacerbation. Moreover, symptoms of OSA can also be more difficult to identify or interpreted as menopausal manifestations, leading to misdiagnosis [45]. Therefore, these gender differences could mostly be explained by anatomic and physiologic variabilities, a difficulty identifying and categorizing non-specific symptoms, and underdiagnosis due to physician biases [48].

2.4. Smoking and Alcohol

Several studies have cited smoking and alcohol as risk factors for OSA. This could be explained by a general decrease in sleep latency, difficulty in initiating sleep, and irregular sleeping patterns after smoking or drinking [52]. The chemicals consumed during smoking can also result in local inflammation and fluid retention in the upper airway, which could exacerbate these symptoms. Where many studies have observed a positive correlation between smoking and OSA, such as in a study by Kashyap et al., who reported the occurrence of OSA to be approximately twice as likely in current smokers as compared to previous smokers and non-smokers combined [53], many other studies have shown the opposite association, or did not observe smoking to be an independent variable for OSA. However, the number of cigarettes consumed per day was still reported to be higher among more severe forms of OSA [54]. Some studies have also reported smoking addiction due to untreated OSA [55]. This variability in results suggests an inconclusive consensus about the role of smoking in OSA progression and severity; thus, further studies are required to elucidate the relevant mechanisms involved.

On the other hand, the role of alcohol seems to be relatively established among OSA patients. The studies show a general consensus that alcohol is positively correlated with an increased risk and severity of OSA by 25% [56]. The likely mechanisms include the relaxation of muscles in the neck and throat leading to airway collapse, decreased ventilatory responses to an increase in higher partial pressure of CO2 and lower pressure of oxygen, and reduction in muscle activity in the tongue [48]. While we know that alcohol can increase the risk of OSA, it would be beneficial to understand if these effects are impacted by individual race, metabolic and immune status, number of drinks consumed per week, and whether individuals are suffering from any other comorbidities. Answers to these questions will allow for a better public health policy.

2.5. *Inflammation*

Inflammation plays an integral role in the induction, progression, and exacerbation of OSA. Over the years, several inflammatory mediators have been corelated with the pathogenesis of OSA; however, some are more extensively researched and notably reported, including CRP, IL-6, IL-8, IL-33 and its receptor ST2, Pentraxin-3 (PTX-3), procalcitonin (ProCT), and TNF-a [57,58]. These mediators can play an important role as a biomarker to decipher the severity of OSA among patients. Particularly, PTX-3 as a predictor of OSA severity has garnered attention due to its consistent specificity and sensitivity across studies. For example, Sozer et al. reported a specificity and sensitivity of 91.7% for PTX-3 as a predictor of OSA among other inflammatory mediators [57]. They also reported a positive correlation between PTX-3 and BMI, suggesting a potential link between these two variables in the subsequent progression of the disease. Other studies have noted the importance of morning levels of PTX-3 as a sensitive biomarker, as patients with OSA could have a higher hypoxic state during sleep [59]. PTX-3 is essentially from the same family as CRP, an acute phase protein with a role in innate immunity, the regulation of inflammatory reactions, and apoptosis. Its role in the pathogenesis of OSA was further elucidated after treating patients with CPAP. In a study by Kobukai et al., there was a marked reduction in the morning levels of PTX-3 and CRP; however, only PTX-3 levels were shown to be significantly correlated with the severity of OSA using AHI [59].

CRP is another important biomarker that has been extensively researched in the pathogenesis of OSA. However, its association and specificity have been questioned in the last two decades, given the variable results across studies [58,60–62]. A strong relationship between OSA and obesity was established in earlier studies, and perhaps this relationship could distort the data if the patients were not optimally matched for BMI [63].

These inflammatory mediators are subsequent culprits in cardiovascular complications, metabolic dysfunction, and atherosclerosis [58], which is briefly discussed later in this article.

3. Complications of Obstructive Sleep Apnea

While OSA is a major concern on its own for any patient and their healthcare providers, it involves several other complications and sequelae that follows due to OSA pathogenesis and symptomatology. These complications often exacerbate their overall health and further its morbidity and mortality rates. Here, we discuss some of the most pressing complications that are caused by OSA.

3.1. Cardiovascular Diseases

Several studies have established a clear association between OSA and various cardiovascular diseases (CVDs), such as hypertension, stroke, coronary artery disease, and atrial fibrillation [64]. While there is no consensus on a well-adopted mechanism for this association, it is reported to include a heightened sympathetic activation, and in turn release of stress hormones, due to difficulty of breathing [65]. This effect is mediated by the hypothalamic–pituitary–adrenal axis, the activation of which has been shown to be corrected with the use of CPAP, along with a decrease in cortisol levels [65,66]. We know that hypertension is a major risk factor for CVD [67], which is prevalent among OSA patients when oxygen levels are decreased due to the narrowing of the airway. This hypertension that exacerbates the activation of the sympathetic system even causes coronary artery disease. OSA can also cause an abnormal heart rhythm, which is difficult to manage if there are other underlying heart conditions or comorbidities directly, or indirectly, affecting cardiovascular health [68]. It is estimated that approximately 30–50% of OSA patients are also diagnosed with cardiac arrhythmias, including atrial and ventricular premature extrasystoles, ventricular tachycardia, sinus arrest, and atrioventricular conduction block [69]. These acute triggers can be explained by previously studied arrhythmogenic mechanisms, where an alteration in the intrathoracic pressure leads to a stretch in the muscle of the left atrium, which causes distention. This distention in turn gives rise to increased atrial premature beats and QT interval prolongation on an electrocardiogram (EKG) [70]. As a result, ventricular tachycardia and/or sudden cardiac death due to this acute episode can ensue. The triggered activity and automaticity due to enhanced sympathetic discharge, and parasympathetic surge on the other hand during apnea, can cause sinus nodal disease and atrial fibrillation (Figure 3) [70,71]. It is important to note that while there is a strong association between OSA and CVD, many trials have failed to establish that these symptoms improve when treating for OSA. Therefore, further studies are required to understand the underlying mechanisms for better targeted therapies.

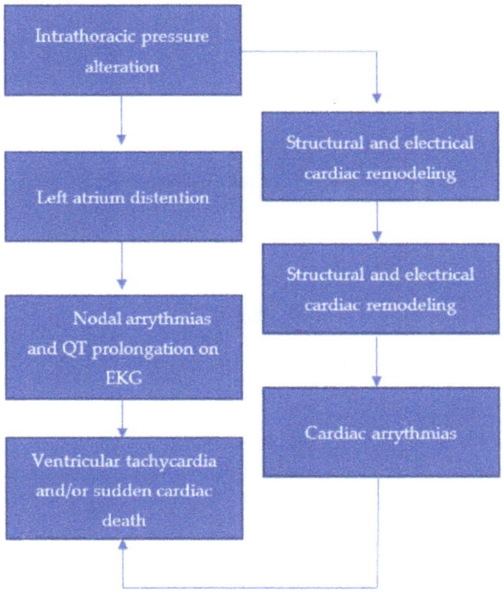

Figure 3. A mechanistic model for the impact of intrathoracic pressure alterations due to OSA on cardiac dysfunction.

Other mechanistic intermediates include a change in intrathoracic pressure, platelet activation due to endothelial damage, and an increase in the levels of inflammatory cytokines, which could all have an impact on CVD [72]. The cytokines involved in these heightened inflammatory reactions are reported to be TNFa, IL-6, IL-8, and C-reactive protein (CRP), which are also associated with excessive daytime sleepiness among OSA patients, along with their action on endothelial damage and myocyte dysfunction [73]. Myocyte hypertrophy due to hypertension can also cause the remodeling of cardiac myocyte, resulting in fibrosis [70]. A similar result could be observed in OSA patients with comorbid obesity, which could enhance the renin–angiotensin–aldosterone system (RAAS) due to sympathetic activation. This in turn can also cause arrhythmias and sudden cardiac death.

3.2. Metabolic Dysfunctions

OSA has been linked to cause, and also manifest, several metabolic derangements, such as insulin resistance, type II diabetes mellitus, metabolic syndrome, and non-alcohol fatty liver disease [45,74]. In a nationwide study of 1,704,905 patients with OSA and an approximately equal number of controls, Mokhlesi et al. attributed a higher prevalence of type II diabetes and ischemic heart disease in men, while hypertension and depression were more prevalent in women as compared to their matched controls [75]. This prevalence of insulin resistance and glucose intolerance among OSA patients was estimated to be anywhere from 20% to approximately 70% [76]. It is important to note that these results are independent of obesity, which could be corroborated as a potential confounding factor. Additionally, while CPAP has been shown to be beneficial in the management of OSA, the results are inconclusive whether it helps in curbing the risks that come with dysfunctional glucose metabolism [76]. This does not only pose a concern for patients with OSA, but also for the healthcare system by further increasing the economic burden of diabetes [77].

As mentioned, not only OSA can cause diabetes, but diabetes could result in OSA as well. This bidirectional association could be explained due to some control of respiration and upper-airway neural reflexes by diabetic neuropathy [78]. Various mechanisms have been proposed to explain the association between insulin resistance and OSA; some of them overlap with CVD complications. For example, sympathetic activation due to hypoxia alters the glucose metabolic cycle, which results in increased cortisol and growth hormone levels, which then can deregulate insulin sensitivity [79]. OSA was also associated with dyslipidemia in several studies due to its association with a state of hypoxia. This was theorized due to the roles of sterol regulatory element-binding protein-1 (SREBP-1) and stearoyl-coenzyme A desaturase-1 (SCD-1). However, these effects were cited due to intermittent hypoxia, and a causative effect of OSA on dyslipidemia is inconclusive [80] (Figure 4).

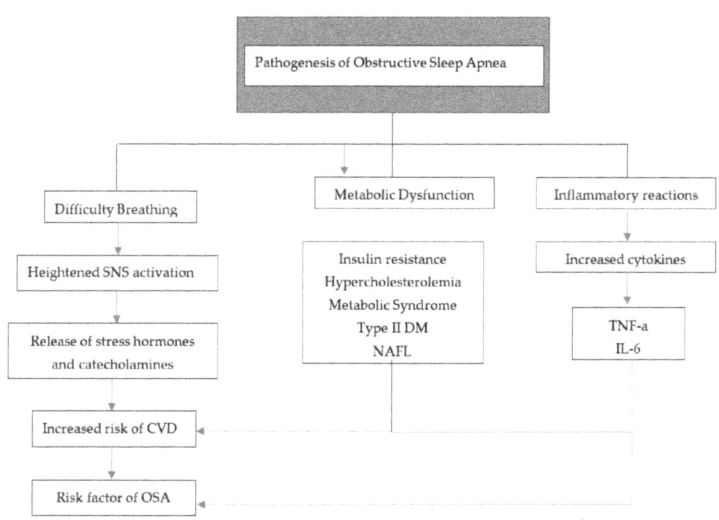

Figure 4. A mechanistic model for pathogenesis of OSA. Abbreviations: SNS = sympathetic nervous system, CVDs: cardiovascular diseases, OSA = obstructive sleep apnea, DM = diabetes mellitus, NAFL = non-alcohol fatty liver, TNF-a = tumor necrosis factor alpha, and IL-6 = interleukin 6.

4. Current Treatment Options

Advances have been made over the years to better manage the symptoms in patients with OSA. Here, we briefly review the available treatments at present and their efficacy.

4.1. Continuous Positive Airway Pressure (CPAP)

CPAP is considered one of the most reliable and effective methods for treating sleep apnea. The use of the term sleep apnea started in late-20th century [9]. However, many changes were made to better accommodate the needs of patients and enhancing its effectiveness. A constant pressure is applied through a tubing system to maintain upper-airway patency during sleep [48]. Many studies have established its efficacy and importance during the course of treatment, and a longer use of CPAP has been shown to be associated with increasing severity. Ravesloot et al. in their study of mathematical function formulas to test its effectiveness reported a 33.3–48.3% reduction in the AHI index upon at least 4 hours per night of CPAP use among patients with moderate OSA [81]. Despite its benefits, the adherence rate is a major concern shared by patients and providers regarding this treatment. Studies suggests that approximately 50–60% of patients discontinue its usage within the first year of their prescription and around 15% discontinue after their very first night of usage [82]. Moreover, the adherence rates were shown to be influenced by severity, body mass index (BMI), AHI, and the oxygen desaturation index (ODI). For example, Jacobsen et al. in their retrospective study of 695 patients reported a higher adherence (89%) in patients with severe OSA as compared to 71% for moderate and 55% for mild OA, and its use was higher among patients with a higher BMI, AHI, and ODI [83]. Having said that, one study showed the importance of formulating a standard protocol, comfortable pressure settings, and offering of mask choice to patients in increasing the adherence to CPAP therapy [84]. While CPAP is widely prescribed, other airway pressure devices might be available depending on individual symptoms.

We believe effective counseling from physicians could greatly impact the rates of adherence. Thus, a proper discussion regarding sleep hygiene and the pros and cons of interrupting their therapies should be thoroughly discussed by clinicians.

4.2. Oral/Dental Devices

Although CPAP is still considered the gold standard when it comes to the treatment of OSA, there are other treatment strategies available if CPAP is not helping or not available to use. Oral devices, most commonly the mandibular advancement device (MAD) and tongue retaining device (TRD) are used, although tongue retainer devices are relatively older devices and are becoming obsolete alongside modern innovations in MADs. However, they are still reported to be an effective alternative treatment option for OSA [85]. MAD is the most commonly used device in patients with sleep apnea, which helps to move the mandible forward, relative to the maxilla [86]. This results in the widening of the airway, which prevents closure and obstruction during sleep. Ultimately, this helps to reduce snoring. MAD is considered as the primary treatment option for mild to moderate cases of OSA, and a secondary option in severe cases for patients having difficulties with CPAP [87]. These devices are available in various designs and can be custom-made or prefabricated. The prefabricated designs are also known as thermoplastic appliances, which are relatively cheaper and the kind that can be bought over the counter. On the other hand, custom-made designs are relatively more sophisticated and expensive because they are produced in a specialized dental laboratory and require the dental imaging of the patient to ensure a good fit [87,88]. While prefabricated devices are simple in design, custom-made designs are more intricate in nature and can consist of multiple and separate parts for the lower and upper jaw. These titratable appliances can be distinguished as middle traction and bilateral thrust devices, which differs in the way they are connected and placed in the oral cavity [87,89]. According to a recent systematic review and metanalysis, there is no clear-cut answer as to which device is superior in terms of alleviating symptoms in patients with mild to moderate

OSA; however, a custom-made MAD was reported to be more superior in terms of comfort and thus a more favorable compliance when compared with a prefabricated MAD (87).

Although many people find this option more comfortable than CPAP, the use of such devices varies. Several unwanted effects of using MADs have been reported, which includes jaw pain, tenderness of denture, and hypersalivation [86]. Moreover, there is no unform guideline for the use of MADs if a patient is already suffering from a pre-existing temporomandibular disorder. However, a relatively recent clinical review reported no contraindication for the use of MADs to treat OSA with a concurrent temporomandibular disorder [90]. TRDs, on the other hand, serve a similar purpose. They appear similar to a large pacifier with a space for the tongue and a defined mandibular protrusion [91]. Similar side effects are also reported for this device. While these devices have shown to be effective, similar to CPAP, the success of this alternative option also relies on patient compliance. The compliance rate pertaining to these devices is anywhere between 30–60%, which is arguably not ideal [92,93]. Most people in these studies cited discomfort, dry mouth, hypersalivation, and other side effects mentioned above to not use it regularly. This also signifies that these devices could be used for short-term periods but are not a long-term solution for these patients. Significant changes in design and comfort are warranted if these options are to be entertained continuously. It is also important to note that those with less propensity for upper-airway collapse and low loop-gain endotypes receive the greatest benefits from oral appliance therapy and upper-airway surgery [94].

4.3. Surgery

Usually, surgery is an option of the last resort, or when a patient insists on choosing this option after expressing frustration and discomfort from the most common treatment options previously discussed. Uvulopalatopharyngoplasty (UPPP) is the most common surgical procedure performed on patients with OSA, where some tissues from the uvula, soft palate, and/or tonsils are resected to open the upper airways [95]. Generally, surgical indications include an AHI of higher than 15, oxyhemoglobin desaturation less than 90%, and cardiac abnormalities associated with OSA. However, it can also be elected if deemed a risk factor for motor vehicle accidents, failed compliance, or intolerance to previous therapies, and polysomnographic parameters of disease [96].

In order to select the patients who are most likely to benefit from this procedure, and to reduce unnecessary harm to patients, a staging system was created, known as the Friedman staging system. In a landmark study, Friedman et al. categorized and scored patients into three different stages based on palate position, tonsil size, and BMI [97]. Based on this scoring system, and a later modification by other scientists, stage 1 patients have a success rate of approximately 81% at present. This success rate is almost decreased by half if a patient is placed in stage 2 using this scoring system. A metanalysis by Choi et al. concluded that while stage 1 is a strong predictor of success after surgery, stage 2 is a negative predictor [98]. This signifies the importance of patient selection and a careful discussion between patients and their physicians to address the efficacy rates and risks involved in this procedure.

4.4. Personalization of the Treatment

Personalized medicine and person-centered care have taken center stage in medicine as of late, and the ailment of OSA is of no exception. The goal of the personalization of treatment in OSA is to carefully evaluate each patient individually, identify their risk factors, and treat them according to their symptoms, keeping in mind their needs, wishes, and values [99]. To provide truly personalized care, it is vital to obtain a detailed patient history surrounding their symptoms and factors of OSA affecting their quality of life. Historically, OSA has been treated according to the severity noted by AHI and generic symptoms [100]. However, at present, we know that OSA has a wide spectrum of symptoms, and, in fact, many patients are presented without any apparent manifestations. Therefore, it is vital to target individual sets of symptoms, which in turn would also motivate treatment adherence among patients. For example, a patient with OSA who is presented with masked cardiovascular manifestations and without daytime sleepiness or snoring at night would be less likely to adhere to CPAP, since there is no perceived immediate reward for that patient. However, someone with OSA who can visibly notice a difference in their sleepiness and snoring would be more motivated to adhere to this regimen [99]. This could be one of the reasons for the low adherence to CPAP among patients with mild OSA, as the perceived burden of using CPAP everyday could outweigh the apparent benefits. This is where patient education can play a vital role, so the patients can remain cognizant of their disease process.

Another important factor in the success of personalized medicine in OSA is actively engaging in the treatment process and self-accountability. As reported for patients with OSA and other chronic diseases, an active pedagogy using telehealth resources and applications to manage CPAP reportedly had better clinical outcomes, reduced progression in disease burden, and resulted in greater self-preventive measures [99,101,102].

The targeted and personalized therapies for OSA can be grossly categorized into two groups: anatomical and non-anatomical. CPAP, oral appliances, weight loss, positional therapy, and upper-airway surgery falls under anatomical therapy, which were previously discussed. Non-anatomical therapy is broadly classified into three groups: muscle function, loop-gain, and arousal threshold [20].

In muscle function therapy, the pharyngeal muscles are targeted since they play a vital role in the patency of the upper airway. Genioglossus is yet another important dilator muscle that can cause upper-airway collapsibility when there is a state-dependent reduction in its activity [103]. It is reported that more than 30% of patients with OSA have minimal muscle responsiveness during sleep, which could further exacerbate the symptoms and collapsibility [20]. To improve patency, hypoglossal nerve stimulation and oropharyngeal muscle training have been suggested as targeted therapies. Both of these personalized interventions have been reported to reduce AHI by more than 50% in patients suffering with OSA and a prior specific reduction in muscle activity [104]. These are categorized under personalized and targeted therapies because the success of these treatment depends on individual's Pcrit, pharyngeal shape, and site of airway collapse [20,104]. Loop-gain therapies include oxygen supplementation and carbonic anhydrase inhibitors. Supplemental oxygen tends to reduce loop-gain and lowers the AHI in selected patients. The mechanism behind this variability for oxygen therapy is unclear at present. Acetazolamide and zonisamide are two carbonic anhydrase inhibitors that have shown promising results by decreasing loop-gains by approximately 40%, while also reducing AHI by half [105,106]. Lastly, therapies for a low arousal threshold include hypnotic agents, such as eszopiclone, zopiclone, and trazodone [20,107]. Without increasing the risk of hypoxemia, these agents can increase the threshold for arousal and reduce the AHI in patients with OSA by approximately 25% to 50% [20].

It is important to realize that these interventions are called targeted for a reason; they seem to have a variable effect on patients with OSA, and several factors can influence the effectiveness of these therapies, which are beyond the scope of this paper.

5. Role of Clinicians

Clinicians play a vital role not only in diagnosing OSA, but also in counseling their patients as lifestyle modifications and prescription compliance play instrumental roles in the management of this medical condition [108]. Counseling can include educating patients about their diet, sleep hygiene, exercise routine, identifying risky behavior, etc. Although OSA is a complex mixture of various intricacies, there are some modifiable risk factors that can be targeted to improve the course of the disease. The major modifiable risk factors of OSA include alcohol, smoking, sleep hygiene, and BMI [109]. For example, weight loss has been reported to decrease the severity of OSA by almost 50% in moderately obese patients with additional benefits in metabolic regulation, such as glycemic control [110]. Similarly, a fixed sleep schedule with the head as elevated and upright as possible could also help reduce symptoms. This is particularly important as sleep hygiene was reported to be indirectly related to daytime sleepiness and depressive symptoms [111]. These modifiable risk factors can also indirectly reduce subsequent cardiovascular complications if targeted early on, since many of these risk factors, such as obesity, exacerbate cardiovascular events, as discussed previously. In a nutshell, counseling should focus on targeting the modifiable risk factors and uplifting patients' quality of life, since several studies have reported a lower quality of life among OSA patients in its most symptomatic forms [112]. The literature suggests that sleep apnea is still widely underdiagnosed [113], which signifies the key role physicians can play in terms of impacting their patients' lives. We suspect that while searching for a correct diagnosis would help narrow the gap in epidemiological variables to devise better public health policies, it could also place an additional burden on the healthcare system, as more patients would consult their physicians and seek appropriate treatment. However, by thoroughly educating their patients about the disease, clinicians cannot only reassure their patients regarding the course of their disease, but also increase the compliant rates pertaining to their treatment. Having said that, it is difficult to predict if a certain healthcare system could handle a large influx of patients with OSA upon an increase in the effectiveness of diagnostics.

6. Conclusions

OSA is increasingly recognized as a prevalent medical condition affecting people globally, making it a pressing public health concern. Although advances have been made over the years, a decent patient population is still undiagnosed due to wide array of reasons, resulting in increasing morbidity and mortality rates related to OSA. This medical condition can manifest itself as several comorbidities, such as cardiovascular dysfunctions, including stroke, hypertension, coronary artery disease, metabolic disorders, chronic inflammation, etc. The most commonly prescribed treatment strategy includes CPAP and oral devices, though patient compliance is a topic of concern. Given the fact that many symptoms are manageable by lifestyle modifications, counseling and education provided by clinicians play important roles. There is a crucial need to understand the many underlying mechanisms concerning OSA and its co-manifestations to better formulate targeted therapies.

Author Contributions: Conceptualization, R.H. and A.S.; methodology, R.H. and A.S.; Writing original draft preparation, R.H. and A.S.; Review and editing, R.H. and A.S. All authors have read and agreed to the published version of the manuscript.

Funding: This research received no external funding.

Institutional Review Board Statement: Not applicable.

Informed Consent Statement: Not applicable.

Data Availability Statement: Not applicable.

Conflicts of Interest: The authors declare no conflict of interest.

References

1. Veasey, S.C.; Rosen, I.M. Obstructive Sleep Apnea in Adults. *N. Engl. J. Med.* **2019**, *380*, 1442–1449. [CrossRef]
2. Strollo, P.J.; Rogers, R.M. Obstructive Sleep Apnea. *N. Engl. J. Med.* **1996**, *334*, 99–104. [CrossRef]
3. Benjafield, A.V.; Ayas, N.T.; Eastwood, P.R.; Heinzer, R.; Ip, M.S.M.; Morrell, M.J.; Nunez, C.M.; Patel, S.R.; Penzel, T.; Pépin, J.-L.; et al. Estimation of the global prevalence and burden of obstructive sleep apnoea: A literature-based analysis. *Lancet Respir. Med.* **2019**, *7*, 687–698. [CrossRef]
4. Finkel, K.J.; Searleman, A.; Tymkew, H.; Tanaka, C.Y.; Saager, L.; Safer-Zadeh, E.; Bottros, M.; Selvidge, J.A.; Jacobsohn, E.; Pulley, D.; et al. Prevalence of undiagnosed obstructive sleep apnea among adult surgical patients in an academic medical center. *Sleep Med.* **2009**, *10*, 753–758. [CrossRef]
5. Dodds, S.; Williams, L.J.; Roguski, A.; Vennelle, M.; Douglas, N.J.; Kotoulas, S.-C.; Riha, R.L. Mortality and morbidity in obstructive sleep apnoea–hypopnoea syndrome: Results from a 30-year prospective cohort study. *ERJ Open Res.* **2020**, *6*, 00057. [CrossRef]

6. Marshall, N.S.; Wong, K.K.H.; Cullen, S.R.; Knuiman, M.; Grunstein, R.R. Sleep Apnea and 20-Year Follow-Up for All-Cause Mortality, Stroke, and Cancer Incidence and Mortality in the Busselton Health Study Cohort. *J. Clin. Sleep Med.* **2014**, *10*, 355–362. [CrossRef]
7. Lavie, P.; Lavie, L.; Herer, P. All-cause mortality in males with sleep apnoea syndrome: Declining mortality rates with age. *Eur. Respir. J.* **2005**, *25*, 514–520. [CrossRef]
8. Economic Burden of Undiagnosed Sleep Apnea in U.S. is Nearly $150B. American Academy of Sleep Medicine—Association for Sleep Clinicians and Researchers. 2016. Available online: https://aasm.org/economic-burden-of-undiagnosed-sleep-apnea-in-u-s-is-nearly-150b-per-year/ (accessed on 17 October 2022).
9. Bahammam, A. Obstructive Sleep Apnea: From Simple Upper Airway Obstruction to Systemic Inflammation. *Ann. Saudi Med.* **2011**, *31*, 1–2. [CrossRef]
10. Littleton, S.W.; Mokhlesi, B. The Pickwickian Syndrome—Obesity Hypoventilation Syndrome. *Clin. Chest Med.* **2009**, *30*, 467–478. [CrossRef]
11. Sleep Eos. A Brief History of Sleep Apnea and Sleep Apnea Signs & Symptoms. Eos Sleep, 2015. Available online: https://www.eossleep.com/2015/05/26/a-brief-history-of-the-causes-of-sleep-apnea/ (accessed on 22 October 2022).
12. Redline, S.; Young, T. Epidemiology and Natural History of Obstructive Sleep Apnea. *Ear Nose Throat J.* **1993**, *72*, 20–26. [CrossRef]
13. Yaremchuk, K.; Garcia-Rodriguez, L. The History of Sleep Surgery. In *Sleep-Related Breathing Disorders*; Karger Publishers: Basel, Switzerland, 2017; Volume 80, pp. 17–21. [CrossRef]
14. Silber, M.H. Tracheostomy Can Fatally Exacerbate Sleep-Disordered Breathing in Multiple System Atrophy. Available online: https://n.neurology.org/content/tracheostomy-can-fatally-exacerbate-sleep-disordered-breathing-multiple-system-atrophy-0 (accessed on 22 October 2022).
15. Kapur, V.K.; Auckley, D.H.; Chowdhuri, S.; Kuhlmann, D.C.; Mehra, R.; Ramar, K.; Harrod, C.G. Clinical Practice Guideline for Diagnostic Testing for Adult Obstructive Sleep Apnea: An American Academy of Sleep Medicine Clinical Practice Guideline. *J. Clin. Sleep Med.* **2017**, *13*, 479–504. [CrossRef]
16. Laratta, C.R.; Ayas, N.T.; Povitz, M.; Pendharkar, S.R. Diagnosis and treatment of obstructive sleep apnea in adults. *Can. Med. Assoc. J.* **2017**, *189*, E1481–E1488. [CrossRef]
17. Finnsson, E.; Ólafsdóttir, G.H.; Loftsdóttir, D.L.; Jónsson, S.Æ.; Helgadóttir, H.; Ágústsson, J.S.; Sands, S.A.; Wellman, A. A scalable method of determining physiological endotypes of sleep apnea from a polysomnographic sleep study. *Sleep* **2021**, *44*, zsaa168. [CrossRef]
18. Wellman, A.; Eckert, D.; Jordan, A.; Edwards, B.; Passaglia, C.; Jackson, A.C.; Gautam, S.; Owens, R.L.; Malhotra, A.; White, D.P. A method for measuring and modeling the physiological traits causing obstructive sleep apnea. *J. Appl. Physiol.* **2011**, *110*, 1627–1637. [CrossRef]
19. Eckert, D.J.; White, D.P.; Jordan, A.S.; Malhotra, A.; Wellman, A. Defining Phenotypic Causes of Obstructive Sleep Apnea. Identification of Novel Therapeutic Targets. *Am. J. Respir. Crit. Care Med.* **2013**, *188*, 996–1004. [CrossRef]
20. Carberry, J.; Amatoury, J.; Eckert, D.J. Personalized Management Approach for OSA. *Chest* **2018**, *153*, 744–755. [CrossRef]
21. Malhotra, A.; Pillar, G.; Fogel, R.; Beauregard, J.; Edwards, J.; White, D.P. Upper-Airway Collapsibility*. *Chest* **2001**, *120*, 156–161. [CrossRef]
22. Malhotra, A.; Jordan, A. The importance of arousal in obstructive sleep apnea—Updates from the American Thoracic Society. *J. Thorac. Dis.* **2016**, *8*, S542–S544. [CrossRef]
23. Subramani, Y.; Singh, M.; Wong, J.; Kushida, C.A.; Malhotra, A.; Chung, F. Understanding Phenotypes of Obstructive Sleep Apnea: Applications in Anesthesia, Surgery, and Perioperative Medicine. *Anesth. Analg.* **2017**, *124*, 179–191. [CrossRef]
24. Vanek, J.; Prasko, J.; Genzor, S.; Ociskova, M.; Kantor, K.; Holubova, M.; Nesnidal, V.; Kolek, A.; Sova, M. Obstructive sleep apnea, depression and cognitive impairment. *Sleep Med.* **2020**, *72*, 50–58. [CrossRef]
25. De Felício, C.M.; da Silva Dias, F.V.; Trawitzki, L.V.V. Obstructive sleep apnea: Focus on myofunctional therapy. *Nat. Sci. Sleep* **2018**, *10*, 271–286. [CrossRef]
26. Mukundan, T.H. Pediatric Obstructive Sleep Apnea and Pediatric Hypersomnia. In *Sleep Disorders in Women: A Guide to Practical Management*; Attarian, H., Viola-Saltzman, M., Eds.; Springer International Publishing: Cham, Switzerland, 2020; pp. 115–128, (Current Clinical Neurology). [CrossRef]
27. Sweetman, A.; Lack, L.; McEvoy, R.D.; Smith, S.; Eckert, D.J.; Osman, A.; Carberry, J.C.; Wallace, D.; Nguyen, P.D.; Catcheside, P. Bi-directional relationships between co-morbid insomnia and sleep apnea (COMISA). *Sleep Med. Rev.* **2021**, *60*, 101519. [CrossRef]
28. Romero-Corral, A.; Caples, S.M.; Lopez-Jimenez, F.; Somers, V.K. Interactions Between Obesity and Obstructive Sleep Apnea: Implications for Treatment. *Chest* **2010**, *137*, 711–719. [CrossRef]
29. Jehan, S.; Zizi, F.; Pandi-Perumal, S.R.; Wall, S.; Auguste, E.; Myers, A.K.; Jean-Louis, G.; McFarlane, S.I. Obstructive Sleep Apnea and Obesity: Implications for Public Health. *Sleep Med. Disord.* **2017**, *1*, 00019.
30. Newman, A.B.; Foster, G.; Givelber, R.; Nieto, F.J.; Redline, S.; Young, T. Progression and regression of sleep-disordered breathing with changes in weight: The Sleep Heart Health Study. *Arch. Intern. Med.* **2005**, *165*, 2408–2413. [CrossRef]
31. Peppard, P.E.; Young, T.; Palta, M.; Dempsey, J.; Skatrud, J. Longitudinal Study of Moderate Weight Change and Sleep-Disordered Breathing. *JAMA* **2000**, *284*, 3015–3021. [CrossRef]
32. Redline, S.; Tishler, P.V.; Tosteson, T.D.; Williamson, J.; Kump, K.; Browner, I.; Ferrette, P.; Krejci, P. The familial aggregation of obstructive sleep apnea. *Am. J. Respir. Crit. Care Med.* **1995**, *151*, 682–687. [CrossRef]

33. Ferini-Strambi, L.; Calori, G.; Oldani, A.; Della Marca, G.; Zucconi, M.; Castronovo, V.; Gallus, G.; Smirne, S. Snoring in twins. *Respir. Med.* **1995**, *89*, 337–340. [CrossRef]
34. Mukherjee, S.; Saxena, R.; Palmer, L.J. The genetics of obstructive sleep apnea. *Respirology* **2018**, *23*, 18–27. [CrossRef]
35. Zhong, A.; Xiong, X.; Xu, H.; Shi, M. An Updated Meta-Analysis of the Association between Tumor Necrosis Factor-α-308G/A Polymorphism and Obstructive Sleep Apnea-Hypopnea Syndrome. *PLoS ONE* **2014**, *9*, e106270. [CrossRef]
36. Patel, S.R.; Goodloe, R.; De, G.; Kowgier, M.; Weng, J.; Buxbaum, S.G.; Cade, B.; Fulop, T.; Gharib, S.A.; Gottlieb, D.J.; et al. Association of genetic loci with sleep apnea in European Americans and African-Americans: The Candidate Gene Association Resource (CARe). *PLoS ONE* **2012**, *7*, e48836. [CrossRef]
37. Glasser, M.; Bailey, N.; McMillan, A.; Goff, E.; Morrell, M. Sleep apnoea in older people. *Breathe* **2011**, *7*, 248–256. [CrossRef]
38. Young, T.; Palta, M.; Dempsey, J.; Skatrud, J.; Weber, S.; Badr, S. The Occurrence of Sleep-Disordered Breathing among Middle-Aged Adults. *N. Engl. J. Med.* **1993**, *328*, 1230–1235. [CrossRef]
39. Hader, C.; Schroeder, A.; Hinz, M.; Micklefield, G.H.; Rasche, K. Sleep disordered breathing in the elderly: Comparison of women and men. *J. Physiol. Pharmacol. Off. J. Pol. Physiol. Soc.* **2005**, *56*, 85–91.
40. Shochat, T.; Pillar, G. Sleep apnoea in the older adult: Pathophysiology, epidemiology, consequences and management. *Drugs Aging* **2003**, *20*, 551–560. [CrossRef]
41. Mant, A.; Saunders, N.A.; Eyland, A.E.; Pond, C.D.; Chancellor, A.H.; Webster, I.W. Sleep-Related Respiratory Disturbance and Dementia in Elderly Females. *J. Gerontol.* **1988**, *43*, M140–M144. [CrossRef]
42. Enright, P.L.; Newman, A.B.; Wahl, P.W.; Manolio, T.A.; Haponik, F.E.; Boyle, P.J.R. Prevalence and Correlates of Snoring and Observed Apneas in 5201 Older Adults. *Sleep* **1996**, *19*, 531–538. [CrossRef]
43. Kim, S.W.; Taranto-Montemurro, L. When do gender differences begin in obstructive sleep apnea patients? *J. Thorac. Dis.* **2019**, *11* (Suppl. 9), S1147–S1149. [CrossRef]
44. Fietze, I.; Laharnar, N.; Obst, A.; Ewert, R.; Felix, S.B.; Garcia, C.; Gläser, S.; Glos, M.; Schmidt, C.O.; Stubbe, B.; et al. Prevalence and association analysis of obstructive sleep apnea with gender and age differences—Results of SHIP-Trend. *J. Sleep Res.* **2018**, *28*, e12770. [CrossRef]
45. Bonsignore, M.R.; Borel, A.-L.; Machan, E.; Grunstein, R. Sleep apnoea and metabolic dysfunction. *Eur. Respir. Rev.* **2013**, *22*, 353–364. [CrossRef]
46. Zhou, X.; Zhou, B.; Li, Z.; Lu, Q.; Li, S.; Pu, Z.; Luo, F. Gender differences of clinical and polysomnographic findings with obstructive sleep apnea syndrome. *Sci. Rep.* **2021**, *11*, 1–6. [CrossRef]
47. Millman, R.P.; Carlisle, C.C.; McGarvey, S.T.; Eveloff, S.E.; Levinson, P.D. Body Fat Distribution and Sleep Apnea Severity in Women. *Chest* **1995**, *107*, 362–366. [CrossRef]
48. Mehra, R.; Gharibeh, T. Obstructive sleep apnea syndrome: Natural history, diagnosis, and emerging treatment options. *Nat. Sci. Sleep* **2010**, *2*, 233–255. [CrossRef]
49. Malhotra, A.; Huang, Y.; Fogel, R.B.; Pillar, G.; Edwards, J.K.; Kikinis, R.; Loring, S.H.; White, D.P. The male predisposition to pharyngeal collapse: Importance of airway length. *Am. J. Respir. Crit. Care Med.* **2002**, *166*, 1388–1395. [CrossRef]
50. Matsumoto, A.M.; Sandblom, R.E.; Schoene, R.B.; Lee, K.A.; Giblin, E.C.; Pierson, D.J.; Bremner, W.J. Testosterone replacement in hypogonadal men: Effects on obstructive sleep apnoea, respiratory drives, and sleep. *Clin. Endocrinol.* **1985**, *22*, 713–721. [CrossRef]
51. Anttalainen, U.; Saaresranta, T.; Aittokallio, J.; Kalleinen, N.; Vahlberg, T.; Virtanen, I.; Polo, O. Impact of menopause on the manifestation and severity of sleep-disordered breathing. *Acta Obstet. Gynecol. Scand.* **2006**, *85*, 1381–1388. [CrossRef]
52. Liao, Y.; Xie, L.; Chen, X.; Kelly, B.C.; Qi, C.; Pan, C.; Yang, M.; Hao, W.; Liu, T.; Tang, J. Sleep quality in cigarette smokers and nonsmokers: Findings from the general population in central China. *BMC Public Health* **2019**, *19*, 808. [CrossRef]
53. Kashyap, R.; Hock, L.M.; Bowman, T.J. Higher prevalence of smoking in patients diagnosed as having obstructive sleep apnea. *Sleep Breath* **2001**, *5*, 167–172. [CrossRef]
54. Ioannidou, D.; Kalamaras, G.; Kotoulas, S.-C.; Pataka, A. Smoking and Obstructive Sleep Apnea: Is There An Association between These Cardiometabolic Risk Factors?—Gender Analysis. *Medicina* **2021**, *57*, 1137. [CrossRef]
55. Krishnan, V.; Dixon-Williams, S.; Thornton, J.D. Where There Is Smoke ... There Is Sleep Apnea: Exploring the relationship between smoking and sleep apnea. *Chest* **2014**, *146*, 1673–1680. [CrossRef]
56. Simou, E.; Britton, J.; Leonardi-Bee, J. Alcohol and the risk of sleep apnoea: A systematic review and meta-analysis. *Sleep Med.* **2018**, *42*, 38–46. [CrossRef]
57. Sozer, V.; Kutnu, M.; Atahan, E.; Ozturk, B.C.; Hysi, E.; Cabuk, C.; Musellim, B.; Simsek, G.; Uzun, H. Changes in inflammatory mediators as a result of intermittent hypoxia in obstructive sleep apnea syndrome. *Clin. Respir. J.* **2017**, *12*, 1615–1622. [CrossRef]
58. McNicholas, W.T. Obstructive Sleep Apnea and Inflammation. *Prog. Cardiovasc. Dis.* **2009**, *51*, 392–399. [CrossRef]
59. Kobukai, Y.; Koyama, T.; Watanabe, H.; Ito, H. Morning pentraxin3 levels reflect obstructive sleep apnea-related acute inflammation. *J. Appl. Physiol.* **2014**, *117*, 1141–1148. [CrossRef]
60. Yokoe, T.; Minoguchi, K.; Matsuo, H.; Oda, N.; Minoguchi, H.; Yoshino, G.; Hirano, T.; Adachi, M. Elevated Levels of C-Reactive Protein and Interleukin-6 in Patients with Obstructive Sleep Apnea Syndrome Are Decreased by Nasal Continuous Positive Airway Pressure. *Circulation* **2003**, *107*, 1129–1134. [CrossRef]
61. Taheri, S.; Austin, D.; Lin, L.; Nieto, F.J.; Young, T.; Mignot, E. Correlates of Serum C-Reactive Protein (CRP)—No Association with Sleep Duration or Sleep Disordered Breathing. *Sleep* **2007**, *30*, 991–996. [CrossRef]

62. Hayashi, M.; Fujimoto, K.; Urushibata, K.; Takamizawa, A.; Kinoshita, O.; Kubo, K. Hypoxia-sensitive molecules may modulate the development of atherosclerosis in sleep apnoea syndrome. *Respirology* **2006**, *11*, 24–31. [CrossRef]
63. Visser, M.; Bouter, L.M.; McQuillan, G.M.; Wener, M.H.; Harris, T.B. Elevated C-reactive protein levels in overweight and obese adults. *JAMA* **1999**, *282*, 2131–2135. [CrossRef]
64. Tietjens, J.R.; Claman, D.; Kezirian, E.J.; De Marco, T.; Mirzayan, A.; Sadroonri, B.; Goldberg, A.N.; Long, C.; Edward, P.; Gerstenfeld Yeghiazarians, Y.; et al. Obstructive Sleep Apnea in Cardiovascular Disease: A Review of the Literature and Proposed Multidisciplinary Clinical Management Strategy. *J. Am. Heart Assoc.* **2019**, *8*, e010440. [CrossRef]
65. Sleep Apnea and its Association with the Stress System, Inflammation, Insulin Resistance and Visceral Obesity—Sleep Medicine Clinics. Available online: https://www.sleep.theclinics.com/article/S1556-407X(07)00048-3/fulltext (accessed on 27 October 2022).
66. Cooper, B.G.; White, J.E.; Ashworth, L.A.; Alberti, K.G.; Gibson, G.J. Hormonal and metabolic profiles in subjects with obstructive sleep apnea syndrome and the acute effects of nasal continuous positive airway pressure (CPAP) treatment. *Sleep* **1995**, *18*, 172–179.
67. Wu, C.-Y.; Hu, H.-Y.; Chou, Y.-J.; Huang, N.; Chou, Y.-C.; Li, C.-P. High Blood Pressure and All-Cause and Cardiovascular Disease Mortalities in Community-Dwelling Older Adults. *Medicine* **2015**, *94*, e2160. [CrossRef]
68. Grimm, W.; Hoffmann, J.; Menz, V.; Köhler, U.; Heitmann, J.; Peter, J.H.; Maisch, B. Electrophysiologic evaluation of sinus node function and atrioventricular conduction in patients with prolonged ventricular asystole during obstructive sleep apnea. *Am. J. Cardiol.* **1996**, *77*, 1310–1314. [CrossRef] [PubMed]
69. Hersi, A.S. Obstructive sleep apnea and cardiac arrhythmias. *Ann. Thorac. Med.* **2010**, *5*, 10–17. [CrossRef] [PubMed]
70. Laczay, B.; Faulx, M.D. Obstructive Sleep Apnea and Cardiac Arrhythmias: A Contemporary Review. *J. Clin. Med.* **2021**, *10*, 3785. [CrossRef] [PubMed]
71. Leung, R.S. Sleep-Disordered Breathing: Autonomic Mechanisms and Arrhythmias. *Prog. Cardiovasc. Dis.* **2009**, *51*, 324–338. [CrossRef]
72. Jean-Louis, G.; Zizi, F.; Clark, L.T.; Brown, C.D.; McFarlane, S.I. Obstructive sleep apnea and cardiovascular disease: Role of the metabolic syndrome and its components. *J. Clin. Sleep Med.* **2008**, *4*, 261–272. [CrossRef]
73. Vgontzas, A.N.; Papanicolaou, D.A.; Bixler, E.O.; Hopper, K.; Lotsikas, A.; Lin, H.M.; Kales, A.; Chrousos, G.P. Sleep apnea and daytime sleepiness and fatigue: Relation to visceral obesity, insulin resistance, and hypercytokinemia. *J. Clin. Endocrinol. Metab.* **2000**, *85*, 1151–1158. [CrossRef]
74. Kostoglou-Athanassiou, I.; Athanassiou, P. Metabolic syndrome and sleep apnea. *Hippokratia* **2008**, *12*, 81–86.
75. Mokhlesi, B.; Ham, S.; Gozal, D. The effect of sex and age on the comorbidity burden of OSA: An observational analysis from a large nationwide US health claims database. *Eur. Respir. J.* **2016**, *47*, 1162–1169. [CrossRef]
76. Pamidi, S.; Tasali, E. Obstructive sleep apnea and type 2 diabetes: Is there a link? *Front Neurol.* **2012**, *3*, 126. [CrossRef]
77. Almendros, I.; García-Río, F. Sleep Apnoea, Insulin Resistance and Diabetes: The First Step Is in the Fat. *Eur. Respir. J.* **2017**, *49*. Available online: https://erj.ersjournals.com/content/49/4/1700179 (accessed on 29 October 2022). [CrossRef] [PubMed]
78. Reutrakul, S.; Mokhlesi, B. Obstructive Sleep Apnea and Diabetes. *Chest* **2017**, *152*, 1070–1086. [CrossRef] [PubMed]
79. Bratel, T.; Wennlund, A.; Carlström, K. Pituitary reactivity, androgens and catecholamines in obstructive sleep apnoea. Effects of continuous positive airway pressure treatment (CPAP). *Respir. Med.* **1999**, *93*, 1–7. [CrossRef] [PubMed]
80. Drager, L.F.; Jun, J.; Polotsky, V.Y. Obstructive sleep apnea and dyslipidemia: Implications for atherosclerosis. *Curr. Opin. Endocrinol. Diabetes* **2010**, *17*, 161–165. [CrossRef]
81. Ravesloot, M.; De Vries, N. Reliable Calculation of the Efficacy of Non-Surgical and Surgical Treatment of Obstructive Sleep Apnea Revisited. *Sleep* **2011**, *34*, 105–110. [CrossRef]
82. Rapelli, G.; Pietrabissa, G.; Manzoni, G.M.; Bastoni, I.; Scarpina, F.; Tovaglieri, I.; Perger, E.; Garbarino, S.; Fanari, P.; Lombardi, C.; et al. Improving CPAP Adherence in Adults with Obstructive Sleep Apnea Syndrome: A Scoping Review of Motivational Interventions. *Front. Psychol.* **2021**, *12*, 705364. [CrossRef]
83. Jacobsen, A.R.; Eriksen, F.; Hansen, R.W.; Erlandsen, M.; Thorup, L.; Damgård, M.B.; Kirkegaard, M.G.; Hansen, K.W. Determinants for adherence to continuous positive airway pressure therapy in obstructive sleep apnea. *PLoS ONE* **2017**, *12*, e0189614. [CrossRef]
84. Morrone, E.; Giordano, A.; Carli, S.; Visca, D.; Rossato, F.; Godio, M.; Paracchini, E.; Rossi, S.; Balbi, B.; Sacco, C.; et al. Something is changing in adherence to CPAP therapy: Real world data after 1 year of treatment in patients with obstructive sleep apnoea. *Eur. Respir. J.* **2019**, *55*, 1901419. [CrossRef]
85. Chang, E.T.; Fernandez-Salvador, C.; Giambo, J.; Nesbitt, N.; Liu, S.Y.C.; Capasso, R.; Kushida, C.A.; Camacho, M. Tongue retaining devices for obstructive sleep apnea: A systematic review and meta-analysis. *Am. J. Otolaryngol.* **2017**, *38*, 272–278. [CrossRef]

86. Jayesh, S.R.; Bhat, W.M. Mandibular advancement device for obstructive sleep apnea: An overview. *J. Pharm. Bioallied. Sci.* **2015**, *7* (Suppl. 1), S223–S225.
87. Venema JA, U.; Rosenmöller, B.R.; De Vries, N.; de Lange, J.; Aarab, G.; Lobbezoo, F.; Hoekema, A. Mandibular advancement device design: A systematic review on outcomes in obstructive sleep apnea treatment. *Sleep Med. Rev.* **2021**, *60*, 101557. [CrossRef] [PubMed]
88. Segù, M.; Campagnoli, G.; Di Blasio, M.; Santagostini, A.; Pollis, M.; Levrini, L. Pilot Study of a New Mandibular Advancement Device. *Dent. J.* **2022**, *10*, 99. [CrossRef] [PubMed]
89. Almeida, F.R.; Lowe, A.A. Principles of Oral Appliance Therapy for the Management of Snoring and Sleep Disordered Breathing. *Oral Maxillofac. Surg. Clin. N. Am.* **2009**, *21*, 413–420. [CrossRef] [PubMed]
90. Bonetti, A.A.; Bortolotti, F.; Moreno-Hay, I.; Michelotti, A.; Cordaro, M.; Bonetti, G.A.; Okeson, J.P. Effects of mandibular advancement device for obstructive sleep apnea on temporomandibular disorders: A systematic review and meta-analysis. *Sleep Med. Rev.* **2019**, *48*, 101211. [CrossRef]
91. Lazard, D.S.; Blumen, M.; Lévy, P.; Chauvin, P.; Fragny, D.; Buchet, I.; Chabolle, F. The Tongue-Retaining Device: Efficacy and Side Effects in Obstructive Sleep Apnea Syndrome. *J. Clin. Sleep Med.* **2009**, *5*, 431–438. [CrossRef]
92. Skalna, M.; Novak, V.; Buzga, M.; Skalny, P.; Hybaskova, J.; Stransky, J.; Stembirek, J. Oral Appliance Effectiveness and Patient Satisfaction with Obstructive Sleep Apnea Treatment in Adults. *Experiment* **2019**, *25*, 516–524. [CrossRef]
93. Saglam-Aydinatay, B.; Taner, T. Oral appliance therapy in obstructive sleep apnea: Long-term adherence and patients' experiences. *Med. Oral Patol. Oral Y Cir. Bucal* **2017**, *23*, e72–e77. [CrossRef]
94. Edwards, B.A.; Redline, S.; Sands, S.A.; Owens, R.L. More Than the Sum of the Respiratory Events: Personalized Medicine Approaches for Obstructive Sleep Apnea. *Am. J. Respir. Crit. Care Med.* **2019**, *200*, 691–703. [CrossRef]
95. Khan, A.; Ramar, K.; Maddirala, S.; Friedman, O.; Pallanch, J.F.; Olson, E.J. Uvulopalatopharyngoplasty in the Management of Obstructive Sleep Apnea: The Mayo Clinic Experience. In *Mayo Clinic Proceedings*; Elsevier: Amsterdam, The Netherlands, 2009; Volume 84, pp. 795–800.
96. MacKay, S.G.; Lewis, R.; McEvoy, D.; Joosten, S.; Holt, N.R. Surgical management of obstructive sleep apnoea: A position statement of the Australasian Sleep Association. *Respirology* **2020**, *25*, 1292–1308. [CrossRef]
97. Friedman, M.; Ibrahim, H.; Bass, L. Clinical staging for sleep-disordered breathing. *Otolaryngol.-Head Neck Surg.* **2002**, *127*, 13–21. [CrossRef]
98. Choi, J.H.; Cho, S.H.; Kim, S.N.; Suh, J.D.; Cho, J.H. Predicting Outcomes after Uvulopalatopharyngoplasty for Adult Obstructive Sleep Apnea: A Meta-analysis. *Otolaryngol.–Head Neck Surg.* **2016**, *155*, 904–913. [CrossRef]
99. Watach, A.J.; Hwang, D.; Sawyer, A.M. Personalized and Patient-Centered Strategies to Improve Positive Airway Pressure Adherence in Patients with Obstructive Sleep Apnea. *Patient Prefer. Adherence* **2021**, *15*, 1557–1570. [CrossRef] [PubMed]
100. Lim, D.C.; Sutherland, K.; Cistulli, P.A.; Pack, A.I. P4 medicine approach to obstructive sleep apnoea. *Respirology* **2017**, *22*, 849–860. [CrossRef] [PubMed]
101. Greene, J.; Hibbard, J.H.; Sacks, R.; Overton, V.; Parrotta, C.D. When Patient Activation Levels Change, Health Outcomes And Costs Change, Too. *Healrh Aff.* **2015**, *34*, 431–437. [CrossRef]
102. Hibbard, J.H.; Greene, J.; Sacks, R.M.; Overton, V.; Parrotta, C. Improving Population Health Management Strategies: Identifying Patients Who Are More Likely to Be Users of Avoidable Costly Care and Those More Likely to Develop a New Chronic Disease. *Health Serv. Res.* **2016**, *52*, 1297–1309. [CrossRef] [PubMed]
103. Shaik, L.; Cheema, M.S.; Subramanian, S.; Kashyap, R.; Surani, S.R. Sleep and Safety among Healthcare Workers: The Effect of Obstructive Sleep Apnea and Sleep Deprivation on Safety. *Medicina* **2022**, *58*, 1723. [CrossRef] [PubMed]
104. Camacho, M.; Certal, V.; Abdullatif, J.; Zaghi, S.; Ruoff, C.M.; Capasso, R.; Kushida, C.A. Myofunctional Therapy to Treat Obstructive Sleep Apnea: A Systematic Review and Meta-analysis. *Sleep* **2015**, *38*, 669–675. [CrossRef] [PubMed]
105. Edwards, B.A.; Sands, S.A.; Eckert, D.J.; White, D.P.; Butler, J.P.; Owens, R.L.; Malhotra, A.; Wellman, A. Acetazolamide improves loop gain but not the other physiological traits causing obstructive sleep apnoea. *J. Physiol.* **2012**, *590*, 1199–1211. [CrossRef]
106. Edwards, B.A.; Connolly, J.G.; Campana, L.M.; Sands, S.A.; Trinder, J.A.; White, D.P.; Wellman, A.; Malhotra, A. Acetazolamide Attenuates the Ventilatory Response to Arousal in Patients with Obstructive Sleep Apnea. *Sleep* **2013**, *36*, 281–285. [CrossRef]
107. Eckert, D.J.; Malhotra, A.; Wellman, A.; White, D.P. Trazodone Increases the Respiratory Arousal Threshold in Patients with Obstructive Sleep Apnea and a Low Arousal Threshold. *Sleep* **2014**, *37*, 811–819. [CrossRef]
108. Robbins, R.; Seixas, A.; Jean-Louis, G.; Parthasarathy, S.; Rapoport, D.; Ogedegbe, G.; Ladapo, J.A. National patterns of physician management of sleep apnea and treatment among patients with hypertension. *PLoS ONE* **2018**, *13*, e0196981. [CrossRef]
109. Al-Qattan, H.; Al-Omairah, H.; Al-Hashash, K.; Al-Mutairi, F.; Al-Mutairat, M.; Al-Ajmi, M.; Mohammad, A.; Alterki, A.; Ziyab, A.H. Prevalence, Risk Factors, and Comorbidities of Obstructive Sleep Apnea Risk Among a Working Population in Kuwait: A Cross-Sectional Study. *Front. Neurol.* **2021**, *12*, 620799. [CrossRef]
110. St-Onge, M.-P.; Tasali, E. Weight Loss Is Integral to Obstructive Sleep Apnea Management. Ten-Year Follow-up in Sleep AHEAD. *Am. J. Respir. Crit. Care Med.* **2021**, *203*, 161–162. [CrossRef]
111. Lee, S.-A.; Paek, J.-H.; Han, S.-H. Sleep hygiene and its association with daytime sleepiness, depressive symptoms, and quality of life in patients with mild obstructive sleep apnea. *J. Neurol. Sci.* **2015**, *359*, 445–449. [CrossRef]

112. Stepnowsky, C.; Sarmiento, K.F.; Bujanover, S.; Villa, K.F.; Li, V.W.; Flores, N.M. Comorbidities, Health-Related Quality of Life, and Work Productivity Among People with Obstructive Sleep Apnea With Excessive Sleepiness: Findings From the 2016 US National Health and Wellness Survey. *J. Clin. Sleep Med.* **2019**, *15*, 235–243. [CrossRef]
113. Kramer, N.R.; Cook, T.E.; Carlisle, C.C.; Corwin, R.W.; Millman, R.P. The role of the primary care physician in recognizing obstructive sleep apnea. *Arch. Intern. Med.* **1999**, *159*, 965–968. [CrossRef]

life

MDPI

Review

Adherence to CPAP Treatment: Can Mindfulness Play a Role?

Athanasia Pataka [1,*], Seraphim Chrysovalantis Kotoulas [2], Panagiotis Raphael Gavrilis [3], Alice Karkala [1], Asterios Tzinas [1] and Aimiliza Stefanidou [4]

[1] Respiratory Failure Unit, G. Papanikolaou Hospital Thessaloniki, Aristotle University of Thessaloniki, 57010 Thessaloniki, Greece
[2] ICU, Hippokratio General Hospital, 54942 Thessaloniki, Greece
[3] Medical School, Aristotle University of Thessaloniki, 54124 Thessaloniki, Greece
[4] American College of Thessaloniki (ACT), Vasiliou Sevenidi 17, Pilea, 55535 Thessaloniki, Greece
* Correspondence: patakath@yahoo.gr

Abstract: Obstructive sleep apnea (OSA) is considered a chronic disease that requires long-term multidisciplinary management for effective treatment. Continuous Positive Airway Pressure (CPAP) is still considered the gold standard of therapy. However, CPAP effectiveness is limited due to poor patients' adherence, as almost 50% of patients discontinue treatment after a year. Several interventions have been used in order to increase CPAP adherence. Mindfulness-based therapies have been applied in other sleep disorders such as insomnia but little evidence exists for their application on OSA patients. This review aims to focus on the current data on whether mindfulness interventions may be used in order to increase CPAP adherence and improve the sleep quality of OSA patients. Even though controlled trials of mindfulness and CPAP compliance remain to be performed, this review supports the hypothesis that mindfulness may be used as an adjunct method in order to increase CPAP adherence in OSA patients.

Keywords: obstructive sleep apnea; OSA; CPAP; adherence; mindfulness; cognitive therapy

Citation: Pataka, A.; Kotoulas, S.C.; Gavrilis, P.R.; Karkala, A.; Tzinas, A.; Stefanidou, A. Adherence to CPAP Treatment: Can Mindfulness Play a Role? *Life* **2023**, *13*, 296. https://doi.org/10.3390/life13020296

Academic Editors: Milan Štengl and David Slouka

Received: 18 December 2022
Revised: 10 January 2023
Accepted: 17 January 2023
Published: 20 January 2023

1. Introduction

Obstructive sleep apnea (OSA) is the most prevalent sleep breathing disorder (SDB) caused by complete or partial upper airway occlusion during sleep. The incidence of OSA is higher than previously believed, with almost 20% of adult males and 10% of women suffering from the moderate-to-severe disease. Obesity is considered to be one of the most important risk factors for OSA [1]. Apart from obesity, increased neck circumference, male sex, older age, upper airway, and craniofacial abnormalities are also considered significant clinical risk factors of the disease [2]. Several cluster analysis studies have found that the classical phenotype of OSA, i.e., the obese sleepy male, represents only a part of the patients, and have identified other clinical phenotypes, with atypical symptoms, such as insomnia, gender-specific, with different co-morbidities and polysomnographic findings [3].

OSA interrupts the physiological sleep structure leading in sleep fragmentation causing excessive daytime sleepiness (EDS), and impaired vigilance, resulting in an increased risk of work and motor vehicle accidents [4]. Undiagnosed and untreated OSA has been associated with increased mortality and has serious health consequences such as hypertension, arrythmias, cardiovascular and cerebrovascular disease, diabetes, impairment of heart failure, and pulmonary hypertension [5]. All the aforementioned consequences have a significant economic burden [6]. However, with early identification and treatment, the negative implications of OSA can be significantly decreased.

OSA is considered a chronic disease that requires long-term multidisciplinary management for effective treatment. Continuous Positive Airway Pressure (CPAP) is still considered the gold standard of therapy, even though there are several treatment options such as mandibular advancement devices, weight loss, lifestyle interventions, positional therapy, hypoglossal nerve stimulation, and surgical operations. CPAP is recommended as the first choice for patients with moderate to severe disease and those with mild and clinical symptoms, such as EDS, or co-morbidities [7]. CPAP provides a stream of pressurized air constantly during inspiration and expiration in order to maintain the upper airways open. The application of CPAP resolves obstructive respiratory events, improves oxygen desaturations, resulting in improved daytime sleepiness, cognitive function, and mood [8]. Additionally, treatment with CPAP has beneficial cardiovascular effects as it reduces arterial blood pressure, especially in patients with severe disease; it improves pulmonary hypertension and left ventricular ejection fraction in patients with heart failure [9,10]. CPAP also reduces mortality and improves the quality of life [11–14]. A dose-response relationship has been found between the improvement in health and CPAP adherence [9,11–14]. However, CPAP effectiveness is limited due to poor patients adherence. It has been shown that almost 50% of patients discontinue CPAP after a year of treatment [15–17], within a range of 29% to 83%, while 8 to 15% of patients reject treatment as early as the first night of application [18,19]. In the comprehensive systematic literature review of Rotenberg et al. [18] that evaluated data from 82 trials regarding CPAP adherence over a twenty-year timeframe, it was found that CPAP adherence remained persistently low, around 34% (30–40%). It was also found that approximately 11% of the participants of the trials were unable to remain on CPAP treatment over the duration of the trial. Discontinuance of CPAP is a global problem despite the different cultural characteristics of the patients.

Mindfulness is defined as being in the moment and aware of one's thoughts and emotions. Living in the present moment through mindfulness practices can be a state or a trait characteristic that is an important element for a healthy life [20–22]. Mindfulness-based therapies have been used in order to improve insomnia and sleep quality, especially in individuals who prefer these types of therapies and those with an expectation of benefit [23]. However, few studies have focused on the effect of mindfulness on OSA treatment, especially as an intervention to increase CPAP adherence and controlled trials remain to be performed. The current review focuses on the evaluation of the hypothesis of whether mindfulness may be used as an alternative method in order to increase CPAP adherence.

2. Practical-Technical Issues Affecting CPAP Adherence

Adherence to CPAP is important for OSA patients. The duration of CPAP use that is required in order to normalize functioning is still unclear, ranging from at least 4 h [24,25] and reaching up to 6–8 h per night of >70% of nights (i.e., >5 nights/week) in different studies [26,27]. Non-adherence to CPAP treatment is attributed to multiple factors including disease severity, possible side effects during the application of the device, psychological factors, and socio-demographic/economic characteristics of the patient [28,29]. The rate of CPAP adherence has been affected by several barriers to successful treatment as mask leaks, skin irritation, conjunctivitis, nasal congestion, dry throat, claustrophobia, or aerophagia [29] (Table 1). CPAP use within the first week is predictive of long-term use [30,31]. For that, it is crucial for the treating physician to assess the possible risk factors for non-adherence early in the application of a treatment, preferably within the first 2 weeks of use [32]. It has been shown that healthcare professionals have different perceptions and knowledge compared with patients regarding CPAP side effects, possible problems, and educational needs. This is important in the design of educational programmes for healthcare professionals and patients in order to increase CPAP adherence [33]. The education that is provided by a knowledgeable and trusted health professional regarding the use of the CPAP device and its expected benefits is important.

Table 1. Problems—practical issues during CPAP use and their solution.

Problem	Solution
Leaks	Better fitting of the mask For the nose mask, use a chinstrap for mouth leaks Try different mask
Skin lesions	Better fitting of the mask Try different mask Topical application of products for skin issues
Rhinitis	If existed previously: • increase treatment (inhaled steroids, antihistamines, very short course of oral steroids) If it did not exist previously: • examine for possible leaks, • examine for persistence of symptoms, • possible allergic test and rhinomanometry • 2 weeks with inhaled steroids, antihistamines, and/or ipratropium bromide • Check again for adequate mask fit (use chinstraps) • If no improvement: humidification • If no improvement: ENT referral, change to oronasal mask
Conjunctivitis	Better fitting of the mask Try different mask
Dry mouth	Better fitting of the mask For nose mask, use a chinstrap for mouth leaks Try oronasal mask Humidification
Noise	Better fitting of the mask
Aerophagia	Better fitting of the mask A transient problem usually
Removal during the night involuntary	Better fitting of the mask to avoid leaks Try different mask Explain that nothing will happen Set an alarm clock in order to put on the mask
Cold air	Humidification
Claustrophobia	Select smaller interfaces such as a nasal pillows or nasal masks Wear CPAP while awake and practice breathing through the mask during the day while reading a book, watching TV Gradually increasing the time of use Select Ramp facility and Expiratory Pressure Relief
Anxiety, phobia, negative social aspects	Psychotherapy Enhance self-efficacy

CPAP: Continuous Positive Airway Pressure.

3. Other Variables Affecting CPAP Acceptance

As the pattern of CPAP adherence is evident during the first weeks of treatment, it may be hypothesized that the patient has already formed perceptions regarding OSA and possible treatment benefits. Based on this hypothesis, the most effective methods in order to promote adherence are based on the perception of the patient [34]. Education is recommended by the American Academy of Sleep Medicine as an important component of adherence [35]. However, there is evidence those educational interventions about OSA and its consequences when untreated, on the different types of devices and masks, or on providing solutions to resolve the different problems, did not result in the complete improvement of adherence [16,17,31–33]. Therefore, adherence to CPAP may depend on other factors such as environmental, motivational, and psychological, and not only technological.

Apart from the various practical issues which have often been considered barriers to treatment adherence (Table 1), psychological variables have also been examined in the prediction of CPAP acceptance. Understanding and assessing patients' prior beliefs regarding their expectations of health care is very important and highly relevant to overall patient satisfaction. According to the Health Belief Model (HBM), it is crucial to communicate effectively to the patients any practical issues and negative experiences which may impact health outcomes [36]. Addressing patient preferences and factors that may cause discomfort across various clinical contexts may achieve a greater likelihood of adherence and avoid intense emotions such as love or hate which are often associated with CPAP use [37]. Patients report a major subjective improvement even the first morning after treatment which could be viewed as an initial quality indicator for future research [38,39].

Patients' level of adherence pattern is strongly associated with their expectations and beliefs [37]. The use of CPAP significantly decreases over 12 months and the decline may be predicted by the experiences of patients with the device early (i.e., at 1 month), making intensive early interventions more feasible to improve long-term compliance [38,39]. Understanding how a person meets or resists expectations could help clinicians identify how some patients adhere to treatment plans and others don't, based on their 'inner' and 'outer' expectations. While most people are not exclusively inner-driven or outer-driven, our tendency influences our behavior. Taking into account the particular psychological factor in understanding CPAP adherence could provide a more holistic framework for offering tailored therapeutic interventions and improving patient engagement [40].

Psychological well-being has also been reported to be severely affected in individuals suffering from OSA and obesity. A core aspect of mental health, psychological well-being, has also been evidenced to be deeply affected in OSA patients who are obese. The study by Scarpina et al. was the first to document the role of OSA in the subjective perception of psychological well-being [41]. In a similar psychosocial direction, social processes, such as personal perception and close relationships should be examined. Sleep may be universal but there are variations in the social context it occurs and social psychological research should investigate such social and cultural disparities. Sleep deprivation has been shown to impair cognitive functioning and heuristic tendencies leading to stereotyping and bias. Bodenhausen has examined the role of stereotypes and biases, especially in a lack of motivation, and has found that "morning people" were more likely to engage in stereotyping at night and "night people" were more likely to engage in stereotyping in the morning [42]. The importance of circadian variations is particularly interesting because they may affect several different processes, i.e., biological, social, and cognitive. The knowledge of this may provide a different perspective on the expectations of CPAP compliance according to circadian variations. It would be rather difficult to expect a 'night' person to apply CPAP early at night than a 'morning' person.

Recent studies have also shown that sleep deprivation impairs empathic responding and overall social experience with reference to interpersonal relationships. Sleep-deprived partners who are not well-rested report more interpersonal conflicts and low frustration tolerance which may set them on a path to an unhappy relationship [43,44]. Beyond any relationship problems, poor sleep can take other social dimensions, such as making patients and partners feel lonely, detached, and eventually withdrawn [22,44]. The emotional interdependence of sleep partners highlights the social component of sleep, and how partners of OSA patients are integral elements contributing to any successful intervention. Marital quality and partner involvement affect adherence to CPAP and has been identified as an important research need to be addressed [45]. Examining the relationship between partner dynamics and sleep, OSA creates a collateral burden to spouses and/or bedtime partners who may often sleep apart from their partners suffering from OSA [46,47]. Partners frequently complain about snoring and sleep interruptions but are also worried about their bed partners who experience various breathing abnormalities during the night [48]. In conclusion, the consequences of OSA itself and its treatment expand beyond the individual that suffers from the disease [45].

Recent research has shown that a patient's personality traits, such as health locus of control and low self-efficacy are other factors for non-adherence. Patients who have a strong internal locus of control and assume personal responsibility for their health are more likely to adhere to treatment [49]. These patients believe that they can change the situation when needed, they are more receptive to consult or following medical advice and are more self-efficacious [50]. Highly empowered patients perceive their condition as urgent, want to take control of their own health, and are more motivated to start CPAP therapy in the first place (self-referring is common), and are more likely to adhere to treatment [50]. On the other hand, patients with Type D personalities are reluctant to follow medical advice and show decreased adherence to CPAP treatment [51]. Individuals with type D personality (D stands for distressed) are characterized by increased negative emotions in different situations and social inhibition as they do not share their emotions with others, because they are afraid of possible disapproval or rejection. Future studies should focus on investigating the causes of low adherence and construct a specific protocol on the basis of personality characteristics and co-morbidity.

Adherence to CPAP may also be influenced by various psychological conditions, especially claustrophobia. Claustrophobia is a type of an anxiety disorder where one can experience anxiety when in a confined space. Claustrophobia is not the same for everyone, it can range from mild anxiety to a panic attack but it is perceived by many patients as one of the most significant obstacles to CPAP therapy [52]. Because of the overall inconvenience and discomfort of the mask and the tube piece, claustrophobia is a common reaction with patients experiencing shortness of breath and feelings of suffocation [52].

4. Interventions to Improve Non-Adherence to CPAP Therapy

During the last decades, several non-pharmacological treatments have been developed in order to help patients with sleep disorders. Sleep medicine combines the work of many health professionals, such as pulmonologists, neurologists, psychiatrists, otolaryngologists, maxillofacial surgeons and psychologists. Due to the multidisciplinary nature of sleep medicine different specialties are required to work together for the effective diagnosis and treatment. Psychology and Sleep medicine are closely related. Non pharmacologic treatment options include cognitive, behavioral, psychosocial, and educational interventions that may help in improving patients' quality of life. In order to improve adherence to CPAP many different interventions have been used (Table 2) [40,49,53]. Behavioral sleep specialists use evidenced-based therapies combining cognitive techniques with behavioral approaches [53]. Cognitive-behavioral treatment is one of the most important behavior change interventions. A recent meta-analysis revealed that motivational interventions were more successful than educational programs and usual care in improving CPAP adherence, even though the results were not always sustained across all the studies [54].

Table 2. Strategies to Enhance self-efficacy for better CPAP adherence [34].

Education	Educational material (leaflets, videos) by one-on-one clinic visits, group meetings, telephone calls, telemedicine interactions, official internet sites
Behavioral Interventions	Cognitive behavioral therapy (CBT) Motivational enhancement therapy (MET),
Telemonitoring	Data on treatment effectiveness and level of adherence. Possible mask leaks, residual respiratory events, CPAP use duration

CPAP: Continuous Positive Airway Pressure.

The clinical observation that even though the therapeutic value of CPAP is undeniable, the percentage of patients that are compliant with treatment is rather low, created the need for educational and behavioral support. Despite the significant technological improvement of masks and devices and telemedicine applications, adherence to CPAP continues to be a major problem [55]. Some patients underestimate the severity of their disease due to its chronicity or some other perceives it as a disability and for that refuse treatment. The continuity of use affects compliance. When used as indicated, CPAP normalizes sleep architecture, reduces daytime sleepiness, cardiovascular risk, and improves health outcomes [56].

One of the most difficult problems to solve is the psychological acceptance of the device. Behavioral change is an important aspect in the acceptance of every treatment and is a complex procedure including not only psychological and motivational, but also socio-environmental aspects [57]. It includes the evaluation of the patient's adherence to a treatment considering the level of awareness of the disease and its health consequences (reasons for change), the eagerness of the patient to change, the readiness of the patient to change, the perceived significance of this change and the spirit in the ability to change [58]. Several behavior change interventions have been used in order to improve adherence to treatment in several chronic conditions including respiratory disease [59] and more specifically for CPAP treatment [31,34]. The most successful intervention over the years for optimizing adherence has been behavioral therapy [60]. The comprehensive explanation to the patient and partner regarding the sleep disorder, its therapy with the function of equipment (mask, humidifier), the early resolution of problems—side effects (Table 1), psychological consultations, and a careful follow-up are the main elements that may increase the compliance [61] (Table 3).

Table 3. Issues that should be discussed during the first visits for CPAP treatment.

Explain about OSA and its impact on patients' health if left untreated
Suggest lifestyle changes such as weight loss, sleep hygiene
Explaining the importance of treatment with CPAP
CPAP device demonstration: different types of masks, humidifier, ramp
Discuss a follow-up plan (short-term and long-term: face-to-face, telephone, telemedicine)
Solve practical issues with CPAP (see Table 1)

Behavioral interventions, such as the use of cognitive-behavioral therapy (CBT) and of motivational enhancement therapy (MET) in order to increase the self-efficacy of the patient, in addition to education, seem to be a promising approach [62]. The goal of CBT is, through the conversational exchange, to correct the patients' beliefs that are incorrect in order to change their behaviors toward treatment [63]. MET applies motivational interviewing through directed interview questions in order to reinforce patients' motivations [64]. A comprehensive program should ideally be multifactorial including the intervention of different specialists such as sleep physicians, technologists, sleep psychologists, and nurses but also partners or caregivers.

OSA and insomnia often coexist. OSA patients present a higher prevalence of insomnia symptoms (40–60%) compared to that of the general population and this has led to the identification of a new disorder named co-morbid insomnia and OSA (COMISA), that has been highly underestimated [65]. The treatment of COMISA should combine positive-airway pressure (PAP) for OSA, together with CBT for insomnia. The combined treatment has been found to have a better patient outcome in comparison to that of every single treatment alone [65].

5. Mindfulness Interventions to Increase CPAP Adherence

Mindfulness, as a quite heterogeneous term in contemporary psychology, is viewed as an umbrella term that can refer to various facets of mindfulness, from a mental state to a personality trait and from a meditation practice to a type of clinical intervention. Mindfulness has been used as a form of meditation emphasizing a nonjudgmental state of complete or heightened awareness of one's thoughts, experiences, or emotions [20,21]. Conceptualizing mindfulness as an art or as a science makes it unique in some way and different backgrounds, disciplines, ideologies, and practices try to achieve 'ownership' of that complicated concept. Depending on the viewing angle, mindfulness can be viewed as a 'state' or 'trait' mindfulness, but it is characterized as both since the practice of mindfulness is linked with the state and trait changes. People may change drastically during their lifetime when experiencing the benefits of mindfulness. It is worth noting that 'state' mindfulness can occur during meditation practices and 'trait' mindfulness is an individual trait that has been associated with being more conscious and aware in everyday life. 'Trait' mindfulness (or sometimes called 'dispositional' mindfulness) can be accessed through several psychometric questionnaires, such as the Mindful Awareness Scale (MAAS) and the Five Facet Mindfulness Questionnaire (FFMQ) and the Cognitive and Affective Mindfulness Scale-Revised (CAMS-R). Mindfulness skills (integration of knowledge and practice) are powerful mind/body life skills that can be applied to a variety of settings and conditions, alleviating the burden of symptoms and increasing psychological well-being [66–68].

A growing body of literature suggests that adding acceptance-based therapies in mindfulness approaches can optimize patient engagement and response to treatment. The idea is for the patient to accept thoughts and feelings (positive or negative) which eventually leads to self-care, a major determinant of outcomes. Mindfulness helps people to accept their experiences and become more compassionate with themselves (self-compassion) and with others as evidenced by enhanced prefrontal activation in imaging studies as fMRI and electrophysiologically in EEG [67,68]. The most widespread protocol used both in the clinical and non-clinical context is the Mindfulness-Based Stress Reduction (MBSR), a rigorous 8-week program that involves formal and informal meditation practices and was originally designed for stress reduction [20,21,68–73]. The aim of MBSR programs is to enhance well-being and coping with stress in diverse populations. MBSR has been proven to address chronic pain, depression, anxiety, and other conditions and overall increase the patient's quality of life yielding significant benefits both in clinical and non-clinical samples [67,70–73].

Mindfulness therapies have been applied to patients suffering from sleep disorders [67–69]. Mindfulness interventions are suggested as a therapeutic option by the American Academy of Sleep Medicine in patients with insomnia, more frequently in a group format [69]. In this group of patients, mindfulness techniques may be also combined with other therapies, such as CBT (sleep restriction therapy, stimulus control, and sleep hygiene) [67,69]. Claustrophobia is highly prevalent among CPAP-treated patients influencing short and longer-term CPAP non-adherence [52]. In an attempt to examine if mindfulness interventions may be effective in improving CPAP adherence of OSA patients, Gawrysiak et al. [71] have structured a detailed protocol targeting claustrophobia (Mindfulness-based Exposure for PAP-associated Claustrophobia, MBE-PC) once per week for eight consecutive weeks in group meetings. The results of this study have not been published yet.

Studies demonstrate that depression, anxiety, and cognitive functions are considered complications of OSA and may be improved after using CPAP [74–77]. For that someone may consider that possibly other treatment interventions targeting psychological distress may be effective in OSA patients [78,79]. Li et al. [80] have evaluated whether mindfulness was associated with CPAP adherence using the MAAS. The authors have concluded that only MAAS and OSA severity were associated with CPAP adherence irrespective of the presence of psychological distress assessed by the Hospital Anxiety and Depression Scale (HADS); even though HADS evaluating depression was found higher in the nonadherent group.

Furthermore, chronic stress can reduce the prefrontal cortex and increase the size of the amygdala making the brain more receptive to stress. Chronic stress can also weaken emotion regulation [81]. Emotion regulation or emotional self-regulation refers to a person's ability to affect one's emotional state. A recent study [82] indicated that fragmented sleep and the reduction of REM sleep, which both characterize the sleep architecture of OSA, were associated with the difficulty of patients to recall details from the past and overall, with poor memory consolidation. In this regard, an embodied emotion regulation framework could be employed to understand how mindfulness, through top-down or bottom-up pathways affects emotion regulation from a cognitive or clinical perspective [79].

In addition, emerging evidence suggests that mobile Health interventions may improve treatment adherence and outcomes. Technological advancements in the digital realm can indeed improve patient compliance. Some mindfulness-related apps have been evaluated for clinical efficacy (e.g., Calm app is one app that specializes in audio and video programs intended to help someone relax before bedtime) and could be a viable option to help patients with OSA reduce self-reported anxiety and get a high-quality sleep [83,84].

6. Conclusions

In order for CPAP therapy to be effective, the patient needs to be committed to treatment using the device every night (or more than 5 nights/weeks) for more than 4 hours/night. The discontinuity of the appropriate CPAP use is reflected in the reduction of the amelioration of symptoms, resulting in a lesser benefit. CPAP treatment is behaviorally based and requires a multidimensional approach. This long-term commitment to treatment regarding CPAP adherence is critical. Technological, educational and behavioral, strategies may be needed in order to target the different disease and patient characteristics, and possible side effects. Personalized medicine should be the future target of treatment individualizing adherence goals by treating patient-specific symptoms (such as excessive daytime sleepiness or insomnia) and reducing the risk of patient-specific consequences (such as cardiovascular) [3]. This review supports the hypothesis that mindfulness can therefore serve as a novel approach to promote CPAP adherence in OSA patients by reducing emotional distress and increasing subjective well-being (Figure 1). As controlled trials have not been performed yet, future research should continue to investigate the role of mindfulness-based interventions in CPAP treatment adherence

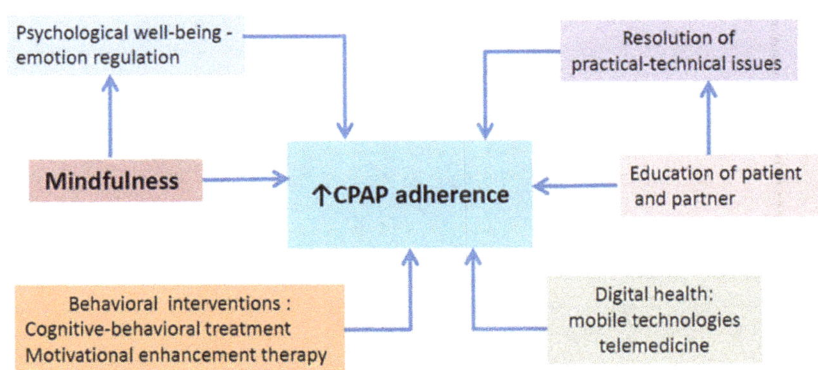

Figure 1. Summary of the different strategies used to increase CPAP adherence. CPAP = Continuous Airway Pressure.

Author Contributions: Conceptualization, A.P. and P.R.G.; Methodology, A.P., S.C.K., A.K., A.T. and A.S.; Investigation, A.P., A.K., A.T. and A.S.; Writing—Original draft preparation, A.P., P.R.G. and A.S.; Writing—review and editing, A.P., A.S. and S.C.K.; Supervision, A.P. and A.S. All authors have read and agreed to the published version of the manuscript.

Funding: This research received no external funding.

Institutional Review Board Statement: Not applicable.

Informed Consent Statement: Not applicable.

Data Availability Statement: Data sharing not applicable: No new data were created or analyzed in this study. Data sharing is not applicable to this article.

Conflicts of Interest: The authors declare no conflict of interest.

References

1. Matsumoto, T.; Murase, K.; Tabara, Y.; Gozal, D.; Smith, D.; Minami, T.; Chin, K. Impact of sleep characteristics and obesity on diabetes and hypertension across genders and menopausal status: The Nagahama Study. *Sleep* **2018**, *41*, zsy071. [CrossRef] [PubMed]
2. Young, T.; Skatrud, J.; Peppard, P.E. Risk factors for obstructive sleep apnea in adult. *JAMA* **2004**, *291*, 2013–2016. [CrossRef] [PubMed]
3. Pevernagie, D.; Bauters, F.A.; Hertegonne, K. The Role of Patient-Reported Outcomes in Sleep Measurements. *Sleep Med. Clin.* **2021**, *16*, 595–606. [CrossRef]
4. Alomri, R.M.; Kennedy, G.A.; Wali, S.O.; Alhejaili, F.; Robinson, S.R. Association between nocturnal activity of the sympathetic nervous systemand cognitive dysfunction in obstructive sleep apnoea. *Sci. Rep.* **2021**, *11*, 11990. [CrossRef] [PubMed]
5. Pépin, J.L.; Jullian-Desayes, I.; Sapène, M.; Treptow, E.; Joyeux-Faure, M.; Benmerad, M.; Bailly, S.; Grillet, Y.; Stach, B.; Richard, P.; et al. Multimodal remote monitoring of high cardiovascular risk patients with OSA initiating CPAP: A randomized trial. *Chest* **2019**, *155*, 730–739. [CrossRef]
6. Jennum, P.; Castro, J.C.; Mettam, S.; Kharkevitch, T.; Cambron-Mellott, M.J. Socioeconomic and humanistic burden of illness of excessive daytime sleepiness severity associated with obstructive sleep apnoea in the European Union 5. *Sleep Med.* **2021**, *84*, 46–55. [CrossRef]
7. Jordan, A.S.; McSharry, D.G.; Malhotra, A. Adult obstructive sleep apnoea. *Lancet* **2014**, *383*, 736–747. [CrossRef]
8. Engleman, H.M.; Martin, S.E.; Deary, I.J.; Douglas, N.J. Effect of continuous positive airway pressure treatment on daytime function in sleep apnoea/hypopnoea syndrome. *Lancet* **1994**, *343*, 572–575. [CrossRef]
9. Marin, J.M.; Carrizo, S.J.; Vicente, E.; Agusti, A.G.N. Long-term cardiovascular outcomes in men with obstructive sleep apnoea-hypopnoea with or without treatment with continuous positive airway pressure: An observational study. *Lancet* **2005**, *365*, 1046–1053. [CrossRef]
10. Dong, J.Y.; Zhang, Y.H.; Qin, L.Q. Obstructive sleep apnea and cardiovascular risk: Meta-analysis of prospective cohort studies. *Atherosclerosis* **2013**, *229*, 489–495. [CrossRef]
11. Martínez-García, M.Á.; Soler-Cataluña, J.J.; Ejarque-Martínez, L.; Soriano, Y.; Román-Sánchez, P.; Illa, F.B.; Canal, J.M.; Duran-Cantolla, J. Continuous positive airway pressure treatment reduces mortality in patients with ischemic stroke and obstructive sleep apnea: A 5-year followup study. *Am. J. Respir. Crit. Care Med.* **2009**, *180*, 36–41. [CrossRef] [PubMed]
12. Gay, P.; Weaver, T.; Loube, D.; Iber, C. Evaluation of positive airway pressure treatment for sleep related breathing disorders in adults. *Sleep* **2006**, *29*, 381–401. [CrossRef] [PubMed]
13. Giles, T.L.; Lasserson, T.J.; Smith, B.; White, J.; Wright, J.J.; Cates, C.J. Continuous positive airways pressure for obstructivesleep apnea in adults. *Cochrane Database Syst. Rev* **2006**, *25*, CD001106.
14. Mashaqi, S.; Gozal, D. The impact of obstructive sleep apnea and PAP therapy on all-cause and cardiovascular mortality based on age and gender—A literature review. *Respir. Investig.* **2020**, *58*, 7–20. [CrossRef] [PubMed]
15. Sawyer, A.M.; Gooneratne, N.S.; Marcus, C.L.; Ofer, D.; Richards, K.C.; Weaver, T.E. A systematic review of CPAP adherence across age groups: Clinical and empiric insights for developing CPAP adherence interventions. *Sleep Med. Rev.* **2011**, *15*, 343–356. [CrossRef] [PubMed]
16. Zozula, R.; Rosen, R. Compliance with continuous positive airway pressure therapy: Assessing and improving treatment outcomes. *Curr. Opin. Pulm. Med.* **2001**, *7*, 391–398. [CrossRef]
17. Weaver, T.; Grunstein, R. Adherence to continuous positive airway pressure therapy. *Proc. Am. Thorac. Soc.* **2008**, *5*, 173–178. [CrossRef]
18. Rotenberg, B.W.; Murariu, D.; Pang, K.P. Trends in CPAP adherence over twenty years of data collection: A flattened curve. *J. Otolaryngol. Head Neck Surg.* **2016**, *45*, 43. [CrossRef]
19. Contal, O.; Poncin, W.; Vaudan, S.; De Lys, A.; Takahashi, H.; Bochet, S.; Grandin, S.; Kehrer, P.; Charbonnier, F. One-year adherence to continuous positive airway pressure with telemonitoring in sleep apnea hypopnea syndrome: A randomized controlled trial. *Front. Med.* **2021**, *8*, 626361. [CrossRef]
20. Kabat-Zinn, J. *Full Catastrophe Living (Revised Version): Using the Wisdom of Your Body and Mind to Face Stress, Pain, and Illness*; Bantam Books: New York, NY, USA, 2013.
21. Kabat-Zinn, J.; Hanh, T.N. *Full Catastrophe Living: Using the Wisdom of Your Body and Mind to Face Stress, Pain, and Illness*; Random House Publishing Group: New York, NY, USA, 2009.
22. Rosa, D.; Amigoni, C.; Rimoldi, E.; Ripa, P.; Ligorio, A.; Fracchiolla, M.; Lombardi, C.; Parati, G.; Perger, E. Obstructive Sleep Apnea and Adherence to Continuous Positive Airway Pressure (CPAP) Treatment: Let's Talk about Partners! *Healthcare* **2022**, *10*, 943. [CrossRef]
23. Rash, J.A.; Kavanagh, V.A.J.; Garland, S.N. A Meta-Analysis of Mindfulness-Based Therapies for Insomnia and Sleep Disturbance: Moving Towards Processes of Change. *Sleep Med. Clin.* **2019**, *14*, 209–233. [CrossRef] [PubMed]
24. Lewis, K.E.; Seale, L.; Bartle, I.E.; Watkins, A.J.; Ebden, P. Early predictors of CPAP use for the treatment of obstructive sleep apnea. *Sleep* **2004**, *27*, 134–138. [CrossRef] [PubMed]
25. Richard, W.; Venker, J.; den Herder, C.; Kox, D.; van den Berg, B.; Laman, M.; van Tinteren, H.; de Vries, N. Acceptance and long-term compliance of nCPAP in obstructive sleep apnea. *Eur. Arch. Oto-Rhino-Laryngol.* **2007**, *264*, 1081–1086. [CrossRef] [PubMed]
26. Zimmerman, E.; Arnedt, J.T.; Stanchina, M.; Millman, R.P.; Aloia, M.S. Normalization of memory performance and positive airway pressure adherence in memory-impaired patients with obstructive sleep apnea. *CHEST* **2006**, *130*, 1772–1778. [CrossRef]

27. Weaver, T.E.; Maislin, G.; Dinges, D.F.; Bloxham, T.; George, C.F.; Greenberg, H.; Kader, G.; Mahowald, M.; Younger, J.; Pack, A.I. Relationship between hours of CPAP use and achieving normal levels of sleepiness and daily functioning. *Sleep* **2007**, *30*, 711–719. [CrossRef]
28. Mehrtash, M.; Bakker, J.P.; Ayas, N. Predictors of Continuous Positive Airway Pressure Adherence in Patients with Obstructive Sleep Apnea. *Lung* **2019**, *197*, 115–121. [CrossRef]
29. Isetta, V.; Ruiz, M.; Farré, R.; Montserrat, J. Supporting patients receiving CPAP treatment: The role of training and telemedicine. *ERS Monogr.* **2015**, *67*, 280–292.
30. Van Ryswyk, E.; Anderson, C.S.; Antic, N.A.; Barbe, F.; Bittencourt, L.; Freed, R.; Heeley, E.; Liu, Z.; Loffler, K.A.; Lorenzi-Filho, G.; et al. Predictors of long-term adherence to continuous positive airway pressure in patients with obstructive sleep apnea and cardiovascular disease. *Sleep* **2019**, *42*, zsz152. [CrossRef]
31. Aloia, M.S.; Arnedt, J.T.; Riggs, R.L.; Hecht, J.; Borrelli, B. Clinical management of poor adherence to CPAP: Motivational enhancement. *Behav. Sleep Med.* **2004**, *2*, 205–222. [CrossRef]
32. Popescu, G.; Latham, M.; Allgar, V.; Elliott, M.W. Continuous positive airway pressure for sleep apnoea/hypopnoea syndrome: Usefulness of a 2 week trial to identify factors associated with long term use. *Thorax* **2001**, *56*, 727–733. [CrossRef]
33. Broström, A.; Strömberg, A.; Ulander, M.; Fridlund, B.; Mårtensson, J.; Svanborg, E. Perceived informational needs, side-effects and their consequences on adherence—A comparison between CPAP treated patients with OSAS and healthcare personnel. *Patient Educ. Couns.* **2009**, *74*, 228–235. [CrossRef] [PubMed]
34. Weaver, T.E. Novel Aspects of CPAP Treatment and Interventions to Improve CPAP Adherence. *J. Clin. Med.* **2019**, *8*, 2220. [CrossRef] [PubMed]
35. Patil, S.; Ayappa, I.A.; Caples, S.M.; Kimoff, R.J.; Patel, S.R. Treatment of adult obstructive sleep apnea with positive airway pressure: An American Academy of Sleep Medicine clinical practice guideline. *J. Clin. Sleep Med.* **2019**, *15*, 335–343. [CrossRef] [PubMed]
36. Smith, S.; Lang, C.; Sullivan, K.; Warren, J. Two new tools for assessing patients' knowledge and beliefs about obstructive sleep apnea and continuous positive airway pressure therapy. *Sleep Med.* **2004**, *5*, 359–367. [CrossRef] [PubMed]
37. Petersen, M.; Kristensen, E.; Berg, S.; Midgren, B. Sexual function in male patients with obstructive sleep apnoea after 1 year of CPAP treatment. *Clin. Respir. J.* **2013**, *7*, 214–219. [CrossRef]
38. Chai-Coetzer, C.L.; Luo, Y.M.; Antic, N.A.; Zhang, X.L.; Chen, B.Y.; He, Q.Y.; Heeley, E.; Huang, S.G.; Anderson, C.; Zhong, N.S.; et al. Predictors of long-term adherence to continuous positive airway pressure therapy in patients with obstructive sleep apnea and cardiovascular disease in the SAVE study. *Sleep* **2013**, *36*, 1929–1937. [CrossRef]
39. Pépin, J.L.; Krieger, J.; Rodenstein, D.; Cornette, A.; Sforza, E.; Delguste, P.; Deschaux, C.; Grillier, V.; Lévy, P. Effective compliance during the first 3 months of continuous positive airway pressure. A European prospective study of 121 patients. *Am. J. Respir. Crit. Care Med.* **1999**, *160*, 1124–1129. [CrossRef]
40. Watach, A.J.; Hwang, D.; Sawyer, A.M. Personalized and Patient-Centered Strategies to Improve Positive Airway Pressure Adherence in Patients with Obstructive Sleep Apnea. *Patient Prefer. Adherence* **2021**, *15*, 1557–1570. [CrossRef]
41. Scarpina, F.; Bastoni, I.; Cappelli, S.; Priano, L.; Giacomotti, E.; Castelnuovo, G.; Molinari, E.; Tovaglieri, I.M.A.; Cornacchia, M.; Fanari, P.; et al. Psychological Well-Being in Obstructive Sleep Apnea Syndrome Associated with Obesity: The Relationship with Personality, Cognitive Functioning, and Subjective and Objective Sleep Quality. *Front. Psychol.* **2021**, *12*, 588767. [CrossRef]
42. Bodenhausen, G.V. Stereotypes as judgmental heuristics: Evidence of circadian variations in discrimination. *Psychol. Sci.* **1990**, *1*, 319–322. [CrossRef]
43. Gordon, A.M.; Chen, S. The Role of Sleep in Interpersonal Conflict Do Sleepless Nights Mean Worse Fights? *Soc. Psychol. Personal. Sci.* **2014**, *5*, 168–175. [CrossRef]
44. Maranges, H.M.; McNulty, J.K. The rested relationship: Sleep benefits marital evaluations. *J. Fam. Psychol.* **2017**, *31*, 117–122. [CrossRef] [PubMed]
45. Ye, L.; Malhotra, A.; Kayser, K.; Willis, D.G.; Horowitz, J.A.; Aloia, M.S.; Weaver, T.E. Spousal involvement and CPAP adherence: A dyadic perspective. *Sleep Med. Rev.* **2015**, *19*, 67–74. [CrossRef] [PubMed]
46. Luyster, F.S. Impact of obstructive sleep apnea and its treatments on partners: A literature review. *J. Clin. Sleep Med.* **2017**, *13*, 467. [CrossRef]
47. Gentina, T.; Bailly, S.; Jounieaux, F.; Verkindre, C.; Broussier, P.M.; Guffroy, D.; Prigent, A.; Gres, J.J.; Kabbani, J.; Kedziora, L.; et al. Marital quality, partner's engagement and continuous positive airway pressure adherence in obstructive sleep apnea. *Sleep Med.* **2019**, *55*, 56–61. [CrossRef] [PubMed]
48. Villa, G.; Mannarini, M.; Della Giovanna, G.; Marzo, E.; Manara, D.F.; Vellone, E. A literature review about self-care on ostomy patients and their caregivers. *Int. J. Urol. Nurs.* **2019**, *13*, 75–80. [CrossRef]
49. Olsen, S.; Smith, S.; Oei, T.P. Adherence to continuous positive airway pressure therapy in obstructive sleep apnoea sufferers: Atheoretical approach to treatment adherence and intervention. *Clin. Psychol. Rev.* **2008**, *28*, 1355–1371. [CrossRef]
50. Wild, M.R.; Engleman, H.M.; Douglas, N.J.; Espie, C.A. Can psychological factors help us to determine adherence to CPAP? A prospective study. *Eur. Respir. J.* **2004**, *24*, 461–465. [CrossRef]
51. Maschauer, E.L.; Fairley, D.M.; Riha, R.L. Does personality play a role in continuous positive airway pressure compliance? *Breathe* **2017**, *13*, 32–43. [CrossRef]

52. Edmonds, J.C.; Yang, H.; King, T.S.; Sawyer, D.A.; Rizzo, A.; Sawyer, A.M. Claustrophobic tendencies and continuous positive airway pressure therapy non-adherence in adults with obstructive sleep apnea. *Heart Lung* **2015**, *44*, 100–106. [CrossRef]
53. Martin, L.R.; Williams, S.L.; Haskard, K.B.; DiMatteo, M.R. The challenge of patient adherence. *Clin. Risk Manag.* **2005**, *1*, 189–199.
54. Rapelli, G.; Pietrabissa, G.; Manzoni, G.M.; Bastoni, I.; Scarpina, F.; Tovaglieri, I.; Perger, E.; Garbarino, S.; Fanari, P.; Lombardi, C.; et al. Improving CPAP Adherence in Adults with Obstructive Sleep Apnea Syndrome: A Scoping Review of Motivational Interventions. *Front. Psychol.* **2021**, *12*, 3266. [CrossRef] [PubMed]
55. Weaver, T.E.; Sawyer, A.M. Adherence to continuous positive airway pressure treatment for obstructive sleep apnoea: Implications for future interventions. *Indian J. Med. Res.* **2010**, *131*, 245–248. [PubMed]
56. Prasad, B.; Steffen, A.D.; Van Dongen, H.; Pack, F.M.; Strakovsky, I.; Staley, B.; Dinges, D.F.; Maislin, G.; Pack, A.I.; Weaver, T.E. Determinants of sleepiness in obstructive sleep apnea. *Sleep* **2018**, *41*, zsx199. [CrossRef]
57. Sawyer, A.M.; Canamucio, A.; Moriarty, H.; Weaver, T.E.; Richards, K.C.; Kuna, S.T. Do cognitive perceptions influence CPAP use? *Patient Educ. Couns.* **2011**, *85*, 85–91. [CrossRef]
58. Ceccarini, M.; Borrello, M.; Pietrabissa, G.; Manzoni, G.M.; Castelnuovo, G. Assessing motivation and readiness to change for weight management and control: An in-depth evaluation of three sets of instruments. *Front. Psychol.* **2015**, *6*, 511. [CrossRef]
59. Shannon, R.; Donovan-Hall, M.; Bruton, A. Motivational interviewing in respiratory therapy:What do clinicians need to make it part of routine care? A qualitative study. *PLoS ONE* **2017**, *12*, e0187335. [CrossRef]
60. Bakker, J.P.; Weaver, T.E.; Parthasarathy, S.; Aloia, M.S. Adherence to CPAP: What Should We Be Aiming For, and How Can We Get There? *Chest* **2019**, *155*, 1272–1287. [CrossRef]
61. Engleman, H.M.; Wild, M.R. Improving CPAP use by patients with the sleep apnoea/hypopnoea syndrome (SAHS). *Sleep Med. Rev.* **2003**, *7*, 81–99. [CrossRef]
62. Richards, D.; Bartlett, D.J.; Wong, K.; Malouff, J.; Grunstein, R.R. Increased adherence to CPAP with a group cognitive-behavioral treatment intervention: A randomized trial. *Sleep* **2007**, *30*, 635–640. [CrossRef]
63. Donovan, L.M.; Boeder, S.; Malhotra, A.; Patel, S.R. New developments in the use of positive airway pressure for obstructive sleep apnea. *J. Thorac. Dis.* **2015**, *7*, 1323–1342.
64. Chaiard, J.; Weaver, T.E. Update on Research and Practices in Major Sleep Disorders: Part I. Obstructive Sleep Apnea Syndrome. *J. Nurs. Scholarsh.* **2019**, *51*, 500–508. [CrossRef] [PubMed]
65. Ragnoli, B.; Pochetti, P.; Raie, A.; Malerba, M. Comorbid Insomnia and Obstructive Sleep Apnea (COMISA): Current Concepts of Patient Management. *Int. J. Environ. Res. Public Health* **2021**, *18*, 9248. [CrossRef] [PubMed]
66. Black, D.S.; O'Reilly, G.A.; Olmstead, R.; Breen, E.C.; Irwin, M.R. Mindfulness meditation and improvement in sleep quality and daytime impairment among older adults with sleep disturbances: A randomized clinical trial. *JAMA Intern. Med.* **2015**, *175*, 494–501. [CrossRef] [PubMed]
67. Xiao, Q.; Yue, C.; He, W.; Yu, J.-Y. The Mindful Self: A Mindfulness-Enlightened Self-view. *Front. Psychol.* **2017**, *8*, 1752. [CrossRef]
68. Lin, Y.; Tang, R.; Braver, T.S. Investigating mindfulness influences on cognitive function: On the promise and potential of converging research strategies. *Psychon. Bull. Rev.* **2022**, *29*, 1198–1222. [CrossRef]
69. Edinger, J.D.; Arnedt, J.T.; Bertisch, S.M.; Carney, C.E.; Harrington, J.J.; Lichstein, K.L.; Sateia, M.J.; Troxel, W.M.; Zhou, E.S.; Kazmi, U.; et al. Behavioral and psychological treatments for chronic insomnia disorder in adults: An American Academy of Sleep Medicine systematic review, meta-analysis, and GRADE assessment. *J. Clin. Sleep Med.* **2021**, *17*, 263–298. [CrossRef] [PubMed]
70. Shonin, E.; Gordon, W.V.; Slade, K.; Griffiths, M.D. Mindfulness and other Buddhist-derived interventions in correctional settings: A systematic review. *Aggress. Violent Behav.* **2013**, *18*, 365–372. [CrossRef]
71. Gawrysiak, M.J.; Baime, M.; King, T.S.; Watach, A.J.; McPhillips, M.V.; Kolanowski, A.; Schutte-Rodin, S.; Sawyer, A.M. Intervention Design and Trial Protocol: Mindfulness-based Exposure for PAP-associated Claustrophobia. *West. J. Nurs. Res.* **2021**, *43*, 261–272. [CrossRef]
72. Jha, A.P.; Denkova, E.; Zanesco, A.P.; Witkin, J.E.; Rooks, J.; Rogers, S.L. Does mindfulness training help working memory 'work' better? *Curr. Opin. Psychol.* **2019**, *28*, 273–278. [CrossRef]
73. Veltri, A.; Scarpellini, P.; Piccinni, A.; Conversano, C.; Giacomelli, C.; Bombardieri, S.; Bazzichi, L.; Dell'Osso, L. Methodological approach to depressive symptoms in fibromyalgia patients. *Clin. Exp. Rheumatol* **2012**, *30*, S136–S142.
74. Bilyukov, R.G.; Nikolov, M.S.; Pencheva, V.P.; Petrova, D.S.; Georgiev, O.B.; Mondeshki, T.L.; Milanova, V.K. Cognitive Impairment and Affective Disorders in Patients with Obstructive Sleep Apnea Syndrome. *Front. Psychiatry* **2018**, *9*, 357. [CrossRef] [PubMed]
75. Haddock, N.; Wells, M.E. The association between treated and untreated obstructive sleep apnea and depression. *Neurodiagn J* **2018**, *58*, 30–39. [CrossRef] [PubMed]
76. Labarca, G.; Saavedra, D.; Dreyse, J.; Jorquera, J.; Barbe, F. Efficacy of CPAP for improvements in sleepiness, cognition, mood, and quality of life in elderly patients with OSA: Systematic review and meta-analysis of randomized controlled trials. *Chest* **2020**, *158*, 751–764. [CrossRef] [PubMed]
77. Carneiro-Barrera, A.; Amaro-Gahete, F.J.; Sáez-Roca, G.; Martín-Carrasco, C.; Ruiz, J.R.; Buela-Casal, G. Anxiety and Depression in Patients with Obstructive Sleep Apnoea before and after Continuous Positive Airway Pressure: The ADIPOSA Study. *J. Clin. Med.* **2019**, *8*, 2099. [CrossRef]
78. Chiesa, A.; Serretti, A.; Jakobsen, J.C. Mindfulness: Top-down or bottom-up emotion regulation strategy? *Clin. Psychol. Rev.* **2013**, *33*, 82–96. [CrossRef]

79. Goyal, M.; Singh, S.; Sibinga, E.M.; Gould, N.F.; Rowland-Seymour, A.; Sharma, R.; Berger, Z.; Sleicher, D.; Maron, D.D.; Shihab, H.M.; et al. Meditation programs for psychological stress and well-being: A systematic review and meta-analysis. *JAMA Intern. Med.* **2014**, *174*, 357–368. [CrossRef]
80. Li, Y.; Huang, X.; Su, J.; Wang, Y. Mindfulness may be a novel factor associated with CPAP adherence in OSAHS patients. *Sleep Breath.* **2020**, *24*, 183–190. [CrossRef]
81. Gross, J.J. The Emerging Field of Emotion Regulation: An Integrative Review. *Rev. Gen. Psychol.* **1998**, *2*, 271–299. [CrossRef]
82. Cunningham, T.J.; Kishore, D.; Guo, M.; Igue, M.; Malhotra, A.; Stickgold, R.; Djonlagic, I. The Effect of Obstructive Sleep Apnea on Sleep-Dependent Emotional Memory Consolidation. *Ann. Am. Thorac. Soc.* **2022**, *10*, 1513. [CrossRef]
83. Huberty, J.L.; Green, J.; Puzia, M.E.; Larkey, L.; Laird, B.; Vranceanu, A.M.; Vlisides-Henry, R.; Irwin, M.R. Testing a mindfulness meditation mobile app for the treatment of sleep-related symptoms in adults with sleep disturbance: A randomized controlled trial. *PLoS ONE* **2021**, *16*, e0244717. [CrossRef] [PubMed]
84. Gao, M.; Roy, A.; Deluty, A.; Sharkey, K.M.; Hoge, E.A.; Liu, T.; Brewer, J.A. Targeting Anxiety to Improve Sleep Disturbance: A Randomized Clinical Trial of App-Based Mindfulness Training. *Psychosom. Med.* **2022**, *84*, 632–642. [CrossRef] [PubMed]

life

Article

The Efficacy of the Partial Glossectomy for Prevention of Airway Volume Reduction in Orthognathic Surgery of Class III Patients

Suyun Seon †, Junho Jung †, Baek-Soo Lee, Yong-Dae Kwon, Byung-Joon Choi and Joo-Young Ohe *

Department of Oral and Maxillofacial Surgery, School of Dentistry, Kyung-Hee University, Seoul 02447, Republic of Korea
* Correspondence: ojyoung81@khu.ac.kr; Tel.: +82-2961-0271
† These authors contributed equally to this work.

Abstract: The aim of this study was to evaluate the effects of a partial glossectomy on volumetric changes of pharyngeal airway space (PAS) in patients with mandibular setback surgery. Overall, 25 patients showing clinical features related to macroglossia treated with mandibular setback surgery were included in this retrospective study. Subjects were divided into two groups: the control group (G1, $n = 13$, with BSSRO) and the study group (G2, $n = 12$, with both BSSRO and partial glossectomy). The PAS volume of both groups was measured by the OnDemand 3D program on CBCT taken shortly before operation (T0), 3 months post-operative (T1), and 6 months post-operative (T2). A paired t-test and repeated analysis of variance (ANOVA) were used for statistical correlation. Total PAS and hypopharyngeal airway space were increased after operation in Group 2 compared to Group 1 ($p < 0.05$), while oropharyngeal airway space showed no significant statistical difference with the tendency of increasing. The combination of partial glossectomy and BSSRO surgical techniques had a significant effect on increasing the hypopharyngeal and total airway space in class III malocclusion patients ($p < 0.05$).

Keywords: pharyngeal airway; glossectomy; mandibular setback surgery

Citation: Seon, S.; Jung, J.; Lee, B.-S.; Kwon, Y.-D.; Choi, B.-J.; Ohe, J.-Y. The Efficacy of the Partial Glossectomy for Prevention of Airway Volume Reduction in Orthognathic Surgery of Class III Patients. *Life* **2023**, *13*, 280. https://doi.org/10.3390/life13020280

Academic Editors: David Slouka and Milan Štengl

Received: 17 October 2022
Revised: 6 January 2023
Accepted: 17 January 2023
Published: 19 January 2023

1. Introduction

It is well known that orthognathic surgery on patients with skeletal malocclusion changes orofacial skeletal and soft tissues. Consideration of not only skeletal changes but also consideration of soft tissue changes are also emphasized after surgery because changes in location of skeletal and soft tissues rely on the mutual relationship between skeletal and neuromuscular components [1,2]. Alternation of soft tissues changes how they look but also changes the location of the tongue, location of the hyoid bone, and the pharyngeal airway space [3]. Among soft tissue components, it is well known that the size and location of the tongue affects the skeletal shape and teeth alignment, especially in those that have the skeletal class III malocclusion. Macroglossia is not only a common cause of open bite and mandibular prognathism but also a cause of increasing risk of post-operative relapse and decreasing skeletal stability after orthognathic surgery [4,5]. Reducing the tongue volume by partial glossectomy is recommended as a treatment for patients with open bite and macroglossia.

The importance of the size and location of the tongue becomes more apparent when mandibular setback is planned because the reduced volume of the oral cavity is occupied more by the tongue even though it is normal size [6]. According to many studies, the partial glossectomy is not a necessary procedure because the hyoid bone and the base of the tongue move downward as a compensation mechanism after orthognathic surgery so that the upper respiratory airway is kept open and the oral cavity is less occupied by the volume of the tongue [7,8]. However, if the hyoid bone and the base of tongue fail to move downward, therefore reducing the patient's oral cavity, the patient is likely to have discomfort. Because the relative increase in size of the tongue causes a higher chance of post-operative relapse, and a lesser amount of overjet and overbite, both the skeletal and soft tissue components must be considered together. This is because the volume of airway space that is determined by the location of the hyoid bone and tongue affects post-operative stability and relapse; therefore, those considerations will improve the stability and prevent the relapse.

Partial glossectomy is known not to cause movement and speech problems of the tongue [9,10]. However, there is no objective criteria for partial glossectomy following orthognathic surgery until recently. Partial glossectomy is used depending on the symptoms of patients and a subjective decision made by surgeons. Up to recently, two-stage surgery is performed when a partial glossectomy is required following mandibular setback surgery because of airway obstruction by the tongue edema and bleeding. However, as the orthognathic surgery technique and method of fixation developed, the problem of post-operative airway obstruction was resolved, and two-stage surgery could be conducted simultaneously. Thus, reducing the size of the tongue by partial glossectomy improves post-operative stability and induces adaptation of the tongue [11,12].

Generally, lateral cephalography is used to measure the volume of airway space, but it has some limitations, such as anatomical structural overlapping and two-dimensional analysis. In contrast, digitalized computed tomography (CT), using a three-dimensional analyzing program, provides more detailed analysis by detecting volumetric changes and avoiding the overlapping problem.

Conventional orthognathic surgery alone can become challenging when the aim is preserving airway space. Thus, additional procedures such as tongue reduction due to macroglossia may be alternatives to a more functional outcome. The purpose of this study was to observe three-dimensional volumetric changes of pharyngeal airway space using Cone-Beam CT (CBCT) and find out the necessity of a partial glossectomy by comparison between the patients who underwent simultaneous mandibular setback surgery and partial glossectomy, and the patients who only underwent mandibular setback surgery.

2. Materials and Methods

2.1. Patients

Twenty-five patients who were diagnosed with skeletal class III malocclusion and treated with maxillary Le Fort I osteotomy and mandibular bilateral sagittal split ramus osteotomy (BSSRO) at the Department of Oral and Maxillofacial surgery, Kyung-Hee University School of Dentistry from 2010 to 2014 were identified. The inclusion criteria were as follows: (1) maxillary advancement surgery by 0–5 mm and mandibular backward movement by 2–15 mm; (2) macroglossia-related clinical features described by Wolford and Cottrell in their study; (3) 6 months follow-up with post-operative CBCT; (4) no obstructive sleep apnea (OSA)-related symptoms. The patients were categorized into 2 groups. Group 1 consisted of 13 patients (mean age of 24 $\pm$ 3.19 years) who underwent Le Fort I osteotomy and mandibular BSSRO. Group 2 included 12 patients (mean age of 25 $\pm$ 6.89 years) who underwent the same surgery as Group 1 but also underwent partial glossectomy simultaneously. Post-operatively, occlusion of those 2 groups was stabilized by splint and they underwent exactly the same post-operative treatment.

2.2. Surgical Procedure

For mandibular surgery, semi-rigid fixation was conducted with a monocortical plate and three miniscrews per side. Following mandibular movement, the miniplate was fixed to maintain the internal gap between bones in order to prevent the displacement of the proximal segment. After making a reference point on the ascending ramus of the mandible, the distance was measured from the reference point to more than 3 points on brackets on the maxillary teeth before surgery to maintain the original position of the proximal segment. Intermaxillary fixation was placed for 2 weeks post-operatively. An opening exercise of the mouth was performed for 4 weeks after the fixation period. In Group 1, maxillary advancement was 0–5 mm (mean 1.9 mm) and mandibular setback was 6–15 mm (mean 8.4 mm), whereas maxillary advancement was 0–3 mm (mean 1.19 mm) and mandibular setback was 2–13 mm (mean 8.5 mm) in Group 2. There were no significant differences in the movement of maxilla and mandible in Group 1 and Group 2.

We performed a T-shaped partial glossectomy that was recommended by Ueyama (9). The horizontal line (Figure 1A—a,b) at the dorsal surface of the tongue begins in front of the vallate papilla, and the tip portion (Figure 1A—e) begins at the dorsal surface of the tongue at least 1 cm from the apex of tongue. The incision is made in a horizontal line down to the upper part of the transverse lingual muscle, and then another incision is made in a V shape to eliminate any in the transverse lingual muscle. In order to eliminate the dead space and gather all parts of tongue in the center, inner muscles and surface tissues of the tongue were sutured. As a result, the length and width of the tongue were reduced. The reduced volume of the tongue was 7–22 cc (mean 9.91 cc). There were no patients complaining about movement, taste, or speech problems.

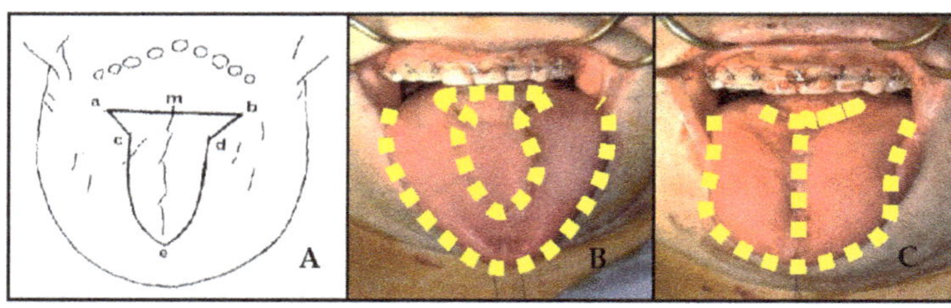

Figure 1. (**A**) Incision used for tongue reduction; (**B**) pre-operative; (**C**) post-operative.

The panoramic mode of CBCT was used to analyze the pre-operative (T0), 3 months post-operative (T1), and 6 months post-operative state (T2). The volume of the airway right after surgery was not measured because the nasopharyngeal airway was used for 3 days to keep the airway open, and it was significantly reduced due to edema of the soft palate, and edema coming from anesthesia and the surgical procedure on the wall of the pharyngeal airway. The CBCT that was used for measurement was the Vega 3030 Dental CT system (Asahi Roentgen Ind. Co., Ltd., Kyoto, Japan). The patient's head was fixed to make the FH plane parallel to the floor and the shot was made in panoramic mode. All pictures were taken at 80 kVp and 5–10 mA with a duration of 17 sec. This study was approved by the Institutional Review Board (IRB) at Kyung-Hee University, School of Dentistry (KHD IRB 1510-2).

2.3. Volumetric Analysis

Analyzation was conducted with a 0.3 mm thickness of Raw-Dicom file, and OnDemand 3D (Cyber Med., Seoul, Republic of Korea) was used to analyze three-dimensionally. The analyzation was conducted by only one person to eliminate measurement errors and collected information was analyzed by OnDemand 3D that was used to measure the volume of airway at T0, T1, and T2.

To measure the volume of the airway space, it was measured in 3 different ways as follows:

(1) Oropharyngeal airway space: the space was defined from the line which was passing through the posterior nasal spine and parallel to the FH plane to the end of the epiglottis, generally the most inferior portion of the soft palate.
(2) Hypopharyngeal airway space: the space was defined below the oropharyngeal airway space and up to the line which was passing through the epiglottis and parallel to the FH plane, generally at the level of the end of the 3rd cervical vertebra.
(3) Total airway was the sum of oropharyngeal airway space and hypopharyngeal airway space.

For easier understanding in this study, oropharyngeal airway space was renamed as "Airway 1" and hypopharyngeal airway space was renamed as "Airway 2" (Figures 2 and 3).

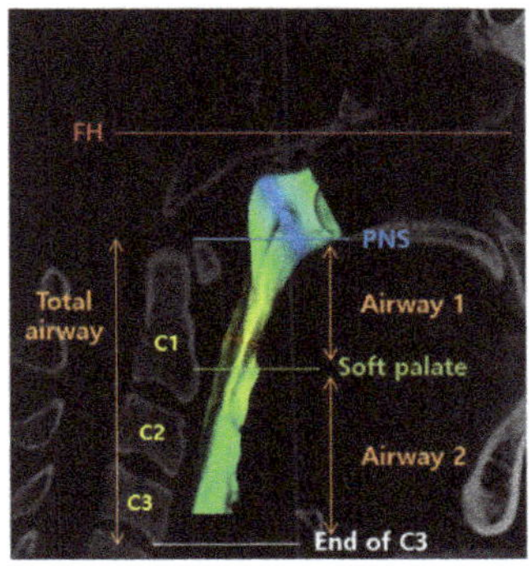

Figure 2. Demonstration of pharyngeal airway section with the volumetric analysis on the CT.

Figure 3. Axial view of airway space shows the increase of airway volume is related with the anterior-posterior width.

2.4. Statistical Analysis

Mean and standard deviation were obtained for each volumetric change in airway at T0, T1, and T2 in both Group 1 and 2. To find out statistical correlation, SPSS v21.0 (IBM Co., Armonk, NY, USA) was used to perform a matched paired t-test. The statistical correlation with and without partial glossectomy was determined using repeated measure analysis of variance (ANOVA). To test their significance, each test was conducted at a significance level of 95%.

3. Results

3.1. Volumetric Changes in Airway 1

Volume of Airway 1 in Group 1 was recorded as 10.33 cc at T0, 9.08 cc at T1, which was slightly reduced from T0 ($p > 0.05$), and 9.67 cc at T2, which showed a tendency of recovering but was still less than volume at T0 ($p > 0.05$). Volume of Airway 1 in Group 2 was recorded as 12.36 cc at T0, 12.91 cc at T1, which showed a tendency of recovering ($p > 0.05$), and 13.55 cc at T2 ($p > 0.05$), which was increased by 1.10 cc compared to volume at T0 (Tables 1 and 2) (Figure 4). When a repeated measure ANOVA was used to find and analyze the volumetric changes in both Group 1 and 2, there was no statistically significant difference depending on time ($p > 0.05$). When the effect of both time and partial glossectomy was analyzed, there was no statistically significant difference ($p > 0.05$).

3.2. Volumetric Changes in Airway 2

The volume of Airway 2 in Group 1 was recorded as 19.00 cc at T0, 17.92 cc at T1, which was slightly reduced ($p > 0.05$), and 17.50 cc at T2 ($p > 0.05$), which was reduced by 1.19 cc compared to volume at T0 but not significantly different ($p > 0.05$). The volume of Airway 2 in Group 2 was measured as 20.27 cc at T0, 17.45 cc at T1, which was reduced ($p > 0.05$), and 22.00 cc at T2, which showed a tendency of recovering ($p < 0.05$) and was increased by 1.65 cc compared to volume at T0 ($p > 0.05$) (Tables 3 and 4) (Figure 5). When the differences in both groups were analyzed using a repeated measure ANOVA, there were statistically significant differences depending on the amount of recovery time after surgery ($p < 0.05$); therefore, the effect of partial glossectomy was proved ($p < 0.05$) (Table 5).

Table 1. Volume of oropharyngeal airway (CC).

	T0	T1	T2
	(Mean ± SD)	**(Mean ± SD)**	**(Mean ± SD)**
Group 1	10.33 ± 3.63	9.08 ± 2.68	9.67 ± 3.08
Group 2	12.36 ± 2.69	12.91 ± 2.88	13.55 ± 3.36

T0: pre-operation; T1: 3 months after operation; T2: 6 months after operation.

Table 2. Change of oropharyngeal airway before and after surgery (CC).

	T0–T1	T1–T2	T2–T0
	(Mean ± SD)	**(Mean ± SD)**	**(Mean ± SD)**
Group 1	1.48 ± 0.74	0.79 ± 1.28	−2.27 ± 1.57
Group 2	−0.54 ± 0.52	0.79 ± 1.28	1.10 ± 0.89

The results of matched paired t-test. T0: pre-operation; T1: 3 months after operation; T2: 6 months after operation.

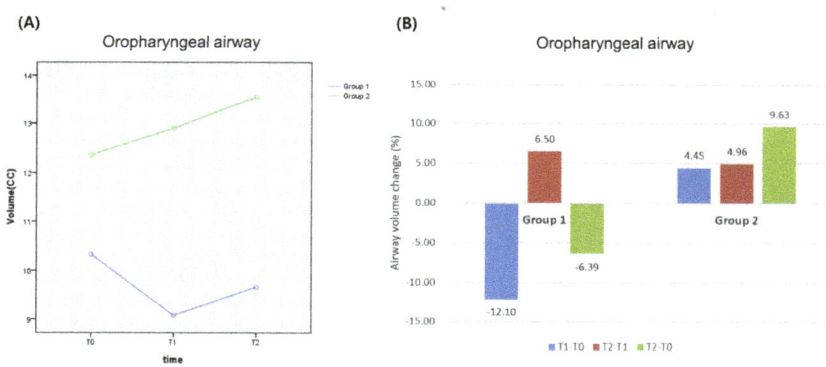

Figure 4. Change of oropharyngeal airway before and after surgery: (**A**) volumetric change; (**B**) percentage change. Group 1: Le Fort I + B-SSRO; Group 2: Le Fort I + B-SSRO + partial glossectomy.

Table 3. Volume of hypopharyngeal airway (CC).

	T0	T1	T2
	(Mean ± SD)	**(Mean ± SD)**	**(Mean ± SD)**
Group 1	19.00 ± 7.92	17.92 ± 5.63	17.50 ± 6.53
Group 2	20.27 ± 7.77	17.45 ± 7.69	22.00 ± 7.25

T0: pre-operation; T1: 3 months after operation; T2: 6 months after operation.

Table 4. Change of hypopharyngeal airway before and after surgery (CC).

	T0–T1	T1–T2	T2–T0
	(Mean ± SD)	**(Mean ± SD)**	**(Mean ± SD)**
Group 1	3.73 ± 1.35	1.79 ± 1.62	−1.19 ± 2.42
Group 2	2.82 ± 1.70	−4.48 ± 1.58 *	1.65 ± 1.69

* Indicates the significant variation of the matched paired t-test ($p < 0.05$). T0: pre-operation; T1: 3 months after operation; T2: 6 months after operation.

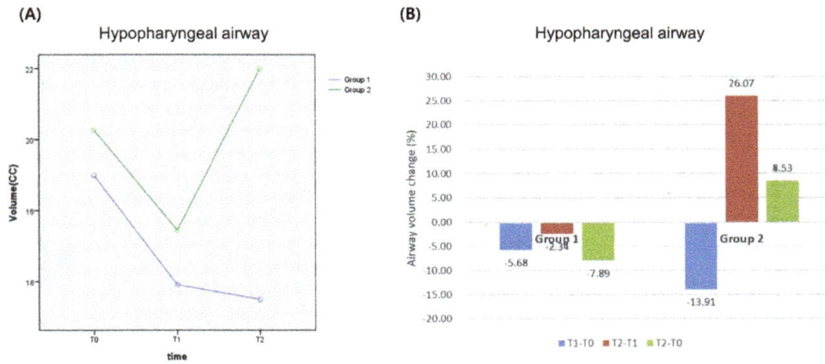

Figure 5. Change of hypopharyngeal airway before and after surgery: (**A**) volumetric change; (**B**) percentage change. Group 1: Le Fort I + B-SSRO; Group 2: Le Fort I + B-SSRO + partial glossectomy.

Table 5. Repeated measure analysis of variance between measurements of hypopharyngeal airway.

Time	Time × Glossectomy
0.012 *	0.013 *

* Indicates the significant variation ($p < 0.05$).

3.3. Volumetric Changes in Total Airway

The volume of the total airway in Group 1 was measured as 29.26 cc at T0, 27.17 cc at T1 ($p > 0.05$), which was reduced from T0 and 27.17 cc at T2, which was not changed from T1. It was reduced by 2.18 cc compared to the volume at T0, and there was no statistical difference ($p > 0.05$). The volume of total airway in Group 2 was recorded as 32.66 cc at T0, 30.66 cc at T1, which was reduced from T0, and 35.36 cc at T2, which showed a tendency of recovering by 2.75 cc. It showed a statistically significant difference compared from volume at T0 to volume at T2 ($p < 0.05$) (Tables 6 and 7) (Figure 6). When the differences in both groups were analyzed by a repeated measure ANOVA, there was statistically significant difference in terms of time; the effect of partial glossectomy was proved ($p < 0.05$). This tendency of change was similar to the tendency of volumetric changes in the hypopharyngeal airway (Table 8).

Table 6. Volume of total airway (CC).

	T0	T1	T2
	(Mean ± SD)	(Mean ± SD)	(Mean ± SD)
Group 1	29.26 ± 11.15	27.17 ± 7.98	27.17 ± 9.06
Group 2	32.66 ± 10.01	30.36 ± 9.86	35.36 ± 9.97

T0: pre-operation; T1: 3 months after operation; T2: 6 months after operation.

Table 7. Change of total airway before and after surgery (CC).

	T0–T1	T1–T2	T2–T0
	(Mean ± SD)	(Mean ± SD)	(Mean ± SD)
Group 1	1.08 ± 1.89	2.58 ± 2.84	−2.18 ± 3.92
Group 2	2.28 ± 2.01	−5.04 ± 2.49 *	2.75 ± 2.37 *

* Indicates the significant variation of the matched paired t-test ($p < 0.05$). T0: pre-operation; T1: 3 months after operation; T2: 6 months after operation.

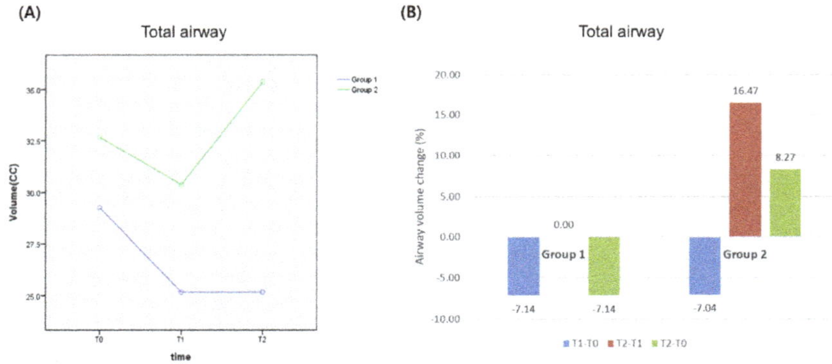

Figure 6. Change of total airway before and after surgery: (**A**) volumetric change; (**B**) percentage change. Group 1: Le Fort I + B-SSRO; Group 2: Le Fort I + B-SSRO + partial glossectomy.

Table 8. Repeated measure analysis of variance between measurements of Total airway.

Time	Time × Glossectomy
0.049 *	0.037 *

* Indicates the significant variation ($p < 0.05$).

4. Discussion

There have been many studies conducted about complications of orthognathic surgery on patients with skeletal class III malocclusion [13,14]. Furthermore, a large number of them are related to the study of changes in airway space after surgery, and it is an actively ongoing subject of research [15,16]. Gu et al. reported that there was a correlation in the location of the hyoid bone, volume of airway, and the position of the head after mandibular setback surgery, and compensation of the volumetric decrease in the airway caused a change in position of the hyoid bone and tongue [17]. Furthermore, Takaki et al. reported that tongue muscles moved downward due to the downward movement of the hyoid bone after mandibular setback surgery. This movement was a type of compensation for reduced airway space because of the downward movement of the tongue. As time passed, the hyoid bone had a tendency to go back to its original position. Moreover, the volume of airway space recovered to its original state [18]. However, these results are contradicted by some other studies [19,20]. These days, there are many studies about obstructive sleep apnea (OSA) after mandibular setback surgery [21,22]. Riley et al. reported that there was a higher risk of getting OSA after orthognathic surgery on patients with mandibular prognathism [23]. Hochban et al. also stated that OSA was induced if pharyngeal airway space after surgery was less than 10 mm. In this case, Hochban mentioned that orthognathic surgery, including maxillary advancement, had to be considered [24].

Especially, patients with mandibular prognathism and pseudomacroglossia have a higher risk of having a post-operative relapse, and this condition is one of factors that decrease skeletal stability [4,5]. Additionally, a decrease in airway space due to the reduced volume of oral cavity increases the chances of having patients feel discomfort after surgery. However, the effect of partial glossectomy on reducing discomfort is still controversial. Wickwire and Sinclare et al. mentioned that partial glossectomy was not required after surgery because pharyngeal airway space was maintained by downward movement of the tongue base and hyoid bone. Furthermore, the volume of the tongue was reduced in the recovering process [7,8]. In contrast, Ingervall et al. mentioned that as the mandible moved backward and the tongue also moved backward, total efficiency of breathing was decreased due to compression of the upper airway space. As the mandible moved downward to compensate for the increased intra-oral cavity pressure, post-operative stability was decreased, inducing anterior open bite [25]. Swanson and Petdachai et al. also insisted that as an enlarged tongue after orthognathic surgery pushed the mandible and anterior teeth of mandible, post-operative relapse could be caused [11,12]. Allison stated that partial glossectomy could help patients to reduce discomfort if the pre-mentioned compensation was not present and if there was an enlarged tongue right after surgery with no compensation yet [11].

Up to recently, there have been a lot of studies conducted about volumetric changes in airway space post-operatively using lateral cephalography and some reference points. However, the analysis was only able to be conducted two-dimensionally on pharyngeal airway space and the hyoid bone, such as the anterior-posterior relationship of airway space and the upward-downward movement of hyoid bone. Therefore, in this study, three-dimensional CBCT was used to analyze the effect of partial glossectomy on hypopharyngeal airway space changes of patients who underwent simultaneous partial glossectomy and mandibular setback surgery. In Group 1 without partial glossectomy, there was no statistically significant decrease in oropharyngeal and hypopharyngeal airway space even though there was a continuous decrease in airway space from T0 to T2. These results supported the study of Wickwire and Athanasious et al. They stated that as the mandible and its surrounding structures moved backward, the airway was expected to be compressed. However, functional and structural adaptation to compensate for airway compression made no significant change in volumetric airway space before and after surgery [7,26].

In Group 2 with partial glossectomy, there was no statistically significant changes in the volume of the oropharyngeal airway, but it showed slightly the tendency of increasing. In Group 1, the hypopharyngeal and total airway were reduced continuously from T0 to T2. In contrast, the hypopharyngeal and total airway were reduced from T0 to T1 but increased from T1 to T2 in Group 2. When the volumetric change of the pharyngeal airway was analyzed dependent upon post-operative time, there was a statistically significant difference in the hypopharyngeal and total airway, and it was thought to be the effect of a partial glossectomy. Since the volumetric change pattern in the hypopharyngeal was similar to the one in the total airway, it was thought to be that partial glossectomy increased hypopharyngeal airway space, and this change increased the total airway space. This result supports the studies of Kwakami that there were differences in the location of the base of the tongue between the patients who received and who did not receive a glossectomy [27]. It was not only the same as the result of Group 1, which was the result of recovering to its original volume of the airway, but also the effect of a partial glossectomy. In other words, decreasing the tongue size caused a decrease in clock-wise rotational movement of the mandible, and a decrease in the amount of invasion into the pharyngeal airway space by the tongue and its surrounding tissues. The reasons of volumetric changes in the hypopharyngeal and total airway of Group 2 from T0 to T2 are that the hyoid bone moves downward to compensate for the decrease in the volume of the oropharyngeal airway due to the backward movement of the mandible after surgery. Then, the movement of the hyoid bone causes a decrease in the volume of the hypopharyngeal airway. However, after 6 months since surgery, the hyoid bone moves back to its original position, and the tongue becomes stable in its size and location. Considering the results of this study, a partial glossectomy is considered to be effective for increasing the airway space with mandibular setback surgery.

Further studies are required on the effect of a partial glossectomy dependent on the shape of the tongue, the size of the tongue, the location of the tongue, the volume of airway space, and the amount of movement of the mandible.

Swallowing and articulation disorders have been reported to be common complications of glossectomy [28]. The factors that influence their development are the extent and location of surgical resection and the flexibility of the residual tongue. Along with many other reports of partial glossectomy resulting in good post-operative tongue function, our post-operative follow-up showed that the function of swallowing and articulation over the long term is generally acceptable after a partial glossectomy.

While this study was focused on the efficacy of the partial glossectomy for prevention of airway reduction, there is currently insufficient direct data to draw any firm conclusions due to the small sample size and lack of post-operative assessment on long-term stability. Furthermore, some potential therapy using low-level lasers and Diode lasers for wound healing and pain relief have been reported [29,30]. Further studies to evaluate the treatment of lasers in prevention of complications of glossectomy in patients with orthognathic surgery are recommended.

5. Conclusions

According to the results of the study, it was concluded that:

1. There was decrease in the volume of airway space in the group without partial glossectomy, but the extent was not statistically significant.

2. In the group with partial glossectomy, their airway spaces were decreased from pre-operation to 3 months post-operative, but it was increased after 3 months. This was because the location of the hyoid bone and tongue muscle was recovered after 3 months post-operative, and partial glossectomy was considered as one of the causes that increased the volume of the oral cavity.

3. Based on the length of time following surgery, the tendency of volumetric changes in the oropharyngeal and total airway space of both groups revealed statistical differences. The reason for this change is that a partial glossectomy increased both hypopharyngeal and the total airway space. The essential clinical factors, such as post-operative stability, relapse prevention, and maintaining an open airway, were all improved by those changes [9,31].

In conclusion, according to the results regarding the tendency of volumetric changes in airway space of patients with skeletal class III malocclusion post-operatively, simultaneous mandibular setback surgery with partial glossectomy was determined to be helpful for patients who need backward movement of the mandible or who have high possibility of having respiratory obstructions such as snoring and obstructive apnea (OSA) after surgery.

Author Contributions: Conceptualization, B.-S.L., Y.-D.K. and B.-J.C.; methodology, S.S.; formal analysis, J.-Y.O.; investigation, S.S.; data curation, J.-Y.O.; writing—original draft preparation, S.S. and J.J.; writing—review and editing, J.-Y.O. All authors have read and agreed to the published version of the manuscript.

Funding: This research received no external funding.

Institutional Review Board Statement: The study was conducted according to the guidelines of the Declaration of Helsinki and approved by the Hungarian Research Ethics Authority (National Institute of Pharmacy and Nutrition, approval reference number: KHD IRB 1510-2).

Informed Consent Statement: Informed consent was obtained from all subjects involved in the study.

Data Availability Statement: Data supporting reported results can be provided upon reasonable request.

Conflicts of Interest: The authors declare no conflict of interest.

References

1. Wu, Y.T.; Sun, J.; Li, Y.L.; Chen, L.Q.; Chen, C. Three dimensional study on change ratios of hard and soft tissue after orthognathic surgery. *Shanghai Kou Qiang Yi Xue* **2019**, *28*, 158–164. [PubMed]
2. Jensen, A.C.; Sinclair, P.M.; Wolford, L.M. Soft tissue changes associated with double jaw surgery. *Am. J. Orthod. Dentofac. Orthop.* **1992**, *101*, 266–275. [CrossRef] [PubMed]
3. Greco, J.M.; Frohberg, U.; Van Sickels, J.E. Long-term airway space changes after mandibular setback using bilateral sagittal split osteotomy. *Int. J. Oral Maxillofac. Surg.* **1990**, *19*, 103–105. [CrossRef]
4. Miyawaki, S.; Oya, S.; Noguchi, H.; Takano-Yamamoto, T. Long-term changes in dentoskeletal pattern in a case with Beckwith-Wiedemann syndrome following tongue reduction and orthodontic treatment. *Angle Orthod.* **2000**, *70*, 326–331. [CrossRef] [PubMed]

5. Wolford, L.M.; Cottrell, D.A. Diagnosis of macroglossia and indications for reduction glossectomy. *Am. J. Orthod. Dentofac. Orthop.* **1996**, *110*, 170–177. [CrossRef]
6. Kawakami, M.; Yamamoto, K.; Fujimoto, M.; Ohgi, K.; Inoue, M.; Kirita, T. Changes in tongue and hyoid positions, and posterior airway space following mandibular setback surgery. *J. Craniomaxillofac. Surg.* **2005**, *33*, 107–110. [CrossRef]
7. Wickwire, N.A.; White, R.P., Jr.; Proffit, W.R. The effect of mandibular osteotomy on tongue position. *J. Oral Surg.* **1972**, *30*, 184–190.
8. Brammer, J.; Finn, R.; Bell, W.H.; Sinn, D.; Reisch, J.; Dana, K. Stability after bimaxillary surgery to correct vertical maxillary excess and mandibular deficiency. *J. Oral Surg.* **1980**, *38*, 664–670.
9. Ueyama, Y.; Mano, T.; Nishiyama, A.; Tsukamoto, G.; Shintani, S.; Matsumura, T. Effects of surgical reduction of the tongue. *Br. J. Oral Maxillofac. Surg.* **1999**, *37*, 490–495. [CrossRef]
10. Tselnik, M.; Pogrel, M.A. Assessment of the pharyngeal airway space after mandibular setback surgery. *J. Oral Maxillofac. Surg.* **2000**, *58*, 282–285. [CrossRef]
11. Petdachai, S.; Inoue, Y.; Inoue, H.; Sakuda, M. Orthognathic surgical approach and partial glossectomy to a skeletal 3 adult open bite. *J. Osaka Univ. Dent. Sch.* **1993**, *33*, 14–20. [PubMed]
12. Swanson, L.T.; Murray, J.E. Partial glossectomy to stabilize occlusion following surgical correction of prognathism. Report of a case. *Oral Surg. Oral Med. Oral Pathol.* **1969**, *27*, 707–715. [CrossRef] [PubMed]
13. Wolford, L.M. Comprehensive Post Orthognathic Surgery Orthodontics: Complications, Misconceptions, and Management. *Oral Maxillofac. Surg. Clin. N. Am.* **2020**, *32*, 135–151. [CrossRef] [PubMed]
14. Panula, K.; Finne, K.; Oikarinen, K. Incidence of complications and problems related to orthognathic surgery: A review of 655 patients. *J. Oral Maxillofac. Surg.* **2001**, *59*, 1128–1136. [CrossRef]
15. Mattos, C.T.; Vilani, G.N.; Sant'Anna, E.F.; Ruellas, A.C.; Maia, L.C. Effects of orthognathic surgery on oropharyngeal airway: A meta-analysis. *Int. J. Oral Maxillofac. Surg.* **2011**, *40*, 1347–1356. [CrossRef]
16. Christovam, I.O.; Lisboa, C.O.; Ferreira, D.M.; Cury-Saramago, A.A.; Mattos, C.T. Upper airway dimensions in patients undergoing orthognathic surgery: A systematic review and meta-analysis. *Int. J. Oral Maxillofac. Surg.* **2016**, *45*, 460–471. [CrossRef]
17. Gu, G.; Gu, G.; Nagata, J.; Suto, M.; Anraku, Y.; Nakamura, K.; Kuroe, K.; Ito, G. Hyoid position, pharyngeal airway and head posture in relation to relapse after the mandibular setback in skeletal Class III. *Clin. Orthod. Res.* **2000**, *3*, 67–77. [CrossRef]
18. Takagi, Y.; Gamble, J.W.; Proffit, W.R.; Christiansen, R.L. Postural change of the hyoid bone following osteotomy of the mandible. *Oral Surg. Oral Med. Oral Pathol.* **1967**, *23*, 688–692. [CrossRef]
19. Mehra, P.; Downie, M.; Pita, M.C.; Wolford, L.M. Pharyngeal airway space changes after counterclockwise rotation of the maxillomandibular complex. *Am. J. Orthod. Dentofac. Orthop.* **2001**, *120*, 154–159. [CrossRef]
20. Liukkonen, M.; Vahatalo, K.; Peltomaki, T.; Tiekso, J.; Happonen, R.P. Effect of mandibular setback surgery on the posterior airway size. *Int. J. Adult Orthod. Orthognath. Surg.* **2002**, *17*, 41–46.
21. Demetriades, N.; Chang, D.J.; Laskarides, C.; Papageorge, M. Effects of mandibular retropositioning, with or without maxillary advancement, on the oro-naso-pharyngeal airway and development of sleep-related breathing disorders. *J. Oral Maxillofac. Surg.* **2010**, *68*, 2431–2436. [CrossRef] [PubMed]
22. Jakobsone, G.; Stenvik, A.; Espeland, L. The effect of maxillary advancement and impaction on the upper airway after bimaxillary surgery to correct Class III malocclusion. *Am. J. Orthod. Dentofac. Orthop.* **2011**, *139*, e369–e376. [CrossRef] [PubMed]
23. Riley, R.W.; Powell, N.B.; Guilleminault, C.; Ware, W. Obstructive sleep apnea syndrome following surgery for mandibular prognathism. *J. Oral Maxillofac. Surg.* **1987**, *45*, 450–452. [CrossRef] [PubMed]
24. Hochban, W.; Schurmann, R.; Brandenburg, U.; Conradt, R. Mandibular setback for surgical correction of mandibular hyperplasia—Does it provoke sleep-related breathing disorders? *Int. J. Oral Maxillofac. Surg.* **1996**, *25*, 333–338. [CrossRef] [PubMed]
25. Ingervall, B.; Schmoker, R. Effect of surgical reduction of the tongue on oral stereognosis, oral motor ability, and the rest position of the tongue and mandible. *Am. J. Orthod. Dentofac. Orthop.* **1990**, *97*, 58–65. [CrossRef] [PubMed]
26. Athanasiou, A.E.; Toutountzakis, N.; Mavreas, D.; Ritzau, M.; Wenzel, A. Alterations of hyoid bone position and pharyngeal depth and their relationship after surgical correction of mandibular prognathism. *Am. J. Orthod. Dentofac. Orthop.* **1991**, *100*, 259–265. [CrossRef] [PubMed]
27. Kawakami, M.; Yamamoto, K.; Noshi, T.; Miyawaki, S.; Kirita, T. Effect of surgical reduction of the tongue on dentofacial structure following mandibular setback. *J. Oral Maxillofac. Surg.* **2004**, *62*, 1188–1192. [CrossRef]
28. Lee, D.Y.; Ryu, Y.J.; Hah, J.H.; Kwon, T.K.; Sung, M.W.; Kim, K.H. Long-term subjective tongue function after partial glossectomy. *J. Oral Rehabil.* **2014**, *41*, 754–758. [CrossRef]
29. Matys, J.; Jaszczak, E.; Flieger, R.; Kostrzewska-Kaminiarz, K.; Grzech-Leśniak, K.; Dominiak, M. Effect of ozone and diode laser (635 nm) in reducing orthodontic pain in the maxillary arch—A randomized clinical controlled trial. *Lasers Med. Sci.* **2020**, *35*, 487–496. [CrossRef]

30. Redman, M.G.; Harris, K.; Phillips, B.S. Low-level laser therapy for oral mucositis in children with cancer. *Arch. Dis. Child* **2022**, *107*, 128–133. [CrossRef]
31. Bak, J.Y.; Lee, B.S.; Kwon, Y.D.; Choi, B.J.; Lee, J.W.; Jung, J.H.; Hwang, B.Y.; Ohe, J.Y. Airway changes in patients with skeletal class III malocclusion after orthognathic surgery with partial glossectomy: A three-dimensional cone-beam computed tomography study. In Proceedings of the 57th Congress of The Korean Association of Maxillofacial Plastic and Reconstructive Surgeons, Cheongju, Republic of Korea, 31 October–2 November 2019; p. 82.

life

MDPI

Article

Outcome of Continuous Positive Airway Pressure Adherence Based on Nasal Endoscopy and the Measurement of Nasal Patency—A Prospective Study

Zdeněk Knížek [1,2,*], Miloš Kotulek [1], Pavlína Brothánková [1,3], Eva Pecháčková [3], Pavel Klail [4], Tomáš Kostlivý [4] and Jan Vodička [1,3]

1 Department of Otorhinolaryngology and Head and Neck Surgery, Hospitals of Pardubice Region—Hospital in Pardubice, 530 03 Pardubice, Czech Republic
2 Faculty of Medicine in Hradec Králové, Charles University, 500 03 Hradec Králové, Czech Republic
3 Faculty of Health Studies, University of Pardubice, 532 10 Pardubice, Czech Republic
4 Department of Otorhinolaryngology, University Hospital in Pilsen, Faculty of Medicine, Charles University, 323 00 Pilsen, Czech Republic
* Correspondence: zdenek.knizek@nempk.cz

Abstract: The gold standard for treating obstructive sleep apnea in adults is continuous positive airway pressure (CPAP). However, it can be difficult to convince patients to adhere to this therapy. The aim of this study was to determine the relationship between nasal endoscopy findings/nose patency and CPAP adherence. **Material and methods:** A cohort of 450 consecutive patients suspected of having OSA were prospectively enrolled. For further analyses, 47 OSA patients undergoing CPAP treatment were selected (13 females and 34 males, average age, 65.3 years, BMI 34.1, apnea-hypopnea index. AHI 51.0). The patients were divided into two groups: patients with good CPAP adherence (n = 35) and patients who did not adhere to CPAP therapy (n = 12). The influence of nasal endoscopy and flow measurement on CPAP adherence was explored. **Results:** We found a statistical independence between adherence to CPAP and AHI (p = 0.124), T90 (p = 0.502), endoscopic findings (p = 0.588) and nasal patency measured by a flowmeter (p = 0.498). **Conclusions:** In our studied sample, endoscopic findings and nasal patency measured by a flowmeter were not predictors of CPAP non-adherence in the first year of the treatment. Our data show that while an endoscopic finding in the nasal cavity could indicate that a patient has a severe obstruction, compliance with CPAP therapy is not reduced in these patients and neither is it reduced with a decrease in nasal flow, according to our observation.

Keywords: CPAP; flow measurement; nasal obstruction; OSA

Citation: Knížek, Z.; Kotulek, M.; Brothánková, P.; Pecháčková, E.; Klail, P.; Kostlivý, T.; Vodička, J. Outcome of Continuous Positive Airway Pressure Adherence Based on Nasal Endoscopy and the Measurement of Nasal Patency—A Prospective Study. *Life* **2023**, *13*, 219. https://doi.org/10.3390/life13010219

Academic Editor: Larry D. Sanford

Received: 20 December 2022
Revised: 9 January 2023
Accepted: 10 January 2023
Published: 12 January 2023

1. Introduction

Adult obstructive sleep apnea (OSA) is a sleep-related breathing disorder. Its diagnostic criteria according to the ICSD-3 are the presence of one or more of the following: the patient related daytime or nighttime symptoms (e.g., sleepiness, non-restorative sleep, waking with breath holding etc.) and/or bed-partner observations (snoring, breathing interruptions, etc.) and/or medical condition associated with OSA (hypertension, mood disorder, cognitive dysfunction, coronary artery disease, stroke, congestive heart failure, atrial fibrillation or type 2 diabetes mellitus) and at the same time a polysomnography (PSG) or an out-of-center sleep-testing (OCST) demonstrated five or more predominantly obstructive respiratory events per hour of sleep. Alternatively, the diagnosis is made when a PSG or an OCST demonstrates 15 or more predominantly obstructive respiratory events per hour of sleep [1].

Sleep-related disordered breathing is highly prevalent in our population. The HypnoLaus study which based its data on a cohort of 3043 patients, suggests a prevalence of moderate or severe OSA as high as 23.4% in women and of 49.7% in men [2]. The severity of OSA proportionately raises the risk for OSA-related comorbidities in multiple organ systems. OSA has been shown to contribute to cardiovascular, respiratory and neurologic impairments; notably, OSA and cardiovascular disease are strongly correlated [3,4].

The role nasal obstruction plays in the pathophysiology of OSA is yet to be fully elucidated. A number of pathophysiological mechanisms can potentially explain the role of nasal pathology in OSA. These include: the Starling resistor model, the unstable oral airway, the nasal ventilatory reflex and the role of nitric oxide [5]. Yet, we observe that the more severe the sleep-related breathing disorder, the less the patients tend to breathe through their mouth alone [6].

Continuous positive airway pressure (CPAP) remains the gold standard for the treatment of moderate or severe OSA in adults [7,8]; oral appliances and surgical treatment are preferred for milder forms of OSA [9,10]. Despite the efficacy of CPAP, many patients find it hard to adhere to this form of therapy [11,12]. Adherence to CPAP therapy is defined as using the therapy for at least four hours a day and for at least 70% of the nights in a year [13,14]. Among the most common reasons for CPAP failure are claustrophobia, mask discomfort, difficulties sleeping, an inability to keep the mask on, sensations of suffocating and nasal congestion. Nasal congestion has been reported as the cause of the failure to adhere to the therapy in multiple studies [15–18].

Despite the fact that nasal surgery itself does not reduce obstructive respiratory events, nasal surgery may reduce daytime and nighttime OSA-related symptoms (excessive daytime sleepiness, snoring) and may improve CPAP adherence [19–24].

According to several studies, the objective confirmation of the presence of a nasal obstruction can be used as a predictor of CPAP therapy non-adherence [25,26]. The aim of this study was to determine the impact of clinically significant nasal septal deformities and/or an inferior turbinates' hypertrophy on CPAP therapy adherence.

2. Material and Methods

2.1. Material

A total of 450 patients were enrolled in our prospective, monocentric, analytical study. The study ran from 6/2018 to 3/2021 in the tertiary referral hospital. Finally, a sample of 47 patients fulfilled the inclusion criteria and were included in the study. The study was approved by the Ethics Committee of the Hospitals of Pardubice region (reference number 6/2015). All the subjects signed a consent form before being enrolled in our study.

The inclusion criteria were defined as: suspicion of OSA, sleep monitoring performed by PSG or limited polygraphy (PG), AHI $\geq$ 15, CPAP therapy with a nasal mask, being over 18 years of age. The exclusion criteria were: sleep monitoring performed by tools other than PSG or PG, nasal injury or prior surgery in the upper airways (patients who underwent adenectomy in childhood were not excluded), chronic disease of the paranasal sinuses, chronic pulmonary disease, CPAP therapy with a full-face mask, incomplete data and lack of cooperation. For details, see the flowchart in Figure 1.

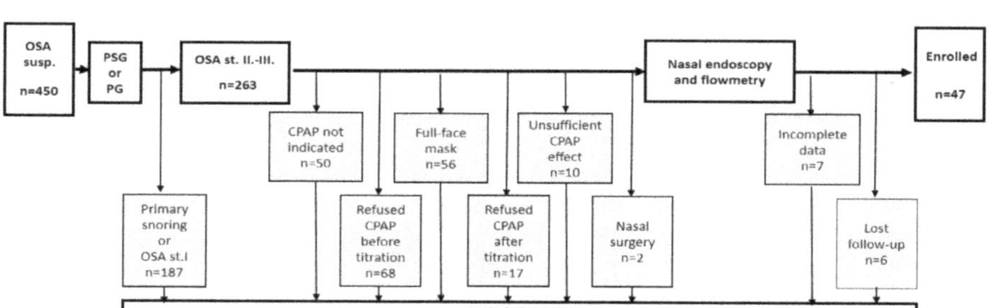

Figure 1. Flowchart of the inclusion/exclusion process.

The selected sample included 34 men and 13 women whose mean age was 56.3 years (for details, see Figure 2), and whose mean BMI was 34.1 kg/m^2.

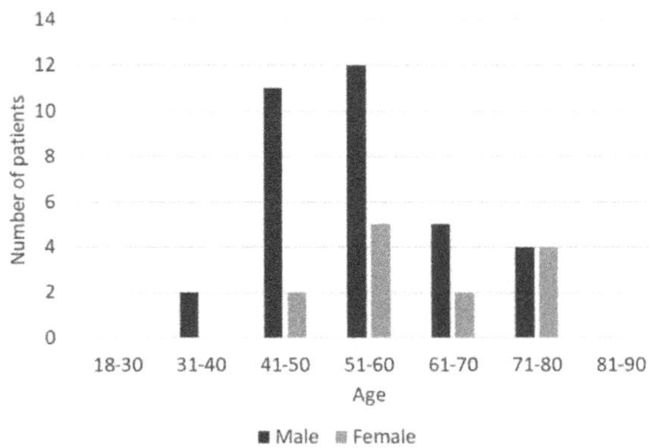

Figure 2. Age and gender distribution of the patients.

Eight patients suffered from moderate OSA, whereas thirty-nine patients suffered from severe OSA, as defined by the AHI. The mean airway pressure of the CPAP they received was 10.3 cm of water pressure (6–18 cm H_2O). For details, see Table 1.

Table 1. Input data of studied sample.

Group	Female					Male					Total				
	n = 13					*n* = 34					*n* = 47				
	Mean	**SD**	**Median**	**IQR**	**Range**	**Mean**	**SD**	**Median**	**IQR**	**Range**	**Mean**	**SD**	**Median**	**IQR**	**Range**
Age (years)	61.6	17.7	61.5	18	49–74	54.2	26.9	55.0	16.8	36–74	56.3	10.7	56	17.0	36–74
BMI (kg/m2)	35.4	5.1	36.5	7.4	27.8–44.4	33.6	7.5	33.5	6.2	25.7–44.7	34.1	7.0	34.3	6.8	25.7–44.7
AHI (n/h)	40.0	34.2	43.4	22.9	18.6–119.2	55.2	26.8	57.9	33.0	18.5–95.8	51.0	23.7	47.4	36.1	18.5–119.2
30 > AHI ≥ 15(n)	4			30.8%		4			11.8%		8			17.0%	
AHI ≥ 30 (n)	9			69.2%		30			88.2%		39			83.0%	
ODI (n/h)	40.3	26.4	31.7	22.6	18.5–118.7	51.9	23.0	55.1	36.3	16.0–116.0	48.6	24.2	45.2	34.0	16.0–118.7
Basal O2 sat. (%)	90.9	3.4	92.0	2.5	83.0–95.0	90.5	16.0	92.0	5.0	79.0–96.0	90.6	13.7	92.0	5.0	79.0–96.0
Average low O2sat. (%)	85.6	4.2	87.0	5.5	77.0–92.0	84.1	15.9	86.0	8.5	65.0–92.0	84.5	13.7	86.5	8.0	65.0–92.0
T 90 (%)	19.8	18.9	10.3	22.4	0.7–63.6	29.7	25.0	26.2	46.6	1.0–69.3	26.9	23.6	14.9	43.0	0.7–69.3
Mean CPAP (cm H2O)	10.5	1.8	10.6	2.0	6.0–13.0	10.1	3.1	10.0	3.75	6.0–18.0	10.3	2.8	10.0	3.0	6.0–18.0

2.2. Methods

The diagnosis of OSA was determined by using the ICSD-3 diagnostic criteria; the sleep testing was performed using a NOX A1 (Resmed Inc., San Diego, CA, U.S.) or a MiniScreen Plus (Saegeling Medizintechnik GmbH, Heidenau, Germany). The CPAP titration was performed using machines from the The AirSense 10 series (Resmed Inc., San Diego, CA, U.S.) or a Philips Dreamstation (Saegeling Medizintechnik GmbH, Heidenau, Germany). The indication for positive airway therapy was determined using the criteria defined by the guidelines of the Czech Sleep Research and Sleep Medicine Society [27]. CPAP therapy is recommended for patients with moderate to severe OSA (AHI/RDI ≥ 15).

Before having the patient use the machine, subjective nasal patency was measured using a visual analogue scale (VAS). The objective nasal patency was measured by a flowmeter (Elmet s.r.o., Přelouč, Czech Republic). Respiration data were analyzed, and the mean value of the inspirational peaks was evaluated. A pathological and clinically significant decrease in flow was defined as a flow of 4.57V or lower [28].

Nasal endoscopy was performed, and anatomical abnormalities that might cause nasal obstruction were recorded; our assessment was based on work published by Mladina et al. in 1987 and 2015 [29,30]. Each side of the nasal cavity was divided into one of six groups (endoscopic score ES6, see Table 2), and an ES6 score of 4 or more was defined as pathological and clinically significant for causing nasal obstruction [28].

Table 2. Endoscopic score ES6.

Group	Septal Deformity	Inferior Turbinate Hypertrophy
1	No	No
2	No	Yes
3	Non-significant (type 1,3,6)	No
4	Non-significant (type 1,3,6)	Yes
5	Significant (type 2,4,5)	No
6	Significant (type 2,4,5)	Yes

After three months of CPAP treatment, the patients filled out a Sinonasal Outcome Test [31,32]. For the purposes of this study, questions targeting sleep and emotions were excluded. A score between 0 and 40 was calculated for each questionnaire. A score of 14.5 or more was taken to indicate a significant nasal intolerance [33].

The data of CPAP therapy adherence were evaluated after 12 months of CPAP treatment.

2.3. Statistics

Statistical analyses were performed using NCSS 2021 Statistical Software (2021), NCSS, LLC. Kaysville, Utah, USA, ncss.com/software/ncss. The groups' characteristics were described. Non-parametric tests were used for the subsequent analyses of the quantitative variables. A comparison of the quantitative values between groups according to compliance was performed. The hypothesis of agreement was tested against the alternative of disagreement. A two-sample t-test and non-parametric Mann–Whitney and Kolmogorov–Smirnov tests were used. The hypothesis of independence was tested in a contingency table against the alternative of dependence to compare genders between the groups. The Fisher's exact test was used. p values < 0.05 were considered statistically significant.

3. Results

Of a total of forty-seven patients, only twelve patients showed low compliance to CPAP therapy after 12 months of treatment. The failure of CPAP therapy was statistically independent of all the measured variables: gender ($p = 0.713$), age ($p = 0.427$), BMI ($p = 0.621$), AHI ($p = 0.124$), ODI ($p = 0.495$), T90 ($p = 0.502$), basal saturation ($p = 0.066$), mean CPAP pressure ($p = 0.057$). For details, see Table 3.

The use of the ES6 yielded a score of 4 or more in at least one side of the nasal cavity in 20 patients. These 20 patients had a mean compliance of 79.0 % and had a mean score of 3.7 points on the questionnaire for CPAP nasal tolerance. The remaining 27 patients were without significant nasal obstruction, as shown by an endoscopy (an ES6 score of less than 4) and had a mean compliance of 73.6 % and a mean score of 1.9 points on the questionnaire for CPAP nasal tolerance. The difference between the groups was not statistically significant for either parameter ($p = 0.498$ for compliance, $p = 0.588$ for the questionnaire). For details, see Table 4 and Figures 3 and 4.

Twenty-one patients had a significant decrease in flow measurement (less than 4.57 V measured with the flowmeter) in at least one side of their nasal cavity. These patients had a mean compliance of 79.7% and a mean of 2.9 (±3.8 SD) points scored on the CPAP nasal tolerance questionnaire. No significant flow drop was observed in twenty-six patients. These patients had a mean compliance of 72.9% and a mean of 2.5 (±4.0 SD) points scored on the CPAP nasal tolerance questionnaire. The difference between the groups was not statistically significant for either parameter (p = 0.754 for compliance, p = 0.657 for questionnaire). For details, see Table 4 and Figures 3 and 4.

Table 3. Comparison of the CPAP compliant and CPAP non-compliant patient groups.

Group	Compliant n = 35					Non-Compliant n = 12					p
	Male		Female			Male		Female			
Gender	28	74.3%		7	25.7%	8	66.7%		4	33.3%	0.713
	Mean	SD	Median	IQR	Range	Mean	SD	Median	IQR	Range	
Age (years)	55.7	11.1	54.0	18	36–74	57.8	9.7	56.5	11.8	36–72	0.427
BMI (kg/m2)	34.3	7.7	34.7	7.6	25.7–44.7	33.5	4.2	32.8	6.3	28.3–43.2	0.621
AHI (n/h)	54.1	24.3	52.0	36.2	18.5–119.2	41.9	19.7	33.6	37.8	18.6–71.8	0.124
30 > AHI ≥ 15(n)	4			11.4%		4			33.3%		
AHI ≥ 30 (n)	31			88.6%		8			66.7%		
ODI (n/h)	50.4	26.4	45.6	37.9	16.0–118.7	43.6	16.4	41.4	23.2	18.6–64.4	0.495
Basal O2 saturation (%)	90.1	15.7	91.0	5.3	79.0–96.0	92.1	2.7	93.0	2.8	86.0–95.0	0.066
Average low O2 sat. (%)	84.2	15.5	86.0	8.0	70.0–92.0	85.5	5.7	87.5	88.0	75.0–92.0	0.501
T 90 (%)	28.4	24.0	23.2	45.3	1.0–69.3	22.6	23.0	11.2	38.8	0.7–65.8	0.502
Mean CPAP (cm H20)	9.9	2.1	10.0	2.4	6.0–13.5	11.4	3.0	10.7	3.8	8.0–18.0	0.057

Table 4. Compliance to CPAP therapy in relation to ES6 and flow measurement.

		≥4 N = 20					<4 N = 27					p
ES6		Mean	SD	Median	IQR	Range	Mean	SD	Median	IQR	Range	
	CPAP compl.	79.0%	23.5	85.7	24.3	16.0–100.0	73.6 %	26.8	86.4	35.6	0.0–100.0	0.498
	questionnaire	3.7	5.0	1.0	6.3	0.0–15.0	1.9	2.7	1.0	3.0	0.0–12.0	0.588
Flow measurement		<4.57 V N = 21					≥4.57 V N = 26					
	CPAP compl.	79.7 %	18.1	87.0	24.4	38.0–100.0	72.9 %	29.9	86.1	41.7	0.0–100.0	0.754
	questionnaire	2.9	3.9	1.0	4.0	0.0–12.0	2.5	4.0	1.0	3.3	0.0–15.0	0.657

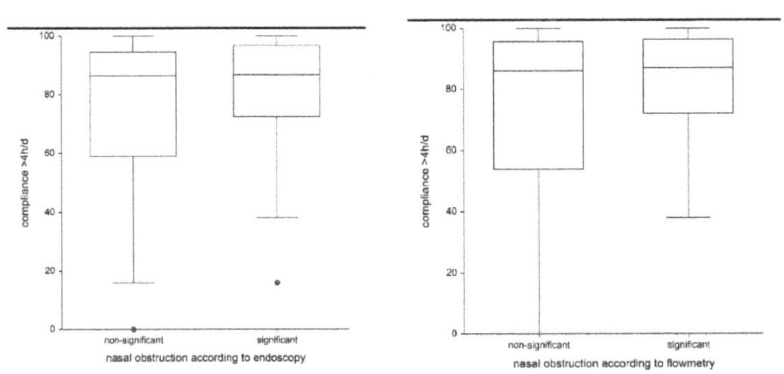

(a) (b)

Figure 3. Relationship between CPAP compliance and nasal obstruction according to the endoscopy (**a**) and flowmetry (**b**) results..

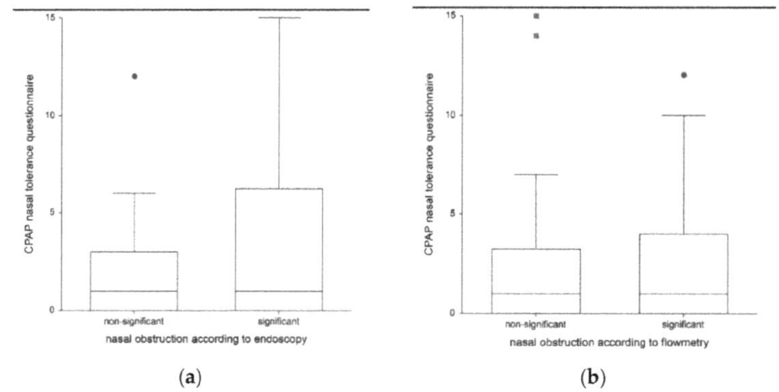

Figure 4. Relationship between the CPAP nasal tolerance questionnaire and nasal obstruction according to the endoscopy (**a**) and flowmetry (**b**) results..

4. Discussion

Unobstructed upper airways are an important condition for an uncomplicated PAP treatment of OSA [15,18,24]. In some nasal obstruction cases, we can provide treatment by applying PAP through a full-face mask, but compliance is higher when the patients use a nasal mask [34–36].

Balsalobre et al. compared patients with nasal polyps to otherwise healthy subjects; both groups underwent CPAP. The control group showed a significant worsening of the nasal obstruction symptoms, as measured by VAS and the NOSE questionnaire ($p < 0.01$), and a significant decrease in nasal patency, as measured by the peak nasal inspiratory flow and acoustic rhinometry ($p < 0.01$) [37].

Our research was aimed at revealing the relationship between CPAP adherence and nasal obstruction, as assessed and evaluated by nasal endoscopy (ES6) and nose cavity flow measurement [28]. We employed a strict exclusion process: only 47 out of 450 patients met the demands of the inclusion criteria. For details, see the flowchart in Figure 1.

The final sample of our study represented the real-world population: 13 female and 34 male patients with mean a BMI of 35.4 and 33.6 and a mean AHI of 40.0 and 55.2, respectively. CPAP failure was associated neither with demographic data such as gender ($p = 0.713$), age (0.427) or BMI ($p = 0.621$), nor with sleep-monitoring data, such as AHI ($p = 0.124$) or T90 ($p = 0.502$).

The guidelines of the surgical division of the Czech Sleep Research and Sleep Medicine Society [38] were complied with based on the premise that nasal surgery does not significantly affect obstructive respiratory events [39,40]. This assumption is based on the fact that only 16.7% of the patients with OSA who undergo nasal surgery meet the Sher criteria [19]. Several works present findings that indicate that a reduction in obstructive respiratory events can be achieved by intranasal corticosteroids application, particularly in children and allergic individuals [41,42]. According to two pooled randomized, placebo-controlled clinical trials, nasal steroids where shown to improve CPAP adherence—the studies recorded an overall 0.4 h per night increase in the usage of the machine. However, this increase did not reach statistical significance ($p = 0.19$). There was no increase in the percentage of nights during which CPAP was used, nor was there a significant difference in nasal symptoms [43].

Nasal surgery can positively influence subjective sleep parameters, e.g., snoring or excessive daytime sleepiness [19,44,45], and nasal patency improvement can be useful for reducing the level of PAP when treating OSA, resulting in better adherence to CPAP therapy [46,47]. However, the positive effect of surgery on CPAP tolerance was not confirmed in all the published literature [48]. Therefore, prudence should be applied when selecting candidates for nasal interventions.

Nasal procedures with the aim of improving nasal patency, e.g., septoplasty, are frequently performed. These operations carry only a low risk with low morbidity and a low rate of complications; we rarely encounter complications such as the formation of nasal septal perforation or nasal synechiae [49,50]. Van Egmond et al. compared the effectiveness of septoplasty combined with turbinate surgery with the efficacy of septoplasty on its own in the treatment of nasal obstruction due to a deviated nasal septum. Subjective and objective outcomes generally appeared to improve after the treatment. However, the additional benefit of turbinate surgery was not evident. Moreover, the subjective benefit was not always accompanied by an objective improvement, and vice versa. Despite the routine application of septoplasty in clinical practice, the body of evidence does not support firm conclusions on its effectiveness [49]. Nasal packing after septoplasty was even more likely to cause adverse events, including respiratory distress, pain, sleep disturbance, crusting, epiphora, dysphagia and adhesion. Routine nasal packing after septoplasty should therefore be avoided [51].

In general, isolated nasal treatments are the least effective in treating OSA, but multi-level surgery may provide an alternative to CPAP treatment [52]. The effect of multi-level surgery (multiple surgeries to include nasal surgery, tonsillectomy, palate surgery, pharyngeal surgery and tongue surgery) on OSA was evaluated by Lin et al., who observed a reduction in the apnea hypopnea index in 1978 patients. They recorded a reduction of 29 events per hour: from 48.0 to 19.0 events/hour (a 60.3% reduction, p-value < 0.0001) [53].

Nasal surgery may be offered to CPAP-intolerant patients—adjunctive nasal surgery may facilitate improved postoperative CPAP adherence due to lower CPAP requirements (average of 2–3 cm H_2O) or improved tolerance of nasal-type masks without the necessity of a chin strap [54].

Kempfle et al. concluded that nasal surgery (septoplasty or turbinate reduction) would be a cost-effective way to increase CPAP adherence. In the short term, septoplasty surgery was not a cost-effective way to improve CPAP adherence in patients who had a great baseline difficulty using CPAP, but over a longer time span of 10 or 15 years, septoplasty became increasingly more cost-effective. The cheaper turbinate reduction would be a cost-effective way to increase CPAP adherence regardless of the time span in question. Notably, perioperative and postoperative surgical complications did not unfavorably influence the cost-effectiveness of either surgery. Therefore, surgical intervention for non-adherent CPAP users, or partially adherent CPAP users, should be considered a part of a multifaceted approach to improve CPAP adherence [55].

CPAP adherence and endoscopic findings assessed by ES6 (modified Sinonasal Outcome Test), as well as the objective assessment of air flow by a flowmeter were shown to be independent of each other in our group.

We compared our results to those of similar studies published in 2019 by Inoue et al. [21] and in 2017 by Park et al. [56]. While Inoue et al. studied a larger sample (n = 543), they reported the same results in their long-term follow up (more than 1 year). Their short-term results (less than 1 year follow up), however, differed. Park et al. used a cohort whose size was similar to that of ours and, like us, they concluded that CPAP non-adherence was independent of sleep parameters (p = 0.671). Their results differed in the evaluation of nose patency influence ($p \leq 0.0001$). Our conclusions concerning the nasal flow measurement are in agreement with a systematic review authored by Brimioulle and Chaidas in 2022 [57] but in conflict with the study by Sugiura et al. from 2007, which was performed on a similar sample (n = 51). Sugiura's work [24] concluded that nasal obstruction is a significant factor for CPAP non-adherence (p = 0.002) but, surprisingly, showed the same association with the AHI (p = 0.003). We attach more reliability to the systematic review by Brimioulle because they evaluated 63 works and pointed out some contradictory results.

The limits of our study are its monocentricity and a strict exclusion process (47 out of 450 patients were enrolled). However, the studied sample represents the real-world population with OSA. For details, see Figure 1 and Table 1.

5. Conclusions

Our study demonstrated independence between CPAP adherence and endoscopic findings/nasal patency. Our data show that although the endoscopic findings in the nasal cavity could indicate an obstruction, compliance to CPAP therapy was not reduced, and neither was it reduced with a decrease in nasal flow, according to our observation.

Author Contributions: Conceptualization, Z.K. and M.K.; methodology, Z.K. and J.V.; investigation, Z.K. and E.P.; resources, Z.K., M.K., P.B., E.P., P.K. and T.K.; data curation, Z.K. and E.P.; writing—original draft preparation, Z.K.; writing—review and editing, Z.K. and J.V.; visualization, Z.K.; supervision, J.V. and P.B.; All authors have read and agreed to the published version of the manuscript.

Funding: This research received no external funding.

Institutional Review Board Statement: The study was conducted according to the guidelines of the Declaration of Helsinki and approved by the Institutional Review Board (or Ethics Committee) of Pardubice Hospital with the reference number 6/2015.

Informed Consent Statement: Informed consent was obtained from all subjects involved in the study.

Data Availability Statement: The data presented in this study are available on request from the corresponding author. The data are not publicly available due to privacy and ethical restrictions.

Conflicts of Interest: The authors declare no conflict of interest.

Abbreviations

AHI	apnea-hypopnea index
BMI	body mass index
CPAP	continuous positive airway pressure
ES6	endoscopic score 6
ICSD-3	international classification of sleep disorders—third edition
OCST	out-of-center sleep testing
ODI	oxygen desaturation index
OSA	obstructive sleep apnea
PAP	positive airway pressure
PG	polygraphy
PSG	polysomnography
VAS	visual analogue scale

References

1. Sateia, M.J. International classification of sleep disorders-third edition: Highlights and modifications. *Chest* **2014**, *146*, 1387–1394. [CrossRef] [PubMed]
2. Heinzer, R.; Vat, S.; Marques-Vidal, P.; Marti-Soler, H.; Andries, D.; Tobback, N.; Mooser, V.; Preisig, M.; Malhotra, A.; Waeber, G.; et al. Prevalence of sleep-disordered breathing in the general population: The HypnoLaus study. *Lancet Respir. Med.* **2015**, *3*, 310–318. [CrossRef]
3. Gilat, H.; Vinker, S.; Buda, I.; Soudry, E.; Shani, M.; Bachar, G. Obstructive sleep apnea and cardiovascular comorbidities: A large epidemiologic study. *Medicine* **2014**, *93*, e45. [CrossRef] [PubMed]
4. Smith, R.; Ronald, J.; Delaive, K.; Walld, R.; Manfreda, J.; Kryger, M.H. What are obstructive sleep apnea patients being treated for prior to this diagnosis? *Chest* **2002**, *121*, 164–172. [CrossRef] [PubMed]
5. de Sousa Michels, D.; da Mota Silveira Rodrigues, A.; Nakanishi, M.; Sampaio, A.L.L.; Venosa, A.R. Nasal involvement in obstructive sleep apnea syndrome. *Int. J. Otolaryngol.* **2014**, *2014*, 717419. [CrossRef]
6. Oeverland, B.; Akre, H.; Skatvedt, O. Oral breathing in patients with sleep-related breathing disorders. *Acta Otolaryngol.* **2002**, *122*, 651–654. [CrossRef]
7. Becker, H.F.; Jerrentrup, A.; Ploch, T.; Grote, L.; Penzel, T.; Sullivan, C.E.; Peter, J.H. Effect of nasal continuous positive airway pressure treatment on blood pressure in patients with obstructive sleep apnea. *Circulation* **2003**, *107*, 68–73. [CrossRef]
8. Slouka, D.; Kucera, R.; Gal, B.; Betka, J.; Skalova, A. Biomarkers—A possibility for monitoring of obstructive sleep apnea syndrome. *Neuro Endocrinol. Lett.* **2019**, *40*, 85–92.
9. Levrini, L.; Sacchi, F.; Milano, F.; Polimeni, A.; Cozza, P.; Bernkopf, E.; Segu, M.; Italian dentist work group about OSAS Collaborators; Zucconi, M.; Vicini, C.; et al. Italian recommendations on dental support in the treatment of adult obstructive sleep apnea syndrome (OSAS). *Ann Stomatol.* **2016**, *6*, 81–86. [CrossRef]

10. Slouka, D.; Honnerova, M.; Hrabe, V.; Matas, A. The prediction of treatment failure of the continuous positive airways pressure. *Bratisl. Lek. Listy* **2014**, *115*, 704–707. [CrossRef]
11. Slouka, D.; Honnerova, M.; Hosek, P.; Matas, A.; Slama, K.; Landsmanova, J.; Kucera, R. Risk factors for failure of continuous positive airway pressure treatment in patients with ostructive sleep apnoea. *Biomed. Pap.* **2018**, *162*, 134–138. [CrossRef] [PubMed]
12. Stepnowsky, C.J.; Moore, P.J. Nasal CPAP treatment for obstructive sleep apnea: Developing a new perspective on dosing strategies and compliance. *J. Psychosom. Res.* **2003**, *54*, 599–605. [CrossRef] [PubMed]
13. Kribbs, N.B.; Pack, A.I.; Kline, L.R.; Schwartz, A.R.; Schubert, N.M.; Redline, S.; Henry, J.N.; Getsy, J.E.; Dinges, D.F. Objective measurement of patterns of nasal CPAP use by patients with obstructive sleep apnea. *Am. Rev. Respir. Dis.* **1993**, *147*, 887–895. [CrossRef] [PubMed]
14. Slouka, D.; Honnerova, M.; Hosek, P.; Gal, B.; Trcka, O.; Kostlivy, T.; Landsmanova, J.; Havel, D.; Baneckova, M.; Kucera, R. Improved prediction of CPAP failure using T90, age and gender. *J. Appl. Biomed.* **2019**, *17*, 81. [CrossRef]
15. Lam, A.S.; Collop, N.A.; Bliwise, D.L.; Dedhia, R.C. Validated Measures of Insomnia, Function, Sleepiness, and Nasal Obstruction in a CPAP Alternatives Clinic Population. *J. Clin. Sleep Med.* **2017**, *13*, 949–957. [CrossRef]
16. Chowdhury, O.; Wedderburn, C.J.; Duffy, D.; Greenough, A. CPAP review. *Eur. J. Pediatr.* **2012**, *171*, 1441–1448. [CrossRef]
17. Cai, Y.; Goldberg, A.N.; Chang, J.L. The Nose and Nasal Breathing in Sleep Apnea. *Otolaryngol. Clin. N. Am.* **2020**, *53*, 385–395. [CrossRef]
18. Balztan, M.A.; Elkholi, O.; Wolkove, N. Evidence of interrelated side effects with reduced compliance in patients treated with nasal continuous positive airway pressure. *Sleep Med.* **2009**, *10*, 198–205. [CrossRef]
19. Li, H.; Wang, P.; Chen, Y.; Lee, L.; Fang, T.; Lin, H. Critical appraisal and meta-analysis of nasal surgery for obstructive sleep apnea. *Am. J. Rhinol. Allergy* **2011**, *25*, 45–49. [CrossRef]
20. Nakata, S.; Noda, A.; Yagi, H.; Yanagi, E.; Mimura, T.; Okada, T.; Misawa, H.; Nakashima, T. Nasal resistance for determinant factor of nasal surgery in CPAP failure patients with obstructive sleep apnea syndrome. *Rhinology* **2005**, *43*, 296–299.
21. Inoue, A.; Chiba, S.; Matsuura, K.; Osafune, H.; Capasso, R.; Wada, K. Nasal function and CPAP compliance. *Auris Nasus Larynx* **2019**, *46*, 548–558. [CrossRef] [PubMed]
22. Elwany, S.; Bahgat, A.Y.; Ibrahim, M.; Bazak, R. Surgical correction of nasal obstruction in obstructive sleep apnea improves CPAP outcomes and compliance. *Ir. J. Med. Sci.* **2022**, *191*, 2723–2728. [CrossRef] [PubMed]
23. Camacho, M.; Riaz, M.; Capasso, R.; Ruoff, C.M.; Guilleminault, C.; Kushida, C.A.; Certal, V. The effect of nasal surgery on continuous positive airway pressure device use and therapeutic treatment pressures: A systematic review and meta-analysis. *Sleep* **2015**, *38*, 279–286. [CrossRef] [PubMed]
24. Sugiura, T.; Noda, A.; Nakata, S.; Yasuda, Y.; Soga, T.; Miyata, S.; Nakai, S.; Koike, Y. Influence of nasal resistance on initial acceptance of continuous positive airway pressure in treatment for obstructive sleep apnea syndrome. *Respiration* **2007**, *74*, 56–60. [CrossRef]
25. Bengtsson, C.; Jonsson, L.; Theorell-Haglow, J.; Holmstrom, M.; Janson, C.; Lindberg, E. Sinonasal outcome test-22 and peak nasal inspiratory flow—Valuable tools in obstructive sleep apnoea. *Rhinology* **2020**, *58*, 341–348. [CrossRef]
26. Awad, M.I.; Kacker, A. Nasal Obstruction Considerations in Sleep Apnea. *Otolaryngol. Clin. N. Am.* **2018**, *51*, 1003–1009. [CrossRef]
27. Pretl, M.; Hobzova, M.; Honnerova, M.; Lnenicka, J.; Novak, V.; Sedlak, V.; Vyskocilova, J.; Sonka, K. Indikační kritéria pro léčbu poruch dýchání ve spánku pomocí přetlaku v dýchacích cestách u dospělých. *Neurol. Praxi* **2013**, *14*, 38–41.
28. Knizek, Z.; Vodicka, J.; Jelinek, J.; Rezny, Z. Měření nosní průchodnosti pomocí flowmetrie a klasifikace endoskopického obrazu nosní dutiny. *Otorhinolaryngol. Phoniatr. Otorinolaryngol. A Foniatr.* **2019**, *68*, 143–149.
29. Mladina, R. The role of maxillar morphology in the development of pathological septal deformities. *Rhinology* **1987**, *25*, 199–205.
30. Mladina, R.; Skitarelić, N.; Poje, G.; Subaric, M. Clinical Implications of Nasal Septal Deformities. *Balk. Med. J.* **2015**, *32*, 137–146. [CrossRef]
31. Hopkins, C.; Gillett, S.; Slack, R. Psychometric validity of the 22-item Sinonasal Outcome Test. *Clin. Otolaryngol.* **2009**, *34*, 447–1454. [CrossRef] [PubMed]
32. Schalek, P. Česká verze dotazníku kvality života pro pacienty s chronickou rinosinusitidou SNOT-22 (Sino-nasal outcome test). *Otorinolaryngol. Foniatr.* **2010**, *59*, 149–151.
33. Fokkens, W.J.; Lund, V.J.; Hopkins, C.; Hellings, P.W.; Kern, R.; Reitsma, S.; Toppila-Salmi, S.; Bernal-Sprekelsen, M.; Mullol, J.; Alobid, I.; et al. European Position Paper on Rhinosinusitis and Nasal Polyps 2020. *Rhinology* **2020**, *58* (Suppl. S29), 1–464. [CrossRef] [PubMed]
34. Chen, L.Y.; Chen, Y.H.; Hu, S.W.; Lin, M.T.; Lee, P.L.; Chiang, A.A.; Tu, Y.K. In search of a better CPAP interface: A network meta-analysis comparing nasal masks, nasal pillows and oronasal masks. *J. Sleep Res.* **2022**, *31*, e13686. [CrossRef] [PubMed]
35. Andrade, R.G.S.; Piccin, V.S.; Nascimento, J.A.; Viana, F.M.L.; Genta, P.R.; Lorenzi-Filho, G. Impact of the type of mask on the effectiveness of and adherence to continuous positive airway pressure treatment for obstructive sleep apnea. *J. Bras. Pneumol.* **2014**, *40*, 658–668. [CrossRef]
36. Andrade, R.G.S.; Viana, F.M.; Nascimento, J.A.; Drager, L.F.; Moffa, A.; Brunoni, A.R.; Genta, P.R.; Lorenzi-Filho, G. Nasal vs Oronasal CPAP for OSA Treatment: A Meta-Analysis. *Chest* **2018**, *153*, 665–674. [CrossRef]
37. Balsalobre, L.; Pezato, R.; Mangussi-Gomes, J.; Gregorio, L.; Haddad, F.L.M.; Gregorio, L.C.; Fujita, R. What is the Impact of Positive Airway Pressure in Nasal Polyposis? An Experimental Study. *Int. Arch. Otorhinolaryngol.* **2019**, *23*, 147–151. [CrossRef]

38. Lansky, M.; Plzak, J.; Ondrova, M.; Janousek, P.; Kraus, J.; Minarik, R. Doporučený diagnostický a terapeutický postup pro pracoviště zajišťující diagnostiku a chirurgickou léčbu poruch dýchání ve spánku u dospělých pacientů. *Otorinolaryngol. Foniatr.* **2013**, *62*, 48–49.
39. Schoustra, E.; Maanen, P.V.; Haan, C.D.; Ravesloot, M.J.L.; de Vries, N. The Role of Isolated Nasal Surgery in Obstructive Sleep Apnea Therapy-A Systematic Review. *Brain Sci.* **2022**, *12*, 1446. [CrossRef]
40. Rosow, D.E.; Stewart, M.G. Is nasal surgery an effective treatment for obstructive sleep apnea? *Laryngoscope.* **2010**, *120*, 1496–1497. [CrossRef]
41. Liu, H.; Lin, Y.; Kuan, Y.; Huang, Y.; Hou, W.; Liou, T.; Chen, H. Intranasal corticosteroid therapy in the treatment of obstructive sleep apnea: A meta-analysis of randomized controlled trials. *Am. J. Rhinol. Allergy* **2016**, *30*, 215–221. [CrossRef] [PubMed]
42. Kiely, J.L.; Nolan, P.; McNicholas, W.T. Intranasal corticosteroid therapy for obstructive sleep apnoea in patients with co-existing rhinitis. *Thorax* **2004**, *59*, 50–55. [PubMed]
43. Charakorn, N.; Hirunwiwatkul, P.; Chirakalwasan, N.; Chaitusaney, B.; Prakassajjatham, M. The effects of topical nasal steroids on continuous positive airway pressure compliance in patients with obstructive sleep apnea: A systematic review and meta-analysis. *Sleep Breath.* **2017**, *21*, 3–8. [CrossRef] [PubMed]
44. Stapleton, A.L.; Chang, Y.F.; Soose, R.J.; Gillman, G.S. The impact of nasal surgery on sleep quality: A prospective outcomes study. *Otolaryngol. Head Neck Surg.* **2014**, *151*, 868–873. [CrossRef]
45. Li, H.; Lin, Y.; Chen, N.; Lee, L.; Fang, T.; Wang, P. Improvement in quality of life after nasal surgery alone for patients with obstructive sleep apnea and nasal obstruction. *Arch. Otolaryngol. Head Neck Surg.* **2008**, *134*, 429–433. [CrossRef]
46. Iwata, N.; Nakata, S.; Inada, H.; Kimura, A.; Hirata, M.; Yasuma, F. Clinical indication of nasal surgery for the CPAP intolerance in obstructive sleep apnea with nasal obstruction. *Auris Nasus Larynx* **2020**, *47*, 1018–1022. [CrossRef]
47. Poirier, J.; George, C.; Rotenberg, B. The effect of nasal surgery on nasal continuous positive airway pressure compliance. *Laryngoscope* **2014**, *124*, 317–319. [CrossRef]
48. Pniak, T.; Matousek, P.; Strympl, P.; Novak, V.; Kominek, P. Obstrukční spánková apnoe a CPAP—Má význam řešit nosní průchodnost? *Česká A Slov. Neurol. A Neurochir.* **2012**, *2*, 222–226.
49. van Egmond, M.M.H.T.; Rovers, M.M.; Tillema, A.H.J.; van Neerbeek, N. Septoplasty for nasal obstruction due to a deviated nasal septum in adults: A systematic review. *Rhinology* **2018**, *56*, 195–208. [CrossRef]
50. Shin, C.H.; Jang, Y.J. Factors affecting the complication rate of septoplasty: Analysis of 1506 consecutive cases of single surgeon. *Facial Plast. Surg.* 2022; *Online Ahead of Print*. [CrossRef]
51. Titirungruang, C.K.; Charakorn, N.; Chaitusaney, B.; Hirunwiwatkul, P. Is postoperative nasal packing after septoplasty safe? A systematic review and meta-analysis of randomized controlled studies. *Rhinology* **2021**, *59*, 340–351. [CrossRef] [PubMed]
52. Camacho, M.; Chang, E.T.; Neighbors, C.L.P.; Noller, M.W.; Mack, D.; Capasso, R.; Kushida, C.A. Thirty-five alternatives to positive airway pressure therapy for obstructive sleep apnea: An overview of meta-analyses. *Expert Rev. Respir. Med.* **2018**, *12*, 919–929. [CrossRef] [PubMed]
53. Lin, H.C.; Friedman, M.; Chang, H.W.; Gurpinar, B. The efficacy of multilevel surgery of the upper airway in adults with obstructive sleep apnea/hypopnea syndrome. *Laryngoscope* **2008**, *118*, 902–908. [CrossRef] [PubMed]
54. Shelgikar, A.V.; Aronovich, S.; Stanley, J.J. Multidisciplinary Alternatives to CPAP Program for CPAP-Intolerant Patients. *J. Clin. Sleep Med.* **2017**, *13*, 505–510. [CrossRef]
55. Kempfle, J.S.; BuSaba, N.Y.; Dobrowski, J.M.; Westover, M.B.; Bianchi, M.T. A cost-effectiveness analysis of nasal surgery to increase continuous positive airway pressure adherence in sleep apnea patients with nasal obstruction. *Laryngoscope* **2017**, *127*, 977–983. [CrossRef]
56. Park, P.; Kim, J.; Song, Y.; Lim, J.; Cho, S.; Won, T.; Han, D.; Kim, D.; Rhee, C.; Kim, H. Influencing factors on CPAP adherence and anatomic characteristics of upper airway in OSA subjects. *Medicine* **2017**, *96*, e8818. [CrossRef]
57. Brimioulle, M.; Chaidas, K. Nasal function and CPAP use in patients with obstructive sleep apnoea: A systematic review. *Sleep Breath.* **2022**, *26*, 1321–1332. [CrossRef]

life

MDPI

Review

Oromaxillofacial Surgery: Both a Treatment and a Possible Cause of Obstructive Sleep Apnea—A Narrative Review

Dinko Martinovic [1], Daria Tokic [2], Ema Puizina-Mladinic [1], Sanja Kadic [1], Antonella Lesin [1], Slaven Lupi-Ferandin [1], Marko Kumric [3] and Josko Bozic [3,*]

[1] Department of Maxillofacial Surgery, University Hospital of Split, 21000 Split, Croatia
[2] Department of Anesthesiology and Intensive Care, University Hospital of Split, 21000 Split, Croatia
[3] Department of Pathophysiology, University of Split School of Medicine, 21000 Split, Croatia
* Correspondence: josko.bozic@mefst.hr; Tel.: +385-21-557-871

Abstract: Obstructive sleep apnea (OSA) is a chronic, sleep-related breathing disorder. It is characterized by a nocturnal periodic decrease or complete stop in airflow due to partial or total collapse of the oropharyngeal tract. Surgical treatment of OSA is constantly evolving and improving, especially with the implementation of new technologies, and this is needed because of the very heterogeneous reasons for OSA due to the multiple sites of potential airway obstruction. Moreover, all of these surgical methods have advantages and disadvantages; hence, patients should be approached individually, and surgical therapies should be chosen carefully. Furthermore, while it is well-established that oromaxillofacial surgery (OMFS) provides various surgical modalities for treating OSA both in adults and children, a new aspect is emerging regarding the possibility that some of the surgeries from the OMFS domain are also causing OSA. The latest studies are suggesting that surgical treatment in the head and neck region for causes other than OSA could possibly have a major impact on the emergence of newly developed OSA, and this issue is still very scarcely mentioned in the literature. Both oncology, traumatology, and orthognathic surgeries could be potential risk factors for developing OSA. This is an important subject, and this review will focus on both the possibilities of OMFS treatments for OSA and on the OMFS treatments for other causes that could possibly be triggering OSA.

Keywords: obstructive sleep apnea; surgical treatment; oromaxillofacial surgery; head and neck surgery

Citation: Martinovic, D.; Tokic, D.; Puizina-Mladinic, E.; Kadic, S.; Lesin, A.; Lupi-Ferandin, S.; Kumric, M.; Bozic, J. Oromaxillofacial Surgery: Both a Treatment and a Possible Cause of Obstructive Sleep Apnea—A Narrative Review. *Life* **2023**, *13*, 142. https://doi.org/10.3390/life13010142

Academic Editors: David Slouka and Milan Štengl

Received: 1 December 2022
Revised: 28 December 2022
Accepted: 30 December 2022
Published: 4 January 2023

1. Introduction

Obstructive sleep apnea (OSA) is a chronic, sleep-related breathing disorder. It is characterized by a nocturnal periodic decrease or complete stop in airflow due to partial or total collapse of the oropharyngeal tract [1]. These obstructive episodes cause asphyxia, which then stimulates breathing efforts against the collapsed upper airway and subsequent awakening. The etiology of the disorder is still unknown, but it is considered to be multifactorial as a combination of anatomic variations, neuromuscular factors, and genetic predispositions [2]. In adults, OSA is most frequently associated with male gender, obesity, and rising age while in children, it is most commonly associated with enlarged tonsils or adenoids. However, even with today's technology, many patients with OSA are undiagnosed and untreated, which is a major global healthcare problem since it is well-established that OSA is associated with high cardiovascular morbidity and mortality [3]. Moreover, independently from body weight, patients with OSA more frequently develop metabolic disorders, such as insulin resistance and hyperlipidemia [4].

There are several treatment modalities for OSA, both conservative and surgical. Continuous positive airway pressure (CPAP) is considered the "gold standard" for the treatment of OSA due to the fact that it can be used for the improvement of sleep-related symptoms and quality of life [5]. On the other hand, since OSA is related to anatomical disturbances of the upper airway, surgical treatment is aimed at reducing the degree of obstruction in the nasal region, nasopharynx, velopharynx, oropharynx, and hypopharynx (Figure 1).

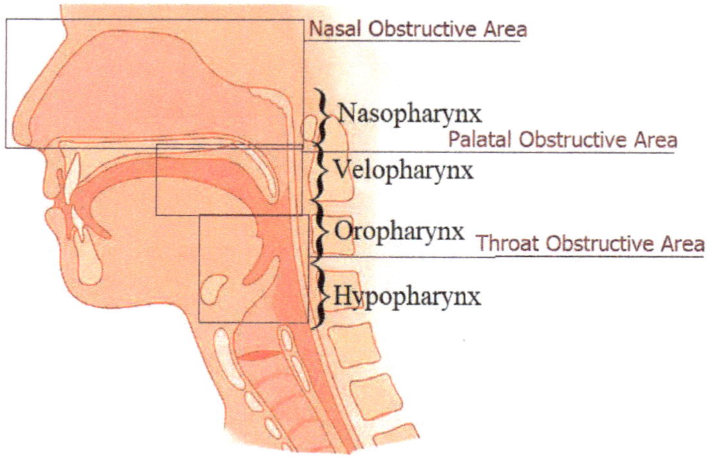

Figure 1. Sagittal view of the upper airway and the places of obstruction in OSA.

While it is well-established that oromaxillofacial surgery (OMFS) provides various surgical modalities for treating OSA both in adults and children; a new aspect is emerging regarding the possibility that some of the surgeries from the OMFS domain are also causing OSA. The latest studies are suggesting that surgical treatment in the head and neck region for causes other than OSA could possibly have a major impact on the emergence of newly developed OSA [6–11]. On one hand, there are oncological surgeries where indications for surgery are of vital importance, but due to tissue loss and reconstruction, they consequently cause anatomical changes and abnormalities. Furthermore, trauma surgery in the OMFS region has its own difficulties, especially regarding maintaining functionality, and in complex fracture cases, it could also be a cause of OSA due to structural alterations. Lastly, there is orthognathic surgery, which represents both a functional and aesthetical component of OMFS with a significant impact on the morphology and anatomy of this specific region. Even though the reasoning behind this possibly higher risk for OSA after OMFS is logical due to anatomical deviations after certain surgical therapies, this subject has become a hot topic only relatively recently. Hence, this review will focus on both the possibilities of maxillofacial surgical treatments for OSA and on the treatments for other causes which could possibly be triggering OSA.

2. Oromaxillofacial Surgery as a Cause of Obstructive Sleep Apnea

Recent studies are suggesting that treatment modalities for cancers in the head and neck region, both surgical and conservative, could be a major risk factor for developing OSA [6]. As aforementioned, surgery in this specific region causes structural alterations that could possibly have severe consequences on airflow through the upper airway while it was also shown that radiotherapy causes edema of the soft tissue and consequent change of anatomic relations [6]. Several studies investigated the development of OSA in patients with head and neck cancers [9,12–14]. The results of these studies greatly vary due to the number of patients enrolled in the study or to the different treatments for cancer. Nevertheless, all of the studies found that surgically treated patients have a higher prevalence of developing OSA. A study by Loth et al. found that 25.49% of the patients enrolled in their study developed OSA after the treatment of cancer regardless of the type of treatment [12]. Furthermore, Friedman et al. estimated the prevalence of OSA at

72.7% [13]. However, this study had a more heterogeneous population and did not exclude pre-existing risk factors. A study conducted by Qian et al. investigated the prevalence of OSA after cancer therapy, and they found that patients treated with surgery have a higher prevalence of developing severe OSA compared to patients on chemo/radiotherapy. Still, this study has limitations, such as a small sample size, incomplete matching in body mass index, and primary tumor location [14]. Furthermore, Ralli et al. found that the prevalence of OSA in head and neck cancer patients is significantly higher when compared to the general population [6]. However, their study was conducted on all treatment modalities, surgical and nonsurgical. Additionally, even though most of the aforementioned studies focused on surgery as a cause of OSA from the cancer excision point of view, a recent study was conducted on patients who were reconstructed with free flaps in the oral region. They found that even the reconstruction technique could possibly influence the emergence of OSA [9].

It is very important to highlight the fact that not only the surgical treatment of cancer in the OMFS region causes OSA, but the tumor itself has a major impact. When growing to the point that it causes anatomical alterations, it is possible that malignancies could influence the development of OSA. Moreover, tumors are a major limitation of the studies regarding OSA development after OMFS oncological surgeries. This limitation is due to the fact that none of the involved participants had a precancer screening of OSA since most of these patients are first screened after the cancer diagnosis; hence, it is impossible to exclude pre-existing OSA and a tumor is a major bias for interpreting the influence of surgery on OSA. Moreover, head and neck tumors as a cause of OSA are very poorly investigated. Payne et al. found a strong association between OSA and oral cavity cancers before the treatment, but they had a small sample of 17 patients included in the study [15]. Moreover, a case report by Gomez-Merino et al. presented a patient with pharyngeal non-Hodgkin's lymphoma whose first symptoms were sleep disorders, such as snoring and breathing pauses during sleep [16]. Another interesting case report by Hockstein et al. showed an interesting finding of retropharyngeal lipoma extended from the nasopharynx to the upper mediastinum which also caused symptoms of OSA [17]. Since the tumor was benign and the patient had several chronic diseases that made him high risk for surgery, the only treatment was CPAP. There are several similar cases describing OSA as the primary symptom of cancer [18–20]. However, this subject is very scarcely represented in the literature, and future larger prospective studies are needed. Nevertheless, healthcare workers involved in the diagnostics and treatment of OSA should always consider both anamnestic details and potential illness as triggers for newly discovered OSA.

When it comes to OMFS trauma, it is often accompanied by injuries to other parts of the body, and according to some studies, it is involved in up to 25% of polytraumatized patients [21]. The most common cause of these injuries is traffic accidents while other more frequent causes are assaults and falls. Furthermore, injuries in this region can range from simple soft-tissue lacerations to complex bone fractures [22]. Fractures in OMFS have both aesthetic and functional aspects, which makes them even more challenging. As already mentioned, any surgical manipulation in this specific area can lead to both hard and soft tissue impairments and subsequent morphologic alterations, which possibly creates a predisposition for the development of OSA. A recent study conducted by Lupi-Ferandin et al. investigated the prevalence of OSA after surgical treatment of maxillary and zygomatic fractures [11]. They found that 54% of their participants developed OSA in a period of three months after surgery. Furthermore, their study showed that OSA is more common in patients with maxillary fractures (62%) than in patients with zygomatic fractures (38%). Another recent study conducted by El-Anwar et al. also found that OSA is connected with trauma in the OMFS region. Their study investigated mandibular fractures and found that OSA was diagnosed in 35.8% of their patients, all with bilateral fractures [23]. Although both of these studies investigated OSA in OMFS trauma and their results are suggesting a possible association between the development of OSA and OMFS trauma, it must be noted that the data regarding this subject are still too scarce. While it is reasonable to assume that both the bone fractures and consequent surgical therapy could possibly impact OSA, on the other hand, further larger studies are needed to evaluate this issue.

Besides trauma and cancer treatment surgeries, another important segment of OMFS is orthognathic surgery. These OMFS procedures manipulate anatomical relations of the maxilla and mandible in order to achieve certain functional and visual improvements of deformities caused by dentofacial conditions or carried out purely for aesthetic reasons [24]. While these types of surgeries require an advanced level of skills, some effects of these surgeries are expected because of the manipulation of the mandible, anatomical features, and morphology of the upper airway. Several studies found that mandibular setback surgery has an impact on respiratory function, more precisely, a reduction in the retrolingual airway space [25,26]. However, these findings are still controversial. A review by Fernandez-Ferrer et al. analyzed the literature on this topic, and their observations are rather interesting; they found that nasopharyngeal space does not change significantly after maxillary surgery but only after mandibular setback [27]. It was shown that there are significant changes in the volume of space in this area, yet there was no evidence found that confirms the hypothesis that orthognathic surgeries predispose the development of OSA. Another interesting point of view is comparing and analyzing the differences between sexes after orthognathic surgeries. A retrospective study by Dahy et al. found that male patients have a higher risk of developing OSA after mandibular setback surgery while females have a higher risk after bimaxillary surgery [28]. One of the possible explanations for these differences are purely morphological; several studies found that females have smaller pharyngeal airways [29–31]. However, considering that this specific area is in a dynamic interaction between muscular activity and anatomic relations, studies found that the female upper airway is more stable [32–34].

3. Surgical Therapy for Obstructive Sleep Apnea

The anatomical reasons for OSA are very heterogeneous due to the multiple sites of potential airway obstruction. Henceforth, various surgical treatment options have been established, especially aimed at patients who have unsuccessfully undergone positive airway pressure therapy. These surgical treatments aim to alleviate the upper airway obstruction in the nose, nasopharynx, velopharynx, oropharynx, and hypopharynx. Moreover, in the case of a severe OSA, the focus of surgery is on the hypopharyngeal or retrolingual obstruction, which is most commonly associated with an enlarged tongue or maxillar and/or mandibular deficiency (Figure 2). Surgical therapy success is defined as achieving a reduction of the apnea–hypopnea index (AHI) greater than 50% and/or achieving an AHI of less than 20 events per hour [35]. After the procedure, the success rate is based on the polysomnography (PSG) results and the patient-reported quality of life.

While PSG is the main diagnostic tool for confirming OSA, on the other hand, it does not give insight into the anatomical location of the obstruction. Hence, a presurgery patient assessment needs to be conducted to identify the potential sites causing the obstruction. There are several diagnostic methods that are used for this purpose, such as flexible fiberoptic nasopharyngoscopy, lateral cephalogram, 3D cone-beam CT scans, sleep endoscopy, and cine-MRI [36]. The latter two are especially valuable since they give insight into the dynamic aspect of the upper airway while sleeping.

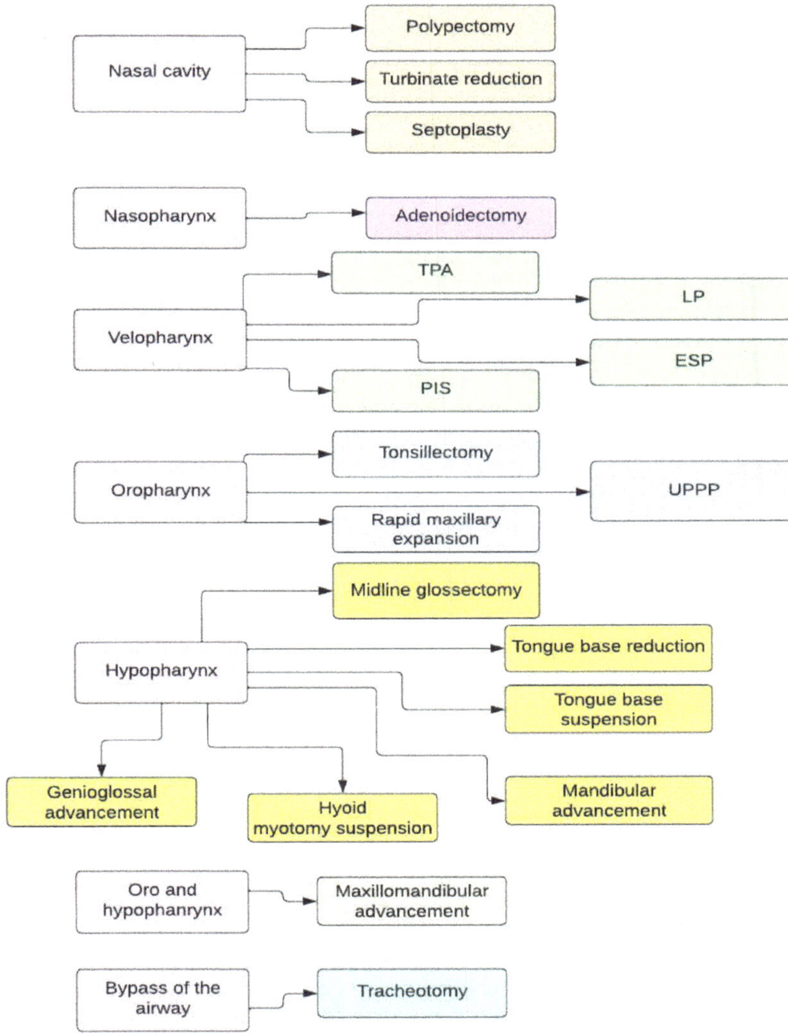

Figure 2. Possible surgical therapies depending on the different level of obstruction. Abbreviations: TPA—trans palatal advancement pharyngoplasty; LP—lateral pharyngoplasty; ESP—expansion sphincter pharyngoplasty; PIS—Pillar implant system; UPPP—Uvulopalatopharyngoplasty.

3.1. Maxillomandibular Advancement

Maxillomandibular advancement (MMA) is typically used for the treatment of refractory or severe OSA or for those with evident and significant maxillomandibular deficiency. The advancement is performed after a Le Fort I osteotomy with a sagittal mandibular split. This consequently enhances the airway space as it pushes forward both the tongue base and the soft palate. MMA is assumed as a highly effective treatment choice for OSA linked with substantial improvements in AHI and RDI [37]. In their meta-analysis, Zaghi et al. reported that in the sample of 518 patients, they found that after MMA, 98.8% had significant improvement in outcomes regarding OSA [38]. There was a significant improvement in postoperative blood oxygen saturation and daytime sleepiness. Moreover, it was found that patients with a more severe preoperative OSA had a lower tendency for surgical success and OSA improvement. John et al. conducted a meta-analysis including details of

462 patients extracted from 20 studies [39]. They showed that, according to AHI and RDI outcomes, the surgical success of MMA was 100%.

Studies show that patients with the highest severity of OSA also have the highest degree of improvement after MMA while, on the other hand, patients with less severe OSA encounter a lesser degree of change in postoperative AHI or RDI, but they have the greatest probability of curing OSA with MMA [38]. Furthermore, patients who have high residual AHI and RDI scores are highly probable to benefit from MMA treatment [40]. MMA causes significant patient morphology change, which can be well-accepted or less favorable, such as an excessively protrusive lower face. Regarding these aesthetic and functional changes, it was advised to modify the MMA surgery technique, which led to the development of the newest surgery adjustments [39]. Modification can consist of a Le Fort I osteotomy with a segmentation or modified step design, counterclockwise or clockwise rotation of the maxillomandibular compound, and reverse T mandibular osteotomy (genioglossal advancement and advancing genioplasty) [41,42]. It is important to highlight that, in general, MMA has a good risk–benefit ratio, and successful outcomes can be achieved with minimal long-term treatment-related adverse outcomes [43]. The advantages of MMA include improved aesthetics, improvement of OSA symptoms, and correcting functional issues that ride along with malocclusion, and the disadvantages are classic surgical adverse effects, such as wound site infection and bleeding, postoperative pain, fixation plate infection, temporomandibular joint pain, limited mouth opening, and sensitive and motor nerve deficit [44].

3.2. Maxillary Expansion and Maxillomandibular Expansion

Patients with a narrow or high-arched hard palate (i.e., transverse maxillary deficiency) are predisposed to OSA and usually have malocclusion and dental crowding, which could be treated with rapid maxillary expansion (RME). One of the characteristics of transverse maxillary deficiency is asymmetric growth of the jaw and consequently discordant size of the maxilla in relation to the mandible. These variations of the skeleton can consequently cause a heightened nasal airflow resistance and a more posterior location of the tongue [45]. As a result of a higher nasal floor, these patients have a higher predisposition for developing nasal airway obstruction, particularly if they also have a septal deviation or enlarged inferior turbinates [46,47].

It was established that RME improves OSA in children [48]. In this population, RME is generally managed without surgery by just using orthodontic apparatus, which often has an expansion screw with numerous arms that administer forces precisely to the maxillary suture through the anchor teeth. RME creates palatal widening, flattening of the palatal arch, inferior movement of the maxilla, and change in the alignment of the mandible [49]. Machado-Júnior et al. conducted a meta-analysis regarding RME for OSA treatment in children [50]. Their outcomes showed that there is a significant decrease in AHI after RME. Moreover, this method improves nasal respiration, increases the maxillary dental arch, and improves the tongue position, and all of these effects can contribute to a reduction in AHI. Additionally, it could be hypothesized that after the broadening of the maxilla, subsequently, there is a rise in the tension of the muscles attached to the palate, which possibly has an alleviating impact on the collapsibility of the upper airway during sleep.

In their randomized controlled trial, Gokce et al. compared different kinds of rapid maxillary expansion appliances on OSA, so they divided patients into three groups: tooth tissue-borne, tooth-borne, and bone-borne expanders [51]. The AHI index did not show any significant difference between the groups; hence, their conclusion was that this is not an effective OSA treatment.

Anatomic and functional factors contribute to the multifactorial etiology of OSA. Amid the soft tissue causes associated with upper airway anatomy, there are skeletal factors described that could have a direct influence on OSA by changing the airway space [52,53].

The treatment modality of transverse maxillary hypoplasia for adults is surgically assisted rapid maxillary expansion (SARME) or distraction osteogenesis maxillary expansion (DOME) [53]. SARME is a combination of orthodontics and surgery that creates an enlargement of the maxillary apical base and the palatal vault, supporting space for the tongue for correct swallowing. Nevertheless, there is a noticeable subjective improvement in nasal breathing associated with the enlargement of the nasal valve. This procedure has been used since 1938 and has become widely accepted as an effective and safe technique for maxillary expansion with minor complications [54,55]. Vinha et al. published an article about the SARME technique for the treatment of OSA in adult patients with good results regarding OSA symptoms, such as decreased rates of respiratory disturbances, desaturation, microarousal, and reduced daytime sleepiness [52]. There are consequently some negative outcomes connected to this surgical method, such as skeletal relapse, different kinds of periodontal problems, lack of movement or failure, gingival recession, pain sensation, and necrosis of the oral mucosa associated with the device [53]

DOME is a method where customized distractors for each patient are first created using 3D cone-beam computed tomography (CBCT). Next, miniscrews are fixed through the palatal roof of both cortical plates as close to the midline as possible. In the last step, a LeFort I osteotomy is conducted. The expander is activated several days after the surgery by using the axial screw, and on average, approximately 8–12 mm of maxillary expansion is created [56]. Recently, Yoon et al. conducted a study on 75 subjects with a narrow maxilla and nasal floor who underwent DOME. The outcomes showed that DOME significantly relieves nasal obstruction, improves the amount of REM sleep, and decreases AHI [56]. Prior to DOME, classic surgically assisted palatal expansion techniques were used and frequently linked to major adverse effects, such as hemorrhage and a high rate of relapse. Compared to that, DOME is less aggressive and has minimal side effects, such as minimal asymmetry (which could be corrected with an orthodontic appliance) and transitory paresthesia of V2 in the anterior maxilla that would resolve in 1–6 months [57].

In their systematic review and meta-analysis, Abdullatif et al. reported six articles regarding maxillary expansion in adults, and they showed that the AHI in these studies has significantly decreased [46]. They also reported outcomes of two studies regarding maxillo-mandibular expansion which also showed a significant reduction in AHI. However, since there were only 39 patients included in these studies, further, larger-scale investigations with a larger sample should be conducted.

3.3. *Septoplasty, Turbinate Reduction, and Polypectomy*

Nasal anomalies, such as alar collapse, nasal valve compromise, bony deformities, nasal polyps, inferior turbinate hypertrophy, and septal deviation, can boost negative intraluminal pressure, initiating airway collapse at the oropharynx or hypopharynx [58]. Procedures used to overcome nasal obstruction include septoplasty, turbinate reduction, and polypectomy.

Septoplasty implicates adjustment of the nasal septum with a lot of different techniques that are used based on the type and the location of the deviation. Even a small reduction of the frontal deviation has been found to cause a substantial improvement of the nasal airway resistance while the restoration of a posterior deviation showed a significantly smaller effect on the airway resistance.

Turbinate reduction is a method that decreases the dimension of the middle or inferior turbinates. The techniques used for turbinate reduction include partial or submucous resection, outfracture, superficial or submucosal cautery, laser treatment, radiofrequency treatment, and endoscopic excision of the concha bullosa [59]. Disregarding the type of nasal surgery, complications are largely the same, such as bleeding, postoperative pain and discomfort, nasal septal perforation, empty nose syndrome, chronic nasal dryness, change in the aesthetic appearance of the nose, etc. [60,61].

In their review, Cai et al. showed that septoplasty, turbinate reduction, rhinoplasty, and sinus surgery subsequently improve OSA-related quality-of-life measures and CPAP tolerance [62]. Takahashi et al. presented the results of patients who had chronic hypertrophic rhinitis and nasal septal deviation and underwent septoplasty and submucous turbinectomy [63]. The postoperative results showed a significant decrease in AHI and the awakening response index while there was also an increase in the mean blood oxygen saturation. However, on the contrary, the recent study by Migueis et al. found that after nasal surgery, there were no differences between baseline and postsurgical values in the frequency of respiratory disturbances, the total apnea time, the distribution of the apnea time within the different apnea types (obstructive and nonobstructive), or the severity of the nocturnal desaturations [64]. Equivalent to these results are outcomes by Kalam, who found that nasal surgery alone, even when nasal obstruction is a predominant symptom, may not be enough to produce a recognizable improvement [65]. These studies suggest that nasal surgeries are possibly ineffective for OSA since they do not improve events during polysomnography.

Furthermore, Hisamatsu et al. described a compound nasal surgery (CNS) method-septoplasty combined with submucosal inferior turbinectomy and posterior nasal neurectomy to assure low nasal resistance during sleep in a total of 45 patients [66]. This surgery method, therefore, inhibits mucosal swelling and nocturnal nasal resistance under a cholinergic dominant action. Regarding the allergy symptoms, it should be emphasized that the outcomes showed improvement of these symptoms in patients who had severe and moderate OSA. Furthermore, it can be concluded that from all nasal surgical procedures, the CNS method of surgery could be the most helpful in patients with severe and moderate OSA since it showed the most significant improvements in PSG findings [66].

Regarding polypectomy, Weder et al., in their literature review and case report, concluded that there are only three articles investigating the association of antrochoanal polyps in children and OSA [67]. In pediatric patients, antrochoanal polyps can be connected to daytime sleepiness and problems with concentration as OSA symptoms and require further investigation and more studies.

3.4. Surgeries of the Oropharynx

The oropharynx of some patients with OSA has excessive tissue that tends to be loose and elongated. Hence, surgery of the oropharynx in the treatment of OSA aims to reduce that excessive amount of the tissue and further stiffen it. The first upper airway surgery that was employed as a treatment method for OSA was the surgery of the soft palate [68,69]. Uvulopalatopharyngoplasty (UPPP) was introduced by Fujita et al., and currently, it is, with or without tonsillectomy, the most commonly implemented surgery for OSA in adults [70]. UPPP increases the upper airway space and alleviates the collapsibility of the pharynx by partial resection of the uvula and the soft palate. It was also shown that UPPP lowers CPAP pressure requirements and consequently improves CPAP compliance in some patients [71]. However, studies showed that the success rates after sole UPPP is highly questionable, and there are several major complications related to UPPP surgery, such as edema of the upper airway, bleeding, velopharyngeal insufficiency, and nasal regurgitation [72–74]. On the other hand, it was established that the combination of UPPP and tonsillectomy is a highly successful surgical treatment for OSA. There are several variations of the UPPP technique as well as similar methods, such as laser-assisted uvulopalatoplasty and upper airway radiofrequency treatment [75]. Complications reported with these procedures were pain, localized edema of the tissues, mucosal sloughing, and oronasal fistulization. Recently, the results of several studies showed that radiofrequency treatment has significant success in the reduction of snoring but limited use for treating OSA [76–79]. However, another study also showed that radiofrequency tissue reduction significantly improves the tolerance and adherence of severe OSA patients using CPAP [80].

Over time, oropharyngeal surgical methods have advanced from excisional techniques towards the more preservative remodeling of the oropharynx, particularly targeting the lateral walls and conducting expansion of the velopharynx space [81]. Among the first of the aforementioned methods was transpalatal advancement pharyngoplasty (TPA), a technique introduced by Woodson et al. who intended it for patients who underwent unsuccessful UPPP and were found to have lasting velopharyngeal collapse [82]. This technique is used to widen the retropalatal space by pulling and fixating the soft palate aponeurosis forward. The results of the meta-analysis by Volner et al. have shown that TPA has significantly improved the AHI in patients with OSA [83]. Furthermore, Cahali introduced lateral pharyngoplasty (LP), a technique in which a suture is placed between the palatoglossal and superior pharyngeal muscle while the latter was previously sectioned in the vertical direction. This method splits the lateral walls of the pharynx and subsequently alleviates the collapsibility of the velopharynx and oropharynx. Furthermore, while preventing excess bleeding and maintaining optimal swallowing function are important when performing any palatopharyngoplasty, on the other hand, this method was reported with the possibility of transient dysphagia [84,85].

More recently, Pang and Woodson described the expansion sphincter pharyngoplasty (ESP), a method that preserves the superior pharyngeal muscle, but the palatopharyngeal muscle is split and fixated to the pterygoid hamulus [86]. This creates tension in the lateral walls of the pharynx and consequently treats both the velopharynx and oropharynx collapse and expands the retropalatal area. ESP is an alternative for patients who are not eligible for CPAP. Recently, a meta-analysis showed that ESP has great results in treating OSA and appears to have the lowest complication rates compared to other techniques [87].

Palate suture suspension is a more recent surgical technique for managing OSA which aims to suspend the soft palate and stabilize the lateral walls of the pharynx by using sutures for fixation. Up until now, there are several similar techniques described in the literature, all of them applying sutures for suspension [88–95]. The main aim of these techniques is pulling the soft palate forward and laterally while also achieving stabilization of lateral walls by fixing the sutures to firm structures without implementing resection, and due to this, they are considered less invasive compared to ESP or UPPP [96].

The Pillar implant system is a newer, minimally invasive technique for managing OSA. There are several variations of this method, but the aim is to induce local stiffness and prevent soft palate collapse in the state of muscle relaxation during sleep. That is achieved by the insertion of artificial implants in the soft palate which significantly alleviates snoring and treats OSA [97]. Disadvantages are foreign body sensation, swallowing difficulties as well as risk for ulceration and implant extrusion. Zhang et al. have described a method in which "U-shaped" titanium mesh and titanium screws are used as implants that are positioned into a soft palate [98]. Its short-term efficacy is good in patients with moderate-to-severe OSA and velopharyngeal obstruction.

3.5. Genioglossal Advancement

Genioglossal advancement (GA) was first described by Riley et al., and it is indicated for patients with a narrow lower pharynx [99]. The procedure involves repositioning the genial tubercle and genioglossus muscle forward, thus enlarging the airway in the anteroposterior dimension. The osteotomy extends over the mandible to incorporate the genioglossus attachment while also leaving the dentoalveolar process intact. The main disadvantages of this procedure were the possibility of mandibular fracture, bony segment necrosis, and a risk of lower incisor teeth damage, so several variations were proposed to improve results and reduce complications [100–103]. They include diverse forms of osteotomies and minor modifications in the transposition of the bony segment. Li et al. presented an alteration of the procedure making a "window" osteotomy by leaving an inferior border of the mandible intact, thus reducing the potential of mandibular fractures [104]. Garcia Vega et al. presented a modified GA technique that reduces muscle damage but is more complicated to perform, especially on smaller mandibles [100]. Nevertheless, the main aim of the technique is the same, and outcomes have shown significant improvement in OSA. A recent systematic review by Kezirian et al. reported a surgical success rate of 67% when undergoing GA as a sole treatment [105]. The recent development of virtual simulation surgeries using 3D technology has helped surgeons to plan more precise osteotomies, thus minimizing the complications of conventional GA procedures [106].

3.6. Hyoid Suspension

Forward movement of the hyoid complex can improve the hypopharyngeal airway due to the excessively low position of the hyoid bone in patients with OSA. Riley et al. described the original method for hyoid myotomy with suspension, and it involved an inferior hyoid myotomy of the infrahyoid musculature followed by suspension of the hyoid to the mandible using fascia lata [107,108]. Due to high morbidity, the technique was later modified to move the hyoid bone forward inferiorly over the thyroid cartilage which is known as hyoidthyroidpexia [109]. The goal of this modification is to increase the retrolingual airway space [110]. Most recent alterations of this treatment include using a stainless steel wire or titanium plates on the thyroid cartilage, thus preventing its damage [111,112]. The use of implantable devices for hyoid expansion was also reported, but no significant improvements in sleep parameters were noted [113]. Serious postoperative complications were rarely reported, and the most common ones were dysfunctions of pronunciation and swallowing [114]. A meta-analysis by Song et al. showed that sole hyoid surgery can reduce OSA severity in adults [115]. Their outcomes showed a significant decrease in sleepiness and that AHI was reduced by 38% whereas for hyothyroidopexy, an AHI reduction of 50.7% was achieved.

3.7. Tongue Base Reduction

Base of tongue (BOT) reduction may be performed through several approaches, such as radiofrequency, surgery, endoscopy, and transoral robotic surgery (TORS) [116]. Radiofrequency surgery has limited indications and is mostly used in cases of moderate tongue base hypertrophy [117]. Midline glossectomy (MLG) involves removing parts of the middle and back of the tongue (usually using a laser) [118]. It is mostly combined with other surgical procedures, such as UPPP, palatoplasty, palatopharyngoplasty, or epiglottidectomy. Submucosal minimally invasive lingual excision (SMILE) is a technique that avoids damaging the mouth floor muscles while undergoing excision of excessive tongue tissue using an endoscope and ultrasound for visualization [119]. A meta-analysis by Murphey et al. showed improvement in outcomes when glossectomy is a part of multilevel surgery in adults with OSA [120]. There are not many reports on the effect of glossectomy as a single treatment, but the study by Suslu et al. showed that coblation MLG is a beneficial procedure when carried out solely with a success rate of 52% [121]. Moreover, it was shown that MLG was very successful when implemented solely in patients who failed UPPP.

With the improvement of new technologies, TORS has become a therapeutic alternative for reducing tongue volume. It is useful due to an endoscope that enables a 3D visualization of the surgical field and a robotic instrument that is handled to reduce the tongue base in an area that is usually very difficult to access through classical surgical methods. The reported complications are bleeding, dehydration, dysphagia, oropharyngeal stenosis, tongue numbness, and dysgeusia [114,122]. In the study by Friedman et al., a comparison of the outcomes between TORS, SMILE, and radiofrequency BOT reduction was conducted. It was shown that TORS in combination with palatopharyngoplasty had a significantly higher reduction of AHI scores [123]. Hwang et al. compared TORS with an endoscope-guided coblation BOT reduction and found that there were no significant differences regarding the outcomes [124]. Moreover, the results of the meta-analysis by Lechien et al. suggest that TORS BOT reduction is associated with improvements in AHI and ESS irrespective of any additional procedures performed [125].

3.8. Tongue Base Suspension

The tongue base suspension (TBS) is a minimally invasive surgical technique whose goal is to reduce the tongue base collapse and enlarge the retrolingual airway [126]. A standard procedure is carried out using a Repose system (Repose Surgical Kit, CKA Air Vance, Medtronic, Inc.). Through a submental incision, a screw is placed into the mandible, and the tongue is stabilized with a suture using a screw as an anchor [127]. Hsin et al. most recently presented a transoral tongue base suspension (TOTS) method [128]. Its main advantage is using an intraoral vestibular incision and no need for screws. Even though there are modifications of the procedure reported, TBS using a Repose system remains a gold standard [129]. It was shown that it has a cure rate of 81.8% when combined with UPPP in patients with multilevel airway obstruction, and it is a good alternative for other more invasive operative procedures, such as GA or various BOT techniques [130]. The most commonly reported complications have been sialadenitis, tongue base swelling, infection, or suture-related problems [131].

3.9. Adenoidectomy and Tonsillectomy

A tonsillectomy is a common surgical treatment used for tonsillar infections and inflammations or tonsillar malignant diseases. However, studies have shown that there is an association between the volume and grade of the tonsils with the AHI and the severity of OSA, but only a few studies have discussed isolated tonsillectomy as a surgical treatment for treating OSA in adults [132–136]. Still, it is important to highlight that the outcomes of several studies imply an association between tonsillar hypertrophy and OSA, and this suggests that tonsillectomy alone is an effective treatment for OSA in adults with tonsil hypertrophy [133,137]. It is considered a simple procedure with few complications, such as temporary pain and dysphagia. Several studies showed that a tonsillectomy considerably reduced the AHI and improved daytime sleepiness and the severity of OSA in these patients [134,138]. Nevertheless, for the management of OSA, tonsillectomy is more commonly applied in combination with UPPP [139,140]. Evidence is showing that the efficacy of UPPP is significantly amplified when joined with a tonsillectomy, and studies established that it leads to an additional improvement of the AHI, snoring, and daytime sleepiness [141]. Regarding pediatric OSA, adenotonsillectomy is the first line of treatment [142]. It is suggested as a safe day-case surgery in carefully selected patients [143]. The results of most secondary analyses suggest that children who underwent adenotonsillectomy experienced the greatest improvements in symptom burden, sleepiness, parent-reported behavior, and quality of life.

3.10. Tracheostomy

A tracheostomy is a procedure that involves creating an opening in the trachea to bypass the upper airway. It was first introduced as a treatment for OSA in 1969, and until then, there were no other surgical options for treating OSA [144]. Since patients with a tracheostomy avoid the whole upper airway, it is the most effective surgical procedure for treating OSA [145]. However, it was shown that patients with concomitant cardiopulmonary diseases or chronic obstructive pulmonary diseases may have residual indispositions [146]. A systematic review by Camacho et al. found that a tracheostomy substantially reduces the AHI and daytime sleepiness while also improving oxygen desaturation index and cardiovascular mortality in patients with OSA [147]. Nevertheless, a permanent tracheostomy tube requires delicate care, and it restricts daily living activities and social interaction. Hence, it should be used as a treatment for OSA only in carefully chosen patients who do not have other surgical options. On the other hand, a temporary tracheostomy can be performed on patients as a preliminary prophylactic measure. Presently, tracheostomy is mostly retained for patients who are not eligible to use CPAP and/or do not have a verifiable upper airway abnormality [148,149].

4. Conclusions

There are a lot of surgical methods and modalities for treating OSA, and they all have advantages and disadvantages. Even though, according to the literature, some surgical methods seem more successful, on the other hand, due to the fact that the anatomical causes of OSA are very heterogeneous, every patient should be approached individually, and the best surgical method should be chosen for his specific needs.

Regarding the possibility that surgical therapy from the OMFS domain could be a trigger for developing OSA, the literature is still too scarce on this subject, and only lately more attention has been drawn to this issue. Nevertheless, the results of several recent studies are implying that oncology, traumatology, and orthognathic surgery could really be impacting OSA. Henceforth, further studies, especially systematic reviews, are needed regarding this important subject.

Author Contributions: Conceptualization, D.M. and J.B.; investigation, D.M., D.T., E.P-M., S.K., A.L. and S.L.-F.; writing—original draft preparation, D.M., D.T., E.P.-M., S.K. and A.L.; writing—review and editing, D.M., D.T., E.P.-M., S.K., A.L. and M.K.; visualization, D.T. and M.K.; supervision, S.L.-F. and J.B.; project administration, S.L.-F. and J.B. All authors have read and agreed to the published version of the manuscript.

Funding: This research received no external funding.

Institutional Review Board Statement: Not applicable.

Informed Consent Statement: Not applicable.

Data Availability Statement: Not applicable.

Conflicts of Interest: The authors declare no conflict of interest.

References

1. Bozic, J.; Galic, T.; Supe-Domic, D.; Ivkovic, N.; Ticinovic Kurir, T.; Valic, Z.; Lesko, J.; Dogas, Z. Morning cortisol levels and glucose metabolism parameters in moderate and severe obstructive sleep apnea patients. *Endocrine* **2016**, *53*, 730–739. [CrossRef] [PubMed]
2. Osman, A.M.; Carter, S.G.; Carberry, J.C.; Eckert, D.J. Obstructive sleep apnea: Current perspectives. *Nat. Sci. Sleep* **2018**, *10*, 21–34. [CrossRef] [PubMed]
3. Borovac, J.A.; Dogas, Z.; Supe-Domic, D.; Galic, T.; Bozic, J. Catestatin serum levels are increased in male patients with obstructive sleep apnea. *Sleep Breath.* **2019**, *23*, 473–481. [CrossRef] [PubMed]
4. Bozic, J.; Borovac, J.A.; Galic, T.; Kurir, T.T.; Supe-Domic, D.; Dogas, Z. Adropin and Inflammation Biomarker Levels in Male Patients with Obstructive Sleep Apnea: A Link With Glucose Metabolism and Sleep Parameters. *J. Clin. Sleep Med.* **2018**, *14*, 1109–1118. [CrossRef]
5. Slouka, D.; Honnerova, M.; Hosek, P.; Gal, B.; Trcka, O.; Kostlivy, T.; Landsmanova, J.; Havel, D.; Baneckova, M.; Kucera, R. Improved prediction of CPAP failure using T90, age and gender. *J. Appl. Biomed.* **2019**, *17*, 10. [CrossRef]
6. Ralli, M.; Campo, F.; Angeletti, D.; Allegra, E.; Minni, A.; Polimeni, A.; Greco, A.; de Vincentiis, M. Obstructive Sleep Apnoea in Patients Treated for Head and Neck Cancer: A Systematic Review of the Literature. *Medicina* **2020**, *56*, 399. [CrossRef]
7. Kim, J.W.; Kwon, T.G. Why most patients do not exhibit obstructive sleep apnea after mandibular setback surgery? *Maxillofac. Plast. Reconstr. Surg.* **2020**, *42*, 7. [CrossRef]
8. Gavidia, R.; Dunietz, G.L.; O'Brien, L.; Shannon, C.; Schuetz, S.; Spector, M.; Swiecicki, P.; Chervin, R.D. Obstructive sleep apnea in patients with head and neck cancer: A systematic review. *J. Clin. Sleep Med.* **2021**, *17*, 1109–1116. [CrossRef]
9. Liao, F.H.; Chang, C.C.; Lu, Y.C.; Lin, C.Y.; Lai, W.S. Impact of free flap reconstruction on obstructive sleep apnea in patients with oral and oropharyngeal cancer. *Asia-Pac. J. Oncol. Nurs.* **2022**, *9*, 100136. [CrossRef]
10. Gavidia, R.; Dunietz, G.L.; O'Brien, L.M.; Schütz, S.G.; Spector, M.E.; Swiecicki, P.L.; Chervin, R.D. Risk of obstructive sleep apnea after treatment of head and neck squamous cell carcinoma: A cross-sectional study. *J. Clin. Sleep Med.* **2022**, *18*, 1681–1686. [CrossRef]
11. Lupi-Ferandin, S.; Galic, T.; Ivkovic, N.; Pecotic, R.; Dogas, Z. Prevalence of obstructive sleep apnea in male patients with surgically treated maxillary and zygomatic fractures. *Can. J. Surg.* **2019**, *62*, 105–110. [CrossRef] [PubMed]
12. Loth, A.; Michel, J.; Giorgi, R.; Santini, L.; Rey, M.; Elbaum, J.M.; Roux, N.; Giovanni, A.; Dessi, P.; Fakhry, N. Prevalence of obstructive sleep apnoea syndrome following oropharyngeal cancer treatment: A prospective cohort study. *Clin. Otolaryngol.* **2017**, *42*, 1281–1288. [CrossRef] [PubMed]
13. Friedman, M.; Landsberg, R.; Pryor, S.; Syed, Z.; Ibrahim, H.; Caldarelli, D.D. The occurrence of sleep-disordered breathing among patients with head and neck cancer. *Laryngoscope* **2001**, *111*, 1917–1919. [CrossRef] [PubMed]
14. Qian, W.; Haight, J.; Poon, I.; Enepekides, D.; Higgins, K.M. Sleep apnea in patients with oral cavity and oropharyngeal cancer after surgery and chemoradiation therapy. *Otolaryngol. Head Neck Surg.* **2010**, *143*, 248–252. [CrossRef]
15. Payne, R.J.; Hier, M.P.; Kost, K.M.; Black, M.J.; Zeitouni, A.G.; Frenkiel, S.; Naor, N.; Kimoff, R.J. High prevalence of obstructive sleep apnea among patients with head and neck cancer. *J. Otolaryngol.* **2005**, *34*, 304–311. [CrossRef]
16. Gómez-Merino, E.; Arriero, J.M.; Chiner, E.; Signes-Costa, J.; Marco, J. Obstructive sleep apnea syndrome as first manifestation of pharyngeal non-Hodgkin's lymphoma. *Respiration* **2003**, *70*, 107–109. [CrossRef]
17. Hockstein, N.G.; Anderson, T.A.; Moonis, G.; Gustafson, K.S.; Mirza, N. Retropharyngeal lipoma causing obstructive sleep apnea: Case report including five-year follow-up. *Laryngoscope* **2002**, *112*, 1603–1605. [CrossRef]

18. Koopmann, C.F., Jr.; Feld, R.A.; Coulthard, S.W. Sleep apnea syndrome associated with a neck mass. *Otolaryngol. Head Neck Surg.* **1981**, *89*, 949–952. [CrossRef]
19. Pellanda, A.; Zagury, S.; Pasche, P. Parapharyngeal lipoma causing obstructive sleep apnea syndrome. *Otolaryngol. Head Neck Surg.* **2003**, *128*, 301–302. [CrossRef]
20. Abdullah, B.J.; Liam, C.K.; Kaur, H.; Mathew, K.M. Parapharyngeal space lipoma causing sleep apnoea. *Br. J. Radiol.* **1997**, *70*, 1063–1065. [CrossRef]
21. Werlinger, F.; Villalón, M.; Duarte, V.; Acevedo, R.; Aguilera, R.; Alcocer, D.; Badillo, O.; Briones, R.; Condal, C.; Del Río, M.; et al. Trends of maxillofacial trauma: An update from the prospective register of a multicenter study in emergency services of Chile. *Med. Oral Patol. Oral Cir. Bucal* **2019**, *24*, e588–e594. [CrossRef] [PubMed]
22. Kaul, R.P.; Sagar, S.; Singhal, M.; Kumar, A.; Jaipuria, J.; Misra, M. Burden of maxillofacial trauma at level 1 trauma center. *Craniomaxillofac. Trauma Reconstr.* **2014**, *7*, 126–130. [CrossRef] [PubMed]
23. El-Anwar, M.W.; Askar, S.; Abou Shab, Y.A.; Abou Sharkh, A.A.M. Could mandibular fractures lead to obstructive sleep apnea? *Cranio* **2021**, *8*, 1–4. [CrossRef] [PubMed]
24. Khechoyan, D.Y. Orthognathic surgery: General considerations. *Semin. Plast. Surg.* **2013**, *27*, 133–136. [CrossRef] [PubMed]
25. Hasebe, D.; Kobayashi, T.; Hasegawa, M.; Iwamoto, T.; Kato, K.; Izumi, N.; Takata, Y.; Saito, C. Changes in oropharyngeal airway and respiratory function during sleep after orthognathic surgery in patients with mandibular prognathism. *Int. J. Oral Maxillofac. Surg.* **2011**, *40*, 584–592. [CrossRef] [PubMed]
26. Kobayashi, T.; Funayama, A.; Hasebe, D.; Kato, Y.; Yoshizawa, M.; Saito, C. Changes in overnight arterial oxygen saturation after mandibular setback. *Br. J. Oral Maxillofac. Surg.* **2013**, *51*, 312–318. [CrossRef]
27. Fernández-Ferrer, L.; Montiel-Company, J.M.; Pinho, T.; Almerich-Silla, J.M.; Bellot-Arcís, C. Effects of mandibular setback surgery on upper airway dimensions and their influence on obstructive sleep apnoea—A systematic review. *J. Craniomaxillofac. Surg.* **2015**, *43*, 248–253. [CrossRef]
28. Dahy, K.; Takahashi, K.; Saito, K.; Kiso, H.; Rezk, I.; Oga, T.; Uozumi, R.; Chin, K.; Bessho, K. Gender differences in morphological and functional outcomes after mandibular setback surgery. *J. Craniomaxillofac. Surg.* **2018**, *46*, 887–892. [CrossRef]
29. Isono, S.; Remmers, J.E.; Tanaka, A.; Sho, Y.; Sato, J.; Nishino, T. Anatomy of pharynx in patients with obstructive sleep apnea and in normal subjects. *J. Appl. Physiol.* **1985**, *82*, 1319–1326. [CrossRef]
30. Martin, S.E.; Mathur, R.; Marshall, I.; Douglas, N.J. The effect of age, sex, obesity and posture on upper airway size. *Eur. Respir. J.* **1997**, *10*, 2087–2090. [CrossRef]
31. Young, T.; Evans, L.; Finn, L.; Palta, M. Estimation of the clinically diagnosed proportion of sleep apnea syndrome in middle-aged men and women. *Sleep* **1997**, *20*, 705–706. [CrossRef] [PubMed]
32. Rowley, J.A.; Zhou, X.; Vergine, I.; Shkoukani, M.A.; Badr, M.S. Influence of gender on upper airway mechanics: Upper airway resistance and Pcrit. *J. Appl. Physiol.* **1985**, *91*, 2248–2254. [CrossRef] [PubMed]
33. Trinder, J.; Kay, A.; Kleiman, J.; Dunai, J. Gender differences in airway resistance during sleep. *J. Appl. Physiol.* **1985**, *83*, 1986–1997. [CrossRef] [PubMed]
34. Popovic, R.M.; White, D.P. Influence of gender on waking genioglossal electromyogram and upper airway resistance. *Am. J. Respir. Crit. Care Med.* **1995**, *152*, 725–731. [CrossRef] [PubMed]
35. Won, C.H.; Li, K.K.; Guilleminault, C. Surgical treatment of obstructive sleep apnea: Upper airway and maxillomandibular surgery. *Proc. Am. Thorac. Soc.* **2008**, *5*, 193–199. [CrossRef]
36. Goyal, M.; Johnson, J. Obstructive Sleep Apnea Diagnosis and Management. *Mo. Med.* **2017**, *114*, 120–124.
37. Boyd, S.B.; Chigurupati, R.; Cillo, J.E., Jr.; Eskes, G.; Goodday, R.; Meisami, T.; Viozzi, C.F.; Waite, P.; Wilson, J. Maxillomandibular Advancement Improves Multiple Health-Related and Functional Outcomes in Patients with Obstructive Sleep Apnea: A Multicenter Study. *J. Oral Maxillofac. Surg.* **2019**, *77*, 352–370. [CrossRef]
38. Zaghi, S.; Holty, J.E.; Certal, V.; Abdullatif, J.; Guilleminault, C.; Powell, N.B.; Riley, R.W.; Camacho, M. Maxillomandibular Advancement for Treatment of Obstructive Sleep Apnea: A Meta-analysis. *JAMA Otolaryngol. Head Neck Surg.* **2016**, *142*, 58–66. [CrossRef]
39. John, C.R.; Gandhi, S.; Sakharia, A.R.; James, T.T. Maxillomandibular advancement is a successful treatment for obstructive sleep apnoea: A systematic review and meta-analysis. *Int. J. Oral Maxillofac. Surg.* **2018**, *47*, 1561–1571. [CrossRef]
40. Friedman, M.; Wilson, M. Re-redefining success in airway surgery for obstructive sleep apnea. *Sleep* **2009**, *32*, 17.
41. Liao, Y.F.; Chiu, Y.T.; Lin, C.H.; Chen, Y.A.; Chen, N.H.; Chen, Y.R. Modified maxillomandibular advancement for obstructive sleep apnoea: Towards a better outcome for Asians. *Int. J. Oral Maxillofac. Surg.* **2015**, *44*, 189–194. [CrossRef] [PubMed]
42. Schendel, S.A.; Broujerdi, J.A.; Jacobson, R.L. Three-dimensional upper-airway changes with maxillomandibular advancement for obstructive sleep apnea treatment. *Am. J. Orthod. Dentofac. Orthop.* **2014**, *146*, 385–393. [CrossRef] [PubMed]
43. Boyd, S.B.; Walters, A.S.; Waite, P.; Harding, S.M.; Song, Y. Long-Term Effectiveness and Safety of Maxillomandibular Advancement for Treatment of Obstructive Sleep Apnea. *J. Clin. Sleep Med.* **2015**, *11*, 699–708. [CrossRef] [PubMed]
44. Blumen, M.B.; Buchet, I.; Meulien, P.; Hausser, H.C.; Neveu, H.; Chabolle, F. Complications/adverse effects of maxillomandibular advancement for the treatment of OSA in regard to outcome. *Otolaryngol. Head Neck Surg.* **2009**, *141*, 591–597. [CrossRef] [PubMed]
45. Vinha, P.P.; Faria, A.C.; Xavier, S.P.; Christino, M.; de Mello-Filho, F.V. Enlargement of the Pharynx Resulting From Surgically Assisted Rapid Maxillary Expansion. *J. Oral Maxillofac. Surg.* **2016**, *74*, 369–379. [CrossRef]

46. Abdullatif, J.; Certal, V.; Zaghi, S.; Song, S.A.; Chang, E.T.; Gillespie, M.B.; Camacho, M. Maxillary expansion and maxillo-mandibular expansion for adult OSA: A systematic review and meta-analysis. *J. Craniomaxillofac. Surg.* **2016**, *44*, 574–578. [CrossRef]
47. Camacho, M.; Zaghi, S.; Certal, V.; Abdullatif, J.; Means, C.; Acevedo, J.; Liu, S.; Brietzke, S.E.; Kushida, C.A.; Capasso, R. Inferior turbinate classification system, grades 1 to 4: Development and validation study. *Laryngoscope* **2015**, *125*, 296–302. [CrossRef]
48. Guilleminault, C.; Quo, S.; Huynh, N.T.; Li, K. Orthodontic expansion treatment and adenotonsillectomy in the treatment of obstructive sleep apnea in prepubertal children. *Sleep* **2008**, *31*, 953–957.
49. Camacho, M.; Chang, E.T.; Song, S.A.; Abdullatif, J.; Zaghi, S.; Pirelli, P.; Certal, V.; Guilleminault, C. Rapid maxillary expansion for pediatric obstructive sleep apnea: A systematic review and meta-analysis. *Laryngoscope* **2017**, *127*, 1712–1719. [CrossRef]
50. Machado-Júnior, A.J.; Zancanella, E.; Crespo, A.N. Rapid maxillary expansion and obstructive sleep apnea: A review and meta-analysis. *Med. Oral Patol. Oral Cir. Bucal* **2016**, *21*, 21073. [CrossRef]
51. Gokce, G.; Basoglu, O.K.; Veli, I. Polygraphic evaluation of the effects of different rapid maxillary expansion appliances on sleep quality: A randomized clinical trial. *Sleep Breath.* **2022**, *5*. [CrossRef] [PubMed]
52. Vinha, P.P.; Eckeli, A.L.; Faria, A.C.; Xavier, S.P.; de Mello-Filho, F.V. Effects of surgically assisted rapid maxillary expansion on obstructive sleep apnea and daytime sleepiness. *Sleep Breath.* **2016**, *20*, 501–508. [CrossRef] [PubMed]
53. Koudstaal, M.J.; Poort, L.J.; van der Wal, K.G.; Wolvius, E.B.; Prahl-Andersen, B.; Schulten, A.J. Surgically assisted rapid maxillary expansion (SARME): A review of the literature. *Int. J. Oral Maxillofac. Surg.* **2005**, *34*, 709–714. [CrossRef] [PubMed]
54. Liu, S.Y.-C.; Guilleminault, C.; Huon, L.-K.; Yoon, A. Distraction Osteogenesis Maxillary Expansion (DOME) for Adult Obstructive Sleep Apnea Patients with High Arched Palate. *Otolaryngol. Head Neck Surg.* **2017**, *157*, 345–348. [CrossRef] [PubMed]
55. Loriato, L.; Ferreira, C.E. Surgically-assisted rapid maxillary expansion (SARME): Indications, planning and treatment of severe maxillary deficiency in an adult patient. *Dent. Press J. Orthod.* **2020**, *25*, 73–84. [CrossRef] [PubMed]
56. Yoon, A.; Guilleminault, C.; Zaghi, S.; Liu, S.Y. Distraction Osteogenesis Maxillary Expansion (DOME) for adult obstructive sleep apnea patients with narrow maxilla and nasal floor. *Sleep Med.* **2020**, *65*, 172–176. [CrossRef] [PubMed]
57. Cakarer, S.; Keskin, B.; Isler, S.C.; Cansiz, E.; Uzun, A.; Keskin, C. Complications associated with surgically assisted rapid palatal expansion without pterygomaxillary separation. *J. Stomatol. Oral Maxillofac. Surg.* **2017**, *118*, 279–282. [CrossRef]
58. Tanna, N.; Smith, B.D.; Zapanta, P.E.; Karanetz, I.; Andrews, B.T.; Urata, M.M.; Bradley, J.P. Surgical Management of Obstructive Sleep Apnea. *Plast. Reconstr. Surg.* **2016**, *137*, 1263–1272. [CrossRef]
59. Mickelson, S.A. Nasal Surgery for Obstructive Sleep Apnea Syndrome. *Otolaryngol. Clin. N. Am.* **2016**, *49*, 1373–1381. [CrossRef]
60. Bloom, J.D.; Kaplan, S.E.; Bleier, B.S.; Goldstein, S.A. Septoplasty complications: Avoidance and management. *Otolaryngol. Clin. N. Am.* **2009**, *42*, 463–481. [CrossRef]
61. Bin Lajdam, G.; Alaryani, K.; Ghaddaf, A.A.; Aljabri, A.; Halawani, A.; Alshareef, M.; Algarni, M.; Al-Hakami, H. Septoplasty versus septoplasty with turbinate reduction for nasal obstruction due to deviated nasal septum: A systematic review and meta-analysis. *Rhinology* **2022**, *60*, 411–420. [CrossRef] [PubMed]
62. Cai, Y.; Goldberg, A.N.; Chang, J.L. The Nose and Nasal Breathing in Sleep Apnea. *Otolaryngol. Clin. N. Am.* **2020**, *53*, 385–395. [CrossRef] [PubMed]
63. Takahashi, R.; Ohbuchi, T.; Hohchi, N.; Takeuchi, S.; Ohkubo, J.; Ikezaki, S.; Suzuki, H. Effect of septoplasty and turbinectomy on obstructive sleep apnea syndrome. *Nihon Jibiinkoka Gakkai Kaiho* **2013**, *116*, 789–792. [CrossRef] [PubMed]
64. Migueis, D.P.; Thuler, L.C.; Lemes, L.N.; Moreira, C.S.; Joffily, L.; Araujo-Melo, M.H. Systematic review: The influence of nasal obstruction on sleep apnea. *Braz. J. Otorhinolaryngol.* **2016**, *82*, 223–231. [CrossRef] [PubMed]
65. Kalam, I. Objective assessment of nasal obstruction in snoring and obstructive sleep apnea patients: Experience of a Police Authority Hospital. *Ann. Saudi Med.* **2002**, *22*, 158–162. [PubMed]
66. Hisamatsu, K.; Kudo, I.; Makiyama, K. The effect of compound nasal surgery on obstructive sleep apnea syndrome. *Am. J. Rhinol. Allergy* **2015**, *29*, e192–e196. [CrossRef]
67. Weder, S.; Landis, B.N.; Banz, Y.; Caversaccio, M.; Dubach, P. Paediatric traffic accident and obstructive sleep apnoea by antrochoanal polyps: Case report and literature review. *Int. J. Pediatr. Otorhinolaryngol.* **2011**, *75*, 1359–1363. [CrossRef]
68. Fujita, S.; Conway, W.; Zorick, F.; Roth, T. Surgical correction of anatomic azbnormalities in obstructive sleep apnea syndrome: Uvulopalatopharyngoplasty. *Otolaryngol. Head Neck Surg.* **1981**, *89*, 923–934. [CrossRef]
69. Holty, J.E.; Guilleminault, C. Surgical options for the treatment of obstructive sleep apnea. *Med. Clin. N. Am.* **2010**, *94*, 479–515. [CrossRef]
70. Collop, N.A. Advances in treatment of obstructive sleep apnea syndrome. *Curr. Treat. Options Neurol.* **2009**, *11*, 340–348. [CrossRef]
71. Chandrashekariah, R.; Shaman, Z.; Auckley, D. Impact of upper airway surgery on CPAP compliance in difficult-to-manage obstructive sleep apnea. *Arch. Otolaryngol. Head Neck Surg.* **2008**, *134*, 926–930. [CrossRef] [PubMed]
72. Pang, K.P.; Vicini, C.; Montevecchi, F.; Piccin, O.; Chandra, S.; Yang, H.C.; Agrawal, V.; Chung, J.C.K.; Chan, Y.H.; Pang, S.B.; et al. Long-term Complications of Palate Surgery: A Multicenter Study of 217 Patients. *Laryngoscope* **2020**, *130*, 2281–2284. [CrossRef] [PubMed]
73. Stuck, B.A.; Ravesloot, M.J.L.; Eschenhagen, T.; de Vet, H.C.W.; Sommer, J.U. Uvulopalatopharyngoplasty with or without tonsillectomy in the treatment of adult obstructive sleep apnea—A systematic review. *Sleep Med.* **2018**, *50*, 152–165. [CrossRef] [PubMed]

74. Friedman, M.; Ibrahim, H.; Bass, L. Clinical staging for sleep-disordered breathing. *Otolaryngol. Head Neck Surg.* **2002**, *127*, 13–21. [CrossRef] [PubMed]
75. Caples, S.M.; Rowley, J.A.; Prinsell, J.R.; Pallanch, J.F.; Elamin, M.B.; Katz, S.G.; Harwick, J.D. Surgical modifications of the upper airway for obstructive sleep apnea in adults: A systematic review and meta-analysis. *Sleep* **2010**, *33*, 1396–1407. [CrossRef] [PubMed]
76. Bäck, L.J.; Hytönen, M.L.; Roine, R.P.; Malmivaara, A.O. Radiofrequency ablation treatment of soft palate for patients with snoring: A systematic review of effectiveness and adverse effects. *Laryngoscope* **2009**, *119*, 1241–1250. [CrossRef]
77. Li, K.K.; Powell, N.B.; Riley, R.W.; Troell, R.J.; Guilleminault, C. Radiofrequency volumetric reduction of the palate: An extended follow-up study. *Otolaryngol. Head Neck Surg.* **2000**, *122*, 410–414. [CrossRef]
78. Finkelstein, Y.; Stein, G.; Ophir, D.; Berger, R.; Berger, G. Laser-assisted uvulopalatoplasty for the management of obstructive sleep apnea: Myths and facts. *Arch. Otolaryngol. Head Neck Surg.* **2002**, *128*, 429–434. [CrossRef]
79. Ryan, C.F.; Love, L.L. Unpredictable results of laser assisted uvulopalatoplasty in the treatment of obstructive sleep apnoea. *Thorax* **2000**, *55*, 399–404. [CrossRef]
80. Emara, T.A.; Ibrahim, H.A.; Elmalt, A.E.; Dahy, K.G.; Rashwan, M.S. Upper airway multilevel radiof-requency under local anesthesia can improve CPAP adherence for severe OSA patients. *Am. J. Otolaryngol.* **2023**, *44*, 103671. [CrossRef]
81. Cammaroto, G.; Stringa, L.M.; Iannella, G.; Meccariello, G.; Zhang, H.; Bahgat, A.Y.; Calvo-Henriquez, C.; Chiesa-Estomba, C.; Lechien, J.R.; Barillari, M.R.; et al. Manipulation of Lateral Pharyngeal Wall Muscles in Sleep Surgery: A Review of the Literature. *Int. J. Environ. Res. Public Health* **2020**, *17*, 5315. [CrossRef] [PubMed]
82. Woodson, B.T.; Toohill, R.J. Transpalatal advancement pharyngoplasty for obstructive sleep apnea. *Laryngoscope* **1993**, *103*, 269–276. [CrossRef] [PubMed]
83. Volner, K.; Dunn, B.; Chang, E.T.; Song, S.A.; Liu, S.Y.; Brietzke, S.E.; O'Connor, P.; Camacho, M. Transpalatal advancement pharyngoplasty for obstructive sleep apnea: A systematic review and meta-analysis. *Eur. Arch. Otorhinolaryngol.* **2017**, *274*, 1197–1203. [CrossRef] [PubMed]
84. Cahali, M.B.; Formigoni, G.G.; Gebrim, E.M.; Miziara, I.D. Lateral pharyngoplasty versus uvulopalatopharyngoplasty: A clinical, polysomnographic and computed tomography measurement comparison. *Sleep* **2004**, *27*, 942–950. [CrossRef]
85. Cahali, M.B. Lateral pharyngoplasty: A new treatment for obstructive sleep apnea hypopnea syndrome. *Laryngoscope* **2003**, *113*, 1961–1968. [CrossRef]
86. Pang, K.P.; Woodson, B.T. Expansion sphincter pharyngoplasty: A new technique for the treatment of obstructive sleep apnea. *Otolaryngol. Head Neck Surg.* **2007**, *137*, 110–114. [CrossRef]
87. Pang, K.P.; Pang, E.B.; Win, M.T.; Pang, K.A.; Woodson, B.T. Expansion sphincter pharyngoplasty for the treatment of OSA: A systemic review and meta-analysis. *Eur. Arch. Otorhinolaryngol.* **2016**, *273*, 2329–2333. [CrossRef]
88. Elbassiouny, A.M. Modified barbed soft palatal posterior pillar webbing flap palatopharyngoplasty. *Sleep Breath.* **2016**, *20*, 829–836. [CrossRef]
89. El-Ahl, M.A.; El-Anwar, M.W. Expansion Pharyngoplasty by New Simple Suspension Sutures without Tonsillectomy. *Otolaryngol. Head Neck Surg.* **2016**, *155*, 1065–1068. [CrossRef]
90. Pianta, L.; Bertazzoni, G.; Morello, R.; Perotti, P.; Nicolai, P. Barbed expansion sphincter pharyngoplasty for the treatment of oropharyngeal collapse in obstructive sleep apnoea syndrome: A retrospective study on 17 patients. *Clin. Otolaryngol.* **2018**, *43*, 696–700. [CrossRef]
91. Salamanca, F.; Costantini, F.; Mantovani, M.; Bianchi, A.; Amaina, T.; Colombo, E.; Zibordi, F. Barbed anterior pharyngoplasty: An evolution of anterior palatoplasty. *Acta Otorhinolaryngol. Ital.* **2014**, *34*, 434–438. [PubMed]
92. Mantovani, M.; Carioli, D.; Torretta, S.; Rinaldi, V.; Ibba, T.; Pignataro, L. Barbed snore surgery for concentric collapse at the velum: The Alianza technique. *J. Craniomaxillofac. Surg.* **2017**, *45*, 1794–1800. [CrossRef] [PubMed]
93. Mantovani, M.; Rinaldi, V.; Torretta, S.; Carioli, D.; Salamanca, F.; Pignataro, L. Barbed Roman blinds technique for the treatment of obstructive sleep apnea: How we do it? *Eur. Arch. Otorhinolaryngol.* **2016**, *273*, 517–523. [CrossRef]
94. Barbieri, M.; Missale, F.; Incandela, F.; Fragale, M.; Barbieri, A.; Roustan, V.; Canevari, F.R.; Peretti, G. Barbed suspension pharyngoplasty for treatment of lateral pharyngeal wall and palatal collapse in patients affected by OSAHS. *Eur. Arch. Otorhinolaryngol.* **2019**, *276*, 1829–1835. [CrossRef]
95. Vicini, C.; Meccariello, G.; Montevecchi, F.; De Vito, A.; Frassineti, S.; Gobbi, R.; Pelucchi, S.; Iannella, G.; Magliulo, G.; Cammaroto, G. Effectiveness of barbed repositioning pharyngoplasty for the treatment of obstructive sleep apnea (OSA): A prospective randomized trial. *Sleep Breath.* **2020**, *24*, 687–694. [CrossRef]
96. Cammaroto, G.; Montevecchi, F.; D'Agostino, G.; Zeccardo, E.; Bellini, C.; Meccariello, G.; Vicini, C. Palatal surgery in a transoral robotic setting (TORS): Preliminary results of a retrospective comparison between uvulopalatopharyngoplasty (UPPP), expansion sphincter pharyngoplasty (ESP) and barbed repositioning pharyngoplasty (BRP). *Acta Otorhinolaryngol. Ital.* **2017**, *37*, 406–409. [CrossRef] [PubMed]
97. Friedman, M.; Schalch, P.; Lin, H.C.; Kakodkar, K.A.; Joseph, N.J.; Mazloom, N. Palatal implants for the treatment of snoring and obstructive sleep apnea/hypopnea syndrome. *Otolaryngol. Head Neck Surg.* **2008**, *138*, 209–216. [CrossRef]
98. Zhang, X.M.; Tham, C.J.; Yin, Y.L.; Sun, Y.Q.; Zhou, X. A novel palatal implant surgery combined with uvulopalatopharyngoplasty and inferior turbinate radiofrequency for the treatment of moderate to severe obstructive sleep apnea: A pilot study. *Eur. Arch. Otorhinolaryngol.* **2015**, *272*, 1195–1202. [CrossRef]

99. Lee, N.R.; Madani, M. Genioglossus muscle advancement techniques for obstructive sleep apnea. *Atlas Oral Maxillofac. Surg. Clin. N. Am.* **2007**, *15*, 179–192. [CrossRef]
100. García Vega, J.R.; de la Plata, M.M.; Galindo, N.; Navarro, M.; Díez, D.; Láncara, F. Genioglossus muscle advancement: A modification of the conventional technique. *J. Craniomaxillofac. Surg.* **2014**, *42*, 239–244. [CrossRef]
101. Emara, T.A.; Omara, T.A.; Shouman, W.M. Modified genioglossus advancement and uvulopalatopharyngoplasty in patients with obstructive sleep apnea. *Otolaryngol. Head Neck Surg.* **2011**, *145*, 865–871. [CrossRef] [PubMed]
102. Emara, T.A.; Elhamshary, A.A.S.; Elkady, A.S.; Elhewity, A.M.M.; Abdelsamee, H.A. Modified genioglossus advancement with radiofrequency tongue base reduction for retroglossal collapse in Obstructive sleep apnea patients. *Am. J. Otolaryngol.* **2022**, *43*, 2. [CrossRef] [PubMed]
103. Goh, Y.H.; Abdullah, V.; Kim, S.W. Genioglossus Advancement and Hyoid Surgery. *Sleep Med. Clin.* **2019**, *14*, 73–81. [CrossRef] [PubMed]
104. Li, K.K. Surgical therapy for obstructive sleep apnea syndrome. *Semin. Respir. Crit. Care Med.* **2005**, *26*, 80–88. [CrossRef] [PubMed]
105. Kezirian, E.J.; Goldberg, A.N. Hypopharyngeal surgery in obstructive sleep apnea: An evidence-based medicine review. *Arch. Otolaryngol. Head Neck Surg.* **2006**, *132*, 206–213. [CrossRef] [PubMed]
106. Lee, J.W.; Lee, D.W.; Ohe, J.Y. Accurate genial tubercle capturing method using computer-assisted virtual surgery for genioglossus advancement. *Br. J. Oral Maxillofac. Surg.* **2017**, *55*, 92–93. [CrossRef]
107. Riley, R.W.; Powell, N.B.; Guilleminault, C.; Nino-Murcia, G. Maxillary, mandibular, and hyoid advancement: An alternative to tracheostomy in obstructive sleep apnea syndrome. *Otolaryngol. Head Neck Surg.* **1986**, *94*, 584–588. [CrossRef]
108. Riley, R.W.; Powell, N.B.; Guilleminault, C. Inferior sagittal osteotomy of the mandible with hyoid myotomy-suspension: A new procedure for obstructive sleep apnea. *Otolaryngol. Head Neck Surg.* **1986**, *94*, 589–593. [CrossRef]
109. den Herder, C.; van Tinteren, H.; de Vries, N. Hyoidthyroidpexia: A surgical treatment for sleep apnea syndrome. *Laryngoscope* **2005**, *115*, 740–745. [CrossRef]
110. Canzi, P.; Berardi, A.; Tinelli, C.; Montevecchi, F.; Pagella, F.; Vicini, C.; Benazzo, M. Thirteen Years of Hyoid Suspension Experience in Multilevel OSAHS Surgery: The Short-Term Results of a Bicentric Study. *Int. J. Otolaryngol.* **2013**, *263043*, 20. [CrossRef]
111. Hormann, K.; Baisch, A. The hyoid suspension. *Laryngoscope* **2004**, *114*, 1677–1679. [CrossRef] [PubMed]
112. Piccin, O.; Scaramuzzino, G.; Martone, C. Modified hyoid suspension technique in the treatment of multilevel related obstructive sleep apnea. *Otolaryngol. Head Neck Surg.* **2014**, *150*, 321–324. [CrossRef]
113. Hamans, E.; Stuck, B.A.; de Vries, N. Hyoid expansion as a treatment for obstructive sleep apnea: A pilot study. *Sleep Breath.* **2013**, *17*, 195–201. [CrossRef] [PubMed]
114. Chen, J.H.; Luo, Z.H.; Yang, R.; Kang, J.; Wang, Y.P.; Yang, X.L.; Zhu, M.W.; Tao, Z.Z. Complications of hyoid suspension with Repose system on obstructive sleep apnea hypopnea syndrome. *Zhonghua Er Bi Yan Hou Tou Jing Wai Ke Za Zhi* **2012**, *47*, 449–453.
115. Song, S.A.; Wei, J.M.; Buttram, J.; Tolisano, A.M.; Chang, E.T.; Liu, S.Y.; Certal, V.; Camacho, M. Hyoid surgery alone for obstructive sleep apnea: A systematic review and meta-analysis. *Laryngoscope* **2016**, *126*, 1702–1708. [CrossRef]
116. Vicini, C.; Dallan, I.; Canzi, P.; Frassineti, S.; La Pietra, M.G.; Montevecchi, F. Transoral robotic tongue base resection in obstructive sleep apnoea-hypopnoea syndrome: A preliminary report. *ORL J. Otorhinolaryngol. Relat. Spec.* **2010**, *72*, 22–27. [CrossRef]
117. Li, K.K.; Powell, N.B.; Riley, R.W.; Guilleminault, C. Temperature-controlled radiofrequency tongue base reduction for sleep-disordered breathing: Long-term outcomes. *Otolaryngol. Head Neck Surg.* **2002**, *127*, 230–234. [CrossRef]
118. Fujita, S.; Woodson, B.T.; Clark, J.L.; Wittig, R. Laser midline glossectomy as a treatment for obstructive sleep apnea. *Laryngoscope* **1991**, *101*, 805–809. [CrossRef]
119. Maturo, S.C.; Mair, E.A. Submucosal minimally invasive lingual excision: An effective, novel surgery for pediatric tongue base reduction. *Ann. Otol. Rhinol. Laryngol.* **2006**, *115*, 624–630. [CrossRef]
120. Murphey, A.W.; Kandl, J.A.; Nguyen, S.A.; Weber, A.C.; Gillespie, M.B. The Effect of Glossectomy for Obstructive Sleep Apnea: A Systematic Review and Meta-analysis. *Otolaryngol. Head Neck Surg.* **2015**, *153*, 334–342. [CrossRef]
121. Süslü, A.E.; Katar, O.; Jafarov, S.; Özer, S.; Önerci, M. Results of coblation midline glossectomy for obstructive sleep apnea. *Auris Nasus Larynx* **2021**, *48*, 697–703. [CrossRef]
122. Hay, A.; Migliacci, J.; Karassawa Zanoni, D.; Boyle, J.O.; Singh, B.; Wong, R.J.; Patel, S.G.; Ganly, I. Complications following transoral robotic surgery (TORS): A detailed institutional review of complications. *Oral Oncol.* **2017**, *67*, 160–166. [CrossRef] [PubMed]
123. Friedman, M.; Soans, R.; Gurpinar, B.; Lin, H.C.; Joseph, N. Evaluation of submucosal minimally invasive lingual excision technique for treatment of obstructive sleep apnea/hypopnea syndrome. *Otolaryngol. Head Neck Surg.* **2008**, *139*, 378–384. [CrossRef] [PubMed]
124. Hwang, C.S.; Kim, J.W.; Lee, E.J.; Kim, C.H.; Yoon, J.H.; Cho, H.J. Comparison of robotic and coblation tongue base resection for obstructive sleep apnoea. *Clin. Otolaryngol.* **2018**, *43*, 249–255. [CrossRef]
125. Lechien, J.R.; Chiesa-Estomba, C.M.; Fakhry, N.; Saussez, S.; Badr, I.; Ayad, T.; Chekkoury-Idrissi, Y.; Melkane, A.E.; Bahgat, A.; Crevier-Buchman, L.; et al. Surgical, clinical, and functional outcomes of transoral robotic surgery used in sleep surgery for obstructive sleep apnea syndrome: A systematic review and meta-analysis. *Head Neck* **2021**, *43*, 2216–2239. [CrossRef] [PubMed]

126. Tsou, Y.A.; Huang, C.W.; Wu, T.F.; Hung, L.W.; Chang, W.D. The effect of tongue base suspension with uvulopalatopharyngoplasty on sleep quality in obstructive sleep apnea. *Sci. Rep.* **2018**, *8*, 8788. [CrossRef]
127. Kocdor, P.; Froymovich, O.; Mickelson, S. Tongue base suspension procedures for obstructive sleep apnea syndrome. *Oper. Tech. Otolaryngol.-Head Neck Surg.* **2015**, *26*, 187–192. [CrossRef]
128. Hsin, L.J.; Lee, Y.C.; Lin, W.N.; Lu, Y.A.; Lee, L.A.; Tsai, M.S.; Cheng, W.N.; Chiang, Y.T.; Li, H.Y. Transoral Tongue Suspension for Obstructive Sleep Apnea-A Preliminary Study. *J. Clin. Med.* **2022**, *11*, 4960. [CrossRef]
129. Toh, S.T.; Hsu, P.P. Robotic Obstructive Sleep Apnea Surgery. *Adv. Otorhinolaryngol.* **2017**, *80*, 125–135.
130. Bostanci, A.; Turhan, M. A systematic review of tongue base suspension techniques as an isolated procedure or combined with uvulopalatopharyngoplasty in obstructive sleep apnea. *Eur. Arch. Otorhinolaryngol.* **2016**, *273*, 2895–2901. [CrossRef]
131. Omur, M.; Ozturan, D.; Elez, F.; Unver, C.; Derman, S. Tongue base suspension combined with UPPP in severe OSA patients. *Otolaryngol. Head Neck Surg.* **2005**, *133*, 218–223. [CrossRef]
132. Cahali, M.B.; Soares, C.F.; Dantas, D.A.; Formigoni, G.G. Tonsil volume, tonsil grade and obstructive sleep apnea: Is there any meaningful correlation? *Clinics* **2011**, *66*, 1347–1352. [CrossRef]
133. Smith, M.M.; Peterson, E.; Yaremchuk, K.L. The Role of Tonsillectomy in Adults with Tonsillar Hypertrophy and Obstructive Sleep Apnea. *Otolaryngol. Head Neck Surg.* **2017**, *157*, 331–335. [CrossRef]
134. Tan, L.T.; Tan, A.K.; Hsu, P.P.; Loh, I.C.; Yuen, H.W.; Chan, Y.H.; Lu, P.K. Effects of tonsillectomy on sleep study parameters in adult patients with obstructive sleep apnea–a prospective study. *Sleep Breath.* **2014**, *18*, 265–268. [CrossRef]
135. Holmlund, T.; Franklin, K.A.; Levring Jäghagen, E.; Lindkvist, M.; Larsson, T.; Sahlin, C.; Berggren, D. Tonsillectomy in adults with obstructive sleep apnea. *Laryngoscope* **2016**, *126*, 2859–2862. [CrossRef]
136. Senchak, A.J.; McKinlay, A.J.; Acevedo, J.; Swain, B.; Tiu, M.C.; Chen, B.S.; Robitschek, J.; Ruhl, D.S.; Williams, L.L.; Camacho, M.; et al. The effect of tonsillectomy alone in adult obstructive sleep apnea. *Otolaryngol. Head Neck Surg.* **2015**, *152*, 969–973. [CrossRef]
137. Verse, T.; Kroker, B.A.; Pirsig, W.; Brosch, S. Tonsillectomy as a treatment of obstructive sleep apnea in adults with tonsillar hypertrophy. *Laryngoscope* **2000**, *110*, 1556–1559. [CrossRef]
138. Camacho, M.; Li, D.; Kawai, M.; Zaghi, S.; Teixeira, J.; Senchak, A.J.; Brietzke, S.E.; Frasier, S.; Certal, V. Tonsillectomy for adult obstructive sleep apnea: A systematic review and meta-analysis. *Laryngoscope* **2016**, *126*, 2176–2186. [CrossRef]
139. Tetter, N.; Tschopp, K. Contribution of Hyoid and Tonsillar Procedures to Outcome in Multilevel Surgery for Obstructive Sleep Apnea Syndrome. *ORL J. Otorhinolaryngol. Relat. Spec.* **2016**, *78*, 353–360. [CrossRef]
140. Sommer, U.J.; Heiser, C.; Gahleitner, C.; Herr, R.M.; Hörmann, K.; Maurer, J.T.; Stuck, B.A. Tonsillectomy with Uvulopalatopharyngoplasty in Obstructive Sleep Apnea. *Dtsch. Arztebl. Int.* **2016**, *113*, 1–8. [CrossRef]
141. Nikisha, G.N.; Mohana Karthikeyan, S. Modified Uvulopalatopharyngoplasty with Tonsil-lectomy in Treatment of Obstructive Sleep Apnoea. *Indian J. Otolaryngol. Head Neck Surg.* **2022**, *74*, 272–278. [CrossRef]
142. Decuzzi, J.; Redline, S.; Isaiah, A. Secondary Analyses of the Childhood Adenotonsillectomy Trial: A Narrative Review. *JAMA Otolaryngol. Head Neck Surg.* **2022**, *148*, 779–784. [CrossRef]
143. Gowda, S.; Leong, W.S.; Edafe, O. Day-case discharge criteria and safety of children undergoing adenoidectomy and tonsillectomy for obstructive symptoms—A systematic review. *Clin. Otolaryngol.* **2022**, *47*, 553–560. [CrossRef]
144. Kuhlo, W.; Doll, E.; Franck, M.C. Successful management of Pickwickian syndrome using long-term tracheostomy. *Dtsch. Med. Wochenschr.* **1969**, *94*, 1286–1290. [CrossRef]
145. Guilleminault, C.; Simmons, F.B.; Motta, J.; Cummiskey, J.; Rosekind, M.; Schroeder, J.S.; Dement, W.C. Obstructive sleep apnea syndrome and tracheostomy. Long-term follow-up experience. *Arch. Intern. Med.* **1981**, *141*, 985–988. [CrossRef]
146. Kim, S.H.; Eisele, D.W.; Smith, P.L.; Schneider, H.; Schwartz, A.R. Evaluation of patients with sleep apnea after tracheotomy. *Arch. Otolaryngol. Head Neck Surg.* **1998**, *124*, 996–1000. [CrossRef]
147. Camacho, M.; Certal, V.; Brietzke, S.E.; Holty, J.E.; Guilleminault, C.; Capasso, R. Tracheostomy as treatment for adult obstructive sleep apnea: A systematic review and meta-analysis. *Laryngoscope* **2014**, *124*, 803–811. [CrossRef]
148. Rizzi, C.J.; Amin, J.D.; Isaiah, A.; Valdez, T.A.; Jeyakumar, A.; Smart, S.E.; Pereira, K.D. Tracheostomy for Severe Pediatric Obstructive Sleep Apnea: Indications and Outcomes. *Otolaryngol. Head Neck Surg.* **2017**, *157*, 309–313. [CrossRef]
149. Haapaniemi, J.J.; Laurikainen, E.A.; Halme, P.; Antila, J. Long-term results of tracheostomy for severe obstructive sleep apnea syndrome. *ORL J. Otorhinolaryngol. Relat. Spec.* **2001**, *63*, 131–136. [CrossRef]

life

MDPI

Review

Effect of Continuous Positive Airway Pressure on Changes of Plasma/Serum Ghrelin and Evaluation of These Changes between Adults with Obstructive Sleep Apnea and Controls: A Meta-Analysis

Amin Golshah [1], Mohammad Moslem Imani [1], Masoud Sadeghi [2], Mozhgan Karami Chalkhooshg [3], Annette Beatrix Brühl [4], Laleh Sadeghi Bahmani [5] and Serge Brand [4,6,7,8,9,10,*]

1 Department of Orthodontics, Kermanshah University of Medical Sciences, Kermanshah 6715847141, Iran
2 Department of Biology, Science and Research Branch, Islamic Azad University, Tehran 1477893855, Iran
3 Students Research Committee, Kermanshah University of Medical Sciences, Kermanshah 6715847141, Iran
4 Center for Affective, Stress and Sleep Disorders (ZASS), Psychiatric University Hospital Basel, 4002 Basel, Switzerland
5 Department of Education and Psychology, Shahid Ashrafi Esfahani University, Esfahan 1461968151, Iran
6 Sleep Disorders Research Center, Kermanshah University of Medical Sciences, Kermanshah 6715847141, Iran
7 Department of Sport, Exercise and Health, Division of Sport Science and Psychosocial Health, University of Basel, 4052 Basel, Switzerland
8 Substance Abuse Prevention Research Center, Kermanshah University of Medical Sciences, Kermanshah 6715847141, Iran
9 School of Medicine, Tehran University of Medical Sciences, Tehran 1417613151, Iran
10 Center for Disaster Psychiatry and Disaster Psychology, Psychiatric University Hospital Basel, 4002 Basel, Switzerland
* Correspondence: serge.brand@upk.ch

Citation: Golshah, A.; Imani, M.M.; Sadeghi, M.; Karami Chalkhooshg, M.; Brühl, A.B.; Sadeghi Bahmani, L.; Brand, S. Effect of Continuous Positive Airway Pressure on Changes of Plasma/Serum Ghrelin and Evaluation of These Changes between Adults with Obstructive Sleep Apnea and Controls: A Meta-Analysis. *Life* **2023**, *13*, 149. https://doi.org/10.3390/life13010149

Academic Editors: David Slouka and Milan Štengl

Received: 18 November 2022
Revised: 15 December 2022
Accepted: 29 December 2022
Published: 4 January 2023

Abstract: Background and objective: Obstructive sleep apnea (OSA) can be related to high ghrelin hormone levels that may encourage additional energy intake. Herein, a new systematic review and meta-analysis were performed to check the changes in serum/plasma levels of ghrelin in adults with OSA compared to controls, as well as before compared after continuous positive airway pressure (CPAP) therapy in adults with OSA. Materials and methods: Four main databases were systematically and comprehensively searched until 17 October 2022, without any restrictions. For assessing the quality, we used the Joanna Briggs Institute (JBI) critical appraisal checklist adapted for case–control studies and the National Institutes of Health (NIH) quality assessment tool for before–after studies. The effect sizes were extracted by the Review Manager 5.3 software for the blood of ghrelin in adults with OSA compared with controls, as well as before and after CPAP therapy. Results: Fifteen articles involving thirteen studies for case–control studies and nine articles for before–after studies were included. The pooled standardized mean differences were 0.30 (95% confidence interval (CI): −0.02, 0.61; p = 0.07; I^2 = 80%) and 0.10 (95% CI: −0.08, 0.27; p = 0.27; I^2 = 42%) for case–control and before–after studies, respectively. For thirteen case–control studies, nine had moderate and four high qualities, whereas for nine before–after studies, five had good and four fair qualities. Based on the trial sequential analysis, more studies are needed to confirm the pooled results of the analyses of blood ghrelin levels in case–control and before–after studies. In addition, the radial plot showed outliers for the analysis of case–control studies that they were significant factors for high heterogeneity. Conclusions: The findings of the present meta-analysis recommended that the blood levels of ghrelin had no significant difference in the adults with OSA compared with the controls, nor did they have significant difference in adults with OSA before compared with after CPAP therapy. The present findings need to be confirmed in additional studies with more cases and higher qualities.

Keywords: sleep apnea syndromes; ghrelin; serum; plasma; meta-analysis

1. Introduction

Obstructive Sleep Apnea (OSA) is a state determined by repeated episodes of partial or complete airway obstruction over sleep [1,2]. Apnea–Hypopnea Index (AHI) or the number of apneas/hypopneas per hour of sleep is the basic metric for identifying OSA that this is evaluated by Polysomnography (PSG) or other shapes of sleep monitoring [3]. The overall prevalence of OSA in adults (AHI $\geq$ 5 events/h) ranged from 9–38% in the public adult population and was more in men [4]. In addition, the OSA prevalence was calculated to be 56.0% in patients with type 2 diabetes [5].

Obesity, smoking, alcohol consumption, higher age, and male gender can be the risk factors for OSA [6]. It has been shown the impact of ethnicity on the prevalence and severity of OSA that this impact can be related to ethnic differences in adipose tissue distributions [7]. Apart from the environmental and demographical factors, the studies reported that genetic [8–11] and blood [12–19] factors could also affect the prevalence or development of OSA.

Ghrelin (a 28 amino acid hormone or orexigenic neuropeptide involving an n-octanoyl group on the serine in position 3) [20] is known as an endocrine pathway in controlling nutrition and energy balance that is secreted by a large number of tissues, but its dominant source is the gastric mucosa [21,22]. Ghrelin is in both acylated and unacylated forms [23]. Acylated ghrelin is the active shape of ghrelin [24] (acylation is essential for the ghrelin binding and function [25]) with some metabolic functions such as appetite stimulation, reduced insulin secretion from pancreatic, elevated growth hormone secretion, reduced body energy consumption, and environmental growth and metabolism, especially carbohydrates and fats [22].

The OSA is related to hormonal features and is illustrated by high levels of ghrelin and leptin hormones that may provoke additional energy intake [26]. In men with OSA, energy expenditure relative to body weight reduces with elevating severity of oxygen desaturation that can contribute to a positive energy balance [27]. The studies [28,29] reported different results for blood levels of ghrelin in adults with OSA in comparison with controls. The relationship between OSA and plasma/serum levels of ghrelin is controversial [30]. Obesity [31], cardiovascular diseases [32], diabetes and metabolic syndrome [33], and hypertention [34] are associated with blood ghrelin levels and on the other hand, OSA is related to these disorders or diseases [35–37]. Therefore, finding a link between ghrelin levels with OSA development can be useful for prediction of related diseases with OSA and possible treatments.

The OSA cases treated by nasal Continuous Positive Airway Pressure (CPAP) require using CPAP therapy to stop the recurrence of symptoms [38]. The changes in energy metabolism accrue after CPAP therapy for OSA [39]. The studies [40–42] reported the impact of CPAP therapy on the blood levels of ghrelin in adults with OSA with different results.

Based on our knowledge of English literature, there was a meta-analysis [30] related to this subject—searching three databases until 2018—with eight case–control and six before–after studies. Therefore, a new systematic review and meta-analysis in four main databases were conducted with more studies (thirteen case–control and nine before–after studies) and additional analyses than the previous meta-analysis for findings potentially effective factors on heterogeneity and bias (radial plot analysis, meta-regression, and trial sequential analysis (TSA)) to check the changes of serum/plasma levels of ghrelin in adults with OSA compared to controls, as well as before compared after CPAP therapy in adults with OSA with more details. In addition to a few new studies, the previous meta-analysis missed several articles before 2018 that could be due of the choices of database or the searching criteria.

2. Materials and Methods

To design of the present meta-analysis, it was followed the PRISMA-P items [43]. The PECO question [44,45] was: Are blood ghrelin levels different in adults with OSA in comparison to controls? (P: human adults with and without OSA, E: OSA disorder, C: adults with OSA compared to controls; O: and the plasma/serum ghrelin level). The clinical PICO (Population, Intervention, Comparator, and Outcome) question was: What is the impact of CPAP therapy on serum/plasma levels of ghrelin in adults with OSA? (P: human adults with OSA, I: CPAP therapy, C: adults with OSA before and after CPAP therapy; O: and the plasma/serum ghrelin level).

2.1. Search Strategy

Four databases (PubMed, Web of Science, Scopus, and Cochrane Library) were systematically and comprehensively searched until 17 October 2022, without any restrictions by one reviewer (M.S.). The search terms were as: ("obstructive sleep apnea" or "sleep apnea" or "OSA" or "obstructive sleep apnea syndrome" or "OSAS" or "obstructive sleep apnea-hypopnea syndrome" or "OSAHS") and ("ghrelin"). The citations of all types of articles linked to the subject and "Google Scholar" were checked to ensure no study was missed.

2.2. Eligibility Criteria

Inclusion criteria: (1) studies including both adults with OSA and controls aged ≥18 years without any treatment or adults with OSA under CPAP therapy, (2) studies reporting plasma/serum ghrelin levels in OSA and controls or adults with OSA before and after CPAP therapy, (3) PSG was applied to diagnose OSA, defined as AHI ≥5 events/h for adult, (4) adults with OSA did not have other systemic diseases (diabetes mellitus, cardiovascular diseases, heart, hepatic, and renal failures, and lung diseases, any malignancy, and infectious diseases, other sleep disorders), (5) controls did not have OSA or systemic disease (see the previous criterion), and (6) venous blood was took in the fasting state on the morning to measure ghrelin. Exclusion criteria: (1) meta-analyses, book chapters, conference papers, the letter to the editor, commentary, and reviews, (2) studies without complete data, (3) studies in the absence of a control group or the control group had AHI was more than 5 events/h, (4) studies including participants aged less than 18 years old, and (5) studies including adults with OSA with any another disease.

2.3. Data Collection

The data were extracted for any study involved in the meta-analysis by two independent reviewers (A.G. and M.S.). The differences between reviewers were resolved by third reviewer (S.B.). Extracted data were the country and ethnicity of participants, the first author, the publication year, ghrelin sampling, the sample size of adults with OSA and controls, quality or quality score, means BMI, age, and AHI the groups, follow-up duration of CPAP therapy, mean AHI before and after CPAP therapy, and mean of blood levels of ghrelin in all groups.

2.4. Quality Assessment

For assessing the quality, we used the Joanna Briggs Institute (JBI) critical appraisal checklist adapted for case–control studies including ten questions or ten scores as Low: 1–4 scores, Moderate: 5–7 scores, High: 8–10 scores [46] and the National Institutes of Health (NIH) quality assessment tool for before–after studies with twelve question or twelve scores as Good: 9–12 scores, Fair: 5–8 scores, Poor: 1–4 scores [47] (See Supplementary File S1). The quality score were performed by two independent reviewers (M.M.I. and M.S.). The differences between reviewers were resolved by third reviewer (M.K.C.).

2.5. Statistical Analyses

The Review Manager 5.3 (RevMan 5.3) software was applied to extract the effect sizes (standardized mean difference (SMD) and 95% confidence interval (CI)) of blood levels of ghrelin amongst adults with OSA and controls, as well as before and after CPAP therapy by one reviewer (M.S.). The p-value (2-sided) of less than 0.05 was considered a significant value. A $P_{heterogeneity} < 0.1(I^2 > 50\%)$ reported a significant heterogeneity that in this state, a random-effects model [48], otherwise, a fixed-effect model [49] was used.

The subgroup and random-effect meta-regression analyses were done based on several variables and evaluating the stability of initial pooled SMDs, both "one-study-removed" and "cumulative" analyses as sensitivity analyses were utilized.

The Begg's funnel plot by Begg's test was applied to test potential publication bias [50] and the Egger's test to report degree of asymmetry [51] that the *p*-values of both tests and the data for sensitivity analyses were extracted by the Comprehensive Meta-Analysis version 2.0 (CMA 2.0) software and a *p*-value (2-sided) less than 0.10 recommended the existence of the publication bias.

To report the potential random error (false-positive and -negative results) in meta-analysis [52], trial sequential analysis (TSA) was accomplished using TSA software (version 0.9.5.10 beta) [53]. The futility threshold can show a no-impact result before attaining the information size. An α-risk of 5%, a β-risk of 20%, and a 2-sided border type reporting the mean difference and variance were based on empirical assumptions created automatically by the software, were used to calculate the required information size (RIS). If the Z-curve reached the RIS line, enough participants were included in the studies and the conclusion was trustworthy or crossed the borderlines the results could be robust. Differently, the volume of information was not large enough and more evidence was needed.

The effect sizes for the studies including the required data just on a graph were extracted from the graph utilizing GetData Graph Digitizer 2.26 software.

3. Results

3.1. Search Strategy

To search in the databases, 362 records were identified and after deleting duplicates and irrelevant records, 28 full-text articles were evaluated (Figure 1). Then, 13 articles were excluded with reasons (one was a meta-analysis, three were reviews, two did not report a control group or adults with OSA under CPAP therapy, two had no relevant data, two included a control group with AHI >5 events/h, one reported geometrical data, and two were reported in children). At last, 15 articles involving 13 studies for case–control studies and 9 articles for before–after studies were included. All studies reported total ghrelin levels except one study [42] that reported acylated ghrelin.

3.2. Characteristics of the Studies

Based on fifteen articles [28,29,39–42,54–62], Tables 1 and 2 display the characteristics of the case–control and before–after studies in the analysis, respectively. With regard to thirteen case–control studies, nine studies were reported in Caucasians, three in Asians, and one in a population with mixed ethnicity, whereas in nine before–after studies, four in Caucasians, four in Asians, and one in a population with mixed ethnicity. In case–control studies, seven studies reported plasma levels of ghrelin and six serum levels, whereas in before–after studies, six plasma levels and three serum levels. Data of other variables such as sample size, mean BMI, mean age, mean AHI, and follow-up duration are reported in Tables 1 and 2.

3.3. Pooled Analyses

Figures 2 and 3 show the forest plot analyses of blood ghrelin levels in adults with OSA in comparison to controls and adults with OSA before and after CPAP therapy, respectively. The pooled SMDs were 0.30 (95% CI: −0.02, 0.61; $p = 0.07$; $I^2 = 80\%$) and 0.10 (95% CI: −0.08, 0.27; $p = 0.27$; $I^2 = 42\%$) in case–control and before–after studies, respectively. Therefore, the results recommended that there were no significant differences between adults with OSA and controls, moreover between adults with OSA before and after CPAP therapy (there was no effect of CPAP therapy on OSA).

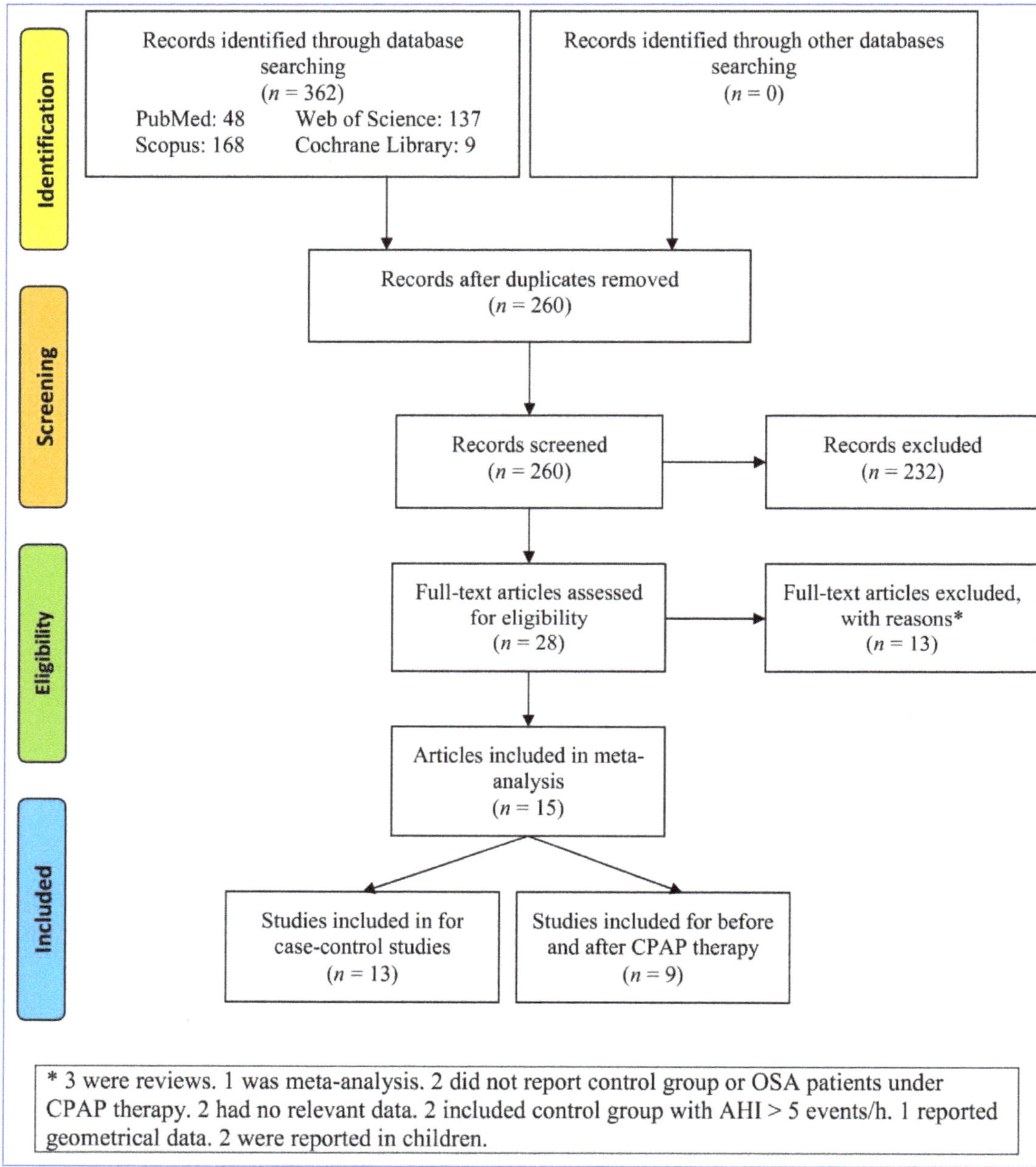

Figure 1. Flowchart of the study selection. CPAP: Continuous positive airways pressure. AHI: Apnea–hypopnea index. OSA: Obstructive sleep apnea.

Table 1. Characteristics of the case–control studies in the meta-analysis.

First Author, Publication Year	Country	Ethnicity	Sample Size (Case/Control)	Mean BMI, kg/m^2		Mean Age, Year		Mean AHI, Events/h		Sample
				Case	Control	Case	Control	Case	Control	
De Santis, 2015 [40]	Italy	Caucasian	26/24	33.0	38.0	41.8	43.7	26.15	1.65	Serum
Liu, 2014 [57]	China	Asian	95/30	28.45	27.85	47.57	45.35	30.28	3.07	Plasma
Ciftci, 2005 [55]	Turkey	Caucasian	30/22	32.12	31.03	Matched	Matched	44.24	1.55	Serum
Yang, 2013 [61]	China	Asian	25/25	27.5	26.22	53	54	25	3	Plasma
Zhang, 2018 [29]	China	Asian	30/20	28.85	27.55	40.73	36.10	61.48	1.93	Plasma
Sánchez-de-la-Torre, 2012 (i) [59]	Spain	Caucasian	10/24	34.34	32.01	46.61	48.7	48.92	2.87	Plasma
Sánchez-de-la-Torre, 2012 (ii) [59]	Spain	Caucasian	21/20	25.02	24.71	49.33	42.9	41.45	3.06	Plasma
Papaioannou, 2011 [58]	UK	Caucasian	33/11	30	28	48	43	30	2	Plasma
Öztürk, 2022 [28]	Turkey	Caucasian	210/62	32.6	30.3	46.4	42.2	31.6	2.8	Serum
Biçer, 2021 [54]	Turkey	Caucasian	75/75	47	32	29.4	25.9	>5	≤5	Plasma
Ursavas, 2010 [60]	Turkey	Caucasian	55/15	51.1	48.4	32.5	31.6	43.5	2.8	Serum
Gharraf, 2019 [56]	Egypt	Caucasian	30/15	41.63	25.09	51	34.27	43.43	<5	Serum
Garbuio, 2009 [41]	Brazil	Mixed	13/13	37	36	29	27	41	2	Serum

AHI: Apnea–hypopnea index. BMI: Body mass index.

Table 2. Characteristics of the studies reporting before and after CPAP therapy in the meta-analysis.

First Author, Publication Year	Country	Ethnicity	Sample Size	Mean BMI, kg/m^2	Mean Age, Year	Mean AHI, Events/h		Sample	Follow-Up Duration
						Before CPAP	After CPAP		
Tachikawa, 2016 (i) [39]	Japan	Asian	63	27.9	60.6	42.2	5.6	Plasma	6 weeks
Tachikawa, 2016 (ii) [39]	Japan	Asian	63	27.9	60.6	42.2	3.9	Plasma	3 months
Takahashi, 2008 [42]	Japan	Asian	21	28.5	53.2	39.4	16.1	Plasma	1 months
Yang, 2013 [62]	China	Asian	22	26.7	60	26	3	Plasma	3 months
Sánchez-de-la-Torre, 2012 (iii) [59]	Italy	Caucasian	21	34.34	46.61	48.92	-	Plasma	3 months
Sánchez-de-la-Torre, 2012 (iv) [59]	Italy	Caucasian	28	25.02	34.34	41.45	-	Plasma	3 months
Garbuio, 2009 [41]	Brazil	Mixed	13	37	29	41	4	Serum	6 months
De Santis, 2015 (i) [40]	Italy	Caucasian	11	-	-	-	2.8	Serum	2 days
De Santis, 2015 (ii) [40]	Italy	Caucasian	11	-	-	-	-	Serum	6 months

CPAP: Continuous positive airway pressure. AHI: Apnea–hypopnea index. BMI: Body mass index.

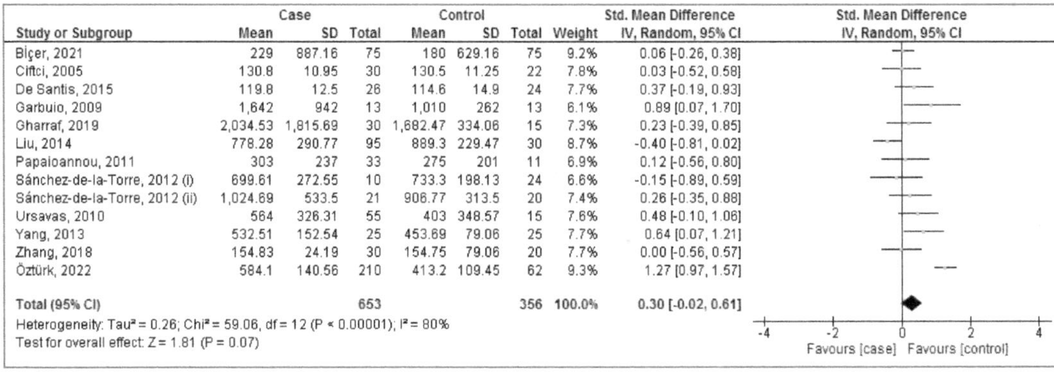

Figure 2. Forest plot analysis of serum/plasma ghrelin levels in adults with obstructive sleep apnea versus controls. The diamond at the bottom of the forest plot represents the result when all the individual studies are combined together and averaged. Names of studies are shown on the left, std. mean differences (green boxes) and confidence intervals (horizontal lines) on the right. The left column shows the first author's names and publication years of studies for twelve articles [28,29,40,41,54–61] included in the analysis. One article [59] included two independent studies marked with i and ii.

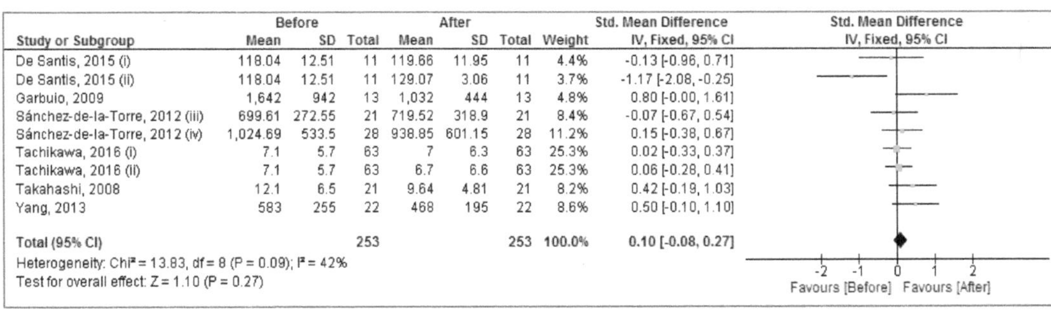

Figure 3. Forest plot analysis of serum/plasma ghrelin levels in adults with obstructive sleep apnea before and after continuous positive airways pressure therapy. The diamond at the bottom of the forest plot represents the result when all the individual studies are combined together and averaged. Names of studies are shown on the left, std. mean differences (green boxes) and confidence intervals (horizontal lines) on the right. The left column shows the first author's names and publication years of studies for six articles [39–42,59,62] included in the analysis. three articles [39,40,59] included two independent studies each one marked with i and ii or iii and iv.

3.4. Quality Scores

Tables 3 and 4 show JBI critical appraisal checklist for case–control studies and NIH quality assessment tool for before–after studies, respectively. The questions of the JBI critical appraisal checklist and the NIH quality assessment tool have been reported in the Supplementary File S1. Of thirteen case–control studies, nine had moderate and four had high qualities. Of nine before–after studies, five had good and four had fair qualities.

Table 3. The Joanna Briggs Institute (JBI) critical appraisal checklist for case–control studies.

First Author, Publication Year	The Joanna Briggs Institute (JBI) Critical Appraisal Checklist										Quality (Total Quality Score)
	Q1	Q2	Q3	Q4	Q5	Q6	Q7	Q8	Q9	Q10	
De Santis, 2015 [40]	No	Yes	Yes	Yes	Yes	No	No	Yes	Yes	Yes	Moderate (7)
Liu, 2014 [57]	No	Yes	Yes	Yes	Yes	No	No	Yes	Yes	Yes	Moderate (7)
Ciftci, 2005 [55]	No	Yes	No	Yes	Yes	No	No	Yes	Yes	Yes	Moderate (6)
Yang, 2013 [61]	Yes	Yes	Yes	Yes	Yes	No	No	Yes	Yes	Yes	High (8)
Zhang, 2018 [29]	Yes	Yes	Yes	Yes	Yes	No	No	Yes	Yes	Yes	High (8)
Sánchez-de-la-Torre, 2012 (i) [59]	No	No	Yes	Yes	Yes	No	No	Yes	Yes	Yes	Moderate (6)
Sánchez-de-la-Torre, 2012 (ii) [59]	No	No	Yes	Yes	Yes	No	No	Yes	Yes	Yes	Moderate (6)
Papaioannou, 2011 [58]	Yes	Yes	Yes	Yes	Yes	No	No	Yes	Yes	Yes	High (8)
Öztürk, 2022 [28]	No	No	Yes	Yes	Yes	No	No	Yes	Yes	Yes	Moderate (6)
Bİçer, 2021 [54]	Yes	No	Yes	Yes	Yes	No	No	Yes	Yes	Yes	Moderate (7)
Ursavas, 2010 [60]	No	Yes	Yes	Yes	Yes	No	No	Yes	Yes	Yes	Moderate (7)
Gharraf, 2019 [56]	No	No	Yes	Yes	Yes	No	No	Yes	Yes	Yes	Moderate (6)
Garbuio, 2009 [41]	Yes	Yes	Yes	Yes	Yes	No	No	Yes	Yes	Yes	High (8)

Low: 1–4 scores, Moderate: 5–7 scores, High: 8–10 scores.

Table 4. The National Institutes of Health (NIH) quality assessment tool for before–after studies.

First Author, Publication Year	The National Institutes of Health (NIH) Quality Assessment Tool												Quality (Total Quality Score)
	Q1	Q2	Q3	Q4	Q5	Q6	Q7	Q8	Q9	Q10	Q11	Q12	
Tachikawa, 2016 (i) [39]	Yes	Yes	NR	Yes	Yes	Yes	Yes	NR	Yes	Yes	Yes	No	Good (9)
Tachikawa, 2016 (ii) [39]	Yes	Yes	NR	Yes	Yes	Yes	Yes	NR	Yes	Yes	Yes	No	Good (9)
Takahashi, 2008 [42]	Yes	Yes	NR	Yes	No	Yes	Yes	NR	Yes	Yes	Yes	No	Fair (8)
Yang, 2013 [62]	Yes	Yes	NR	Yes	No	Yes	Yes	NR	Yes	Yes	Yes	No	Fair (8)
Sánchez-de-la-Torre, 2012 (iii) [59]	Yes	Yes	Yes	Yes	No	Yes	Yes	NR	Yes	Yes	Yes	No	Good (9)
Sánchez-de-la-Torre, 2012 (iv) [59]	Yes	Yes	Yes	Yes	No	Yes	Yes	NR	Yes	Yes	Yes	No	Good (9)
Garbuio, 2009 [41]	Yes	Yes	Yes	Yes	No	Yes	Yes	NR	Yes	Yes	Yes	No	Good (9)
De Santis, 2015 (i) [40]	Yes	Yes	NR	Yes	No	Yes	No	NR	Yes	No	Yes	No	Fair (6)
De Santis, 2015 (ii) [40]	Yes	Yes	NR	Yes	No	Yes	No	NR	Yes	No	Yes	No	Fair (6)

Good: 9–12 scores, Fair: 5–8 scores, Poor: 1–4 scores. NR = not reported.

3.5. Subgroup Analyses

Tables 5 and 6 report the subgroup analyses (for finding probably effective factors for heterogeneity) for the case–control and before–after studies, respectively. The subgroup analysis based on ethnicity, blood sample, sample size, mean BMI of adults and mean age of adults with OSA and controls, mean AHI of adults with OSA, and quality as important factors in adults with OSA and also checked factors in most studies were checked for case–control studies. The results showed that blood sample, sample size, mean age of adults with OSA, and quality were effective factors for the pooled analysis of the blood levels of ghrelin in adults with OSA in comparison to controls, as well as heterogeneity across the studies. In addition, the subgroup analyses based on ethnicity, blood sample, sample size, mean BMI, mean age, mean AHI of adults with OSA before treatment, and quality were checked for before–after studies. The results showed that just mean AHI before treatment was an effective factor for the poled analysis of blood levels of ghrelin in adults with OSA before in comparison to after CPAP therapy.

Table 5. Subgroup analysis of the correlation between blood levels of ghrelin and several variables in the case–control studies in the meta-analysis.

Variable	Subgroup (N)	SMD	95% CI		*p*-Value	I^2, %	$P_{heterogeneity}$
			Min.	Max.			
Ethnicity							
	Caucasian (9)	0.32	−0.06	0.70	0.90	80	**<0.00001**
	Asian (3)	0.06	−0.55	0.66	0.85	76	**0.02**
Sample							
	Serum (6)	0.56	0.09	1.03	0.02	79	**0.0003**
	Plasma (7)	0.03	−0.15	0.22	0.73	36	0.15
Sample size							
	≥100 (3)	0.32	−0.67	1.30	0.53	96	**<0.00001**
	<100 (10)	0.28	0.08	0.47	0.005	0	0.58
Mean BMI of adults with OSA, kg/m^2							
	≥30 (8)	0.19	0.00	0.38	0.05	0	0.54
	<30 (5)	0.36	−0.35	1.08	0.32	91	**<0.00001**
Mean BMI of controls, kg/m^2							
	≥30 (6)	0.20	−0.02	0.41	0.07	16	0.31
	<30 (7)	0.31	−0.23	0.86	0.26	88	**<0.00001**
Mean age of adults with OSA, year							
	≥45 (7)	0.88	−0.26	0.86	0.30	88	**<0.00001**
	<45 (5)	0.22	0.00	0.44	0.05	22	0.27
Mean age of adults with OSA, year							
	≥45 (3)	0.02	−0.65	0.69	0.95	76	**0.02**
	<45 (9)	0.42	0.04	0.79	0.03	79	**<0.00001**
Mean AHI of adults with OSA, events/h							
	≥40 (7)	0.22	−0.02	0.45	0.07	0	0.50
	<40 (5)	0.41	−0.29	1.11	0.25	91	**<0.00001**
Quality							
	Moderate (9)	0.25	−0.16	0.67	0.23	85	**<0.00001**
	High (4)	0.36	0.04	0.68	0.03	34	0.21

SMD: Standardized mean difference. CI: Confidence interval. BMI: Body mass index. AHI: Apnea–hypopnea index. N: number of studies. Bold number means statistically significant ($p < 0.05$).

Table 6. Subgroup analysis of the correlation between blood levels of ghrelin and several variables in the studies reporting before and after continuous positive airways pressure (CPAP) therapy in the meta-analysis.

Variable	Subgroup (N)	SMD	95% CI		*p*-Value	I^2, %	$P_{heterogeneity}$
			Min.	Max.			
Ethnicity							
	Caucasian (4)	−0.13	−0.47	0.20	0.43	50	0.11
	Asian (4)	0.15	−0.07	0.36	0.18	0	0.42
Sample							
	Serum (3)	−0.15	−1.25	0.95	0.80	80	**0.007**
	Plasma (6)	0.13	−0.06	0.31	0.19	0	0.66
Sample size							
	≥20 (6)	0.13	−0.06	0.31	0.19	0	0.66
	<20 (3)	−0.15	−1.25	0.95	0.80	80	0.007
Mean BMI, kg/m^2							
	≥30 (2)	0.33	−0.52	1.17	0.45	65	0.09
	<30 (5)	0.15	−0.05	0.34	0.15	0	0.59
Mean age, year							
	≥45 (5)	0.12	−0.08	0.32	0.24	0	0.52
	<45 (2)	0.34	− 0.10	0.78	0.13	44	0.18
Mean AHI before treatment, events/h							
	≥40 (5)	0.09	−0.11	0.30	0.37	0	0.48
	<40 (2)	0.46	0.03	0.89	0.04	0	0.86
Follow-up duration, month							
	≥3 (6)	0.09	−0.29	0.47	0.63	59	**0.03**
	<3 (3)	0.09	−0.20	0.37	0.55	0	0.46
Quality							
	Good (5)	0.09	−0.11	0.30	0.37	0	0.48
	Fair (4)	−0.02	−0.70	0.65	0.94	71	**0.02**

SMD: Standardized mean difference. CI: Confidence interval. BMI: Body mass index. AHI: Apnea–hypopnea index. N: number of studies. Bold number means statistically significant ($p < 0.05$). OSA: Obstructive sleep apnea.

3.6. Meta-Regression Analyses

Tables 7 and 8 include the data of the meta-regression analyses for the case–control and before–after studies, respectively. The results represented that publication year, sample size, mean AHI of adults with OSA, and quality were confounding factors for the blood levels of ghrelin in adults with OSA versus controls (increasing publication year and sample size, the level of ghrelin significantly increased, but increasing mean AHI of adults with OSA and quality score, the level of ghrelin significantly decreased. Among the factors checked in before–after studies, there was no confounding factor. Therefore, publication year, sample size, mean AHI of adults with OSA, and quality can affect the ghrelin levels and these factors can be probably effective factors for heterogeneity across the studies.

Table 7. Meta-regression analysis of the correlation between blood levels of ghrelin and several variables in the case–control studies in the meta-analysis.

Variable	Point Estimate	Standard Error	Lower Limit	Upper Limit	Z-Value	*p*-Value
Publication year	0.04139	0.01348	0.01497	0.06780	3.07068	**0.00214**
Sample size	0.00524	0.00100	0.00327	0.00721	5.22125	**<0.00001**
Mean BMI of adults with OSA	−0.00468	0.00894	−0.02221	0.01284	−0.52393	0.60033
Mean BMI of controls	0.01264	0.01316	−0.01316	0.03844	0.96033	0.33689
Mean age of adults with OSA	0.01202	0.00886	−0.00534	0.02938	1.35746	0.17463
Mean age of controls	0.01017	0.00863	−0.00674	0.02708	1.17857	0.23850
Mean AHI of adults with OSA	−0.01818	0.00795	−0.03375	−0.00260	−2.28733	**0.02218**
Quality score	−0.31009	0.09887	−0.50387	−0.11630	−3.13616	**0.00171**

BMI: Body mass index. AHI: Apnea–hypopnea index. Bold number means statistically significant ($p < 0.05$).

Table 8. Meta-regression analysis of the correlation between blood levels of ghrelin and several variables in the studies reporting before and after continuous positive airways pressure (CPAP) therapy in the meta-analysis.

Variable	Point Estimate	Standard Error	Lower Limit	Upper Limit	Z-Value	*p*-Value
Publication year	−0.05749	0.03383	−0.12379	0.00881	−1.69959	0.08921
Sample size	−0.00738	0.00456	−0.01631	0.00155	−1.61907	0.10543
Mean BMI	0.01806	0.03137	−0.04342	0.07954	0.57583	0.56473
Mean age	−0.00853	0.00883	−0.02583	0.00877	−0.96615	0.33397
Mean AHI before treatment	−0.02764	0.01775	−0.06242	0.00714	−1.55739	0.00938
Quality score	0.15786	0.10484	−0.04763	0.36334	1.50569	0.13215

AHI: Apnea–hypopnea index. BMI: Body mass index.

3.7. Radial Plots

Figure 4 identifies the radial plots for the case–control and before–after studies, respectively. The radial plot confirmed the high heterogeneity between the case–control studies. Two studies [28,57] were outliers and removing these studies as a sensitivity analysis, there was a lack of heterogeneity. Therefore, outliers are a significant effective factor for high heterogeneity between the studies. The radial plot confirmed that there was no heterogeneity due to outliers for before–after studies.

3.8. Sensitivity Analyses

The sensitivity analyses reported stability of the pooled results for both case–control and before–after studies. There were two outliers [28,57] for the case–control studies that removed them, pooled SMD became 0.22 (95% CI: 0.05, 0.39; $p = 0.010$; $I^2 = 0\%$). The result showed a significantly high level of ghrelin in adults with OSA vs. controls.

3.9. Trial Sequential Analyses (TSAs)

Figure 5 shows TSAs for the case–control and before–after studies. The result of TSA showed that the cumulative Z-curve crossed both the conventional boundary and the trial sequential monitoring boundary, which recommended that the result of the analysis of blood ghrelin level in adults with OSA vs. controls is robust with 1009 cases. Although the actual sample size did not exceed the RIS of 1280 cases, therefore definite result could not be obtained for this analysis, and more studies with sufficient evidence are needed. The results did not confirm the sufficient cases and evidence for the pooled analyses of the blood levels of ghrelin in before–after studies. Because Z-curve did not cross the RIS line or monitored the boundaries in the analysis of before–after studies. Therefore, more studies with sufficient evidence are needed in the future to confirm this result of the analysis of blood ghrelin levels before compared with after CPAP therapy in adults with OSA.

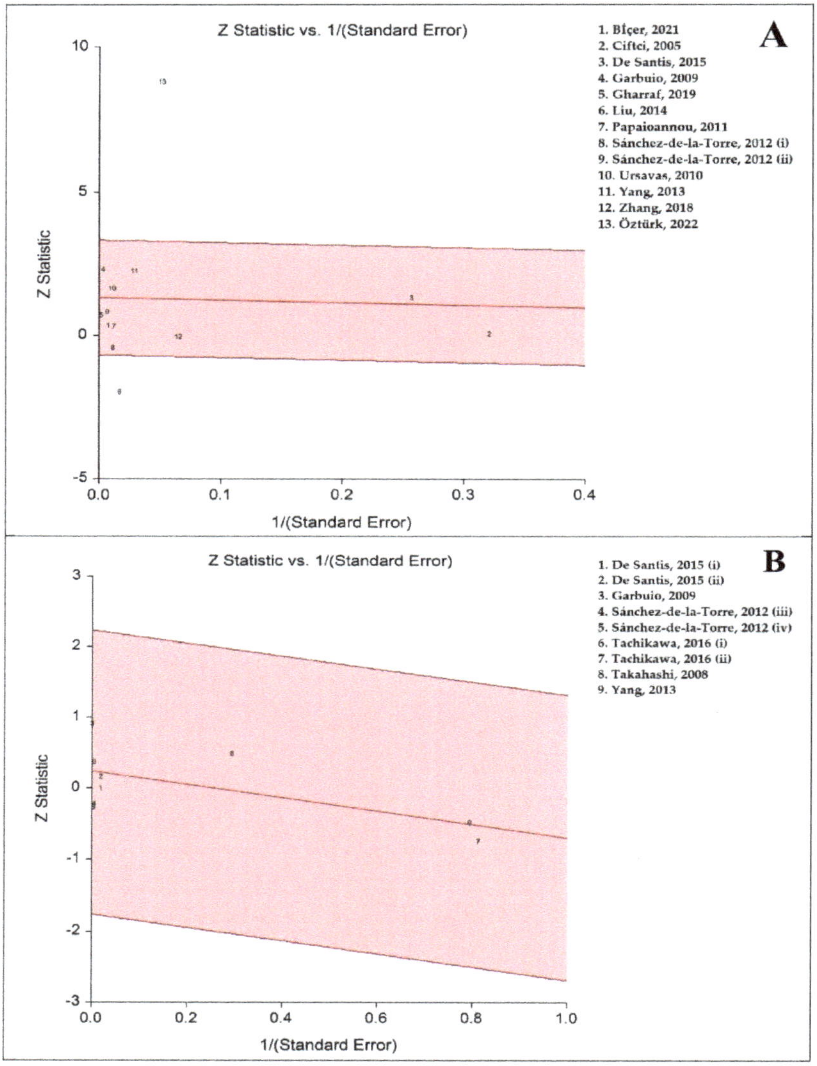

Figure 4. Radial plots of serum/plasma ghrelin levels. (**A**) Adults with obstructive sleep apnea vs. controls. (**B**) Before and after continuous positive airways pressure therapy in adults with OSA. Each number shows one study that the studies with related numbers are represented in right side. Each number in up and right shows one study that (**A**) shows thirteen studies from twelve articles [28,29,40,41,54–61] and (**B**) ahows nine studies from six articles [39–42,59,62]. The articles with two independent studies [39,40,59] have marked with i and ii or iii and iv.

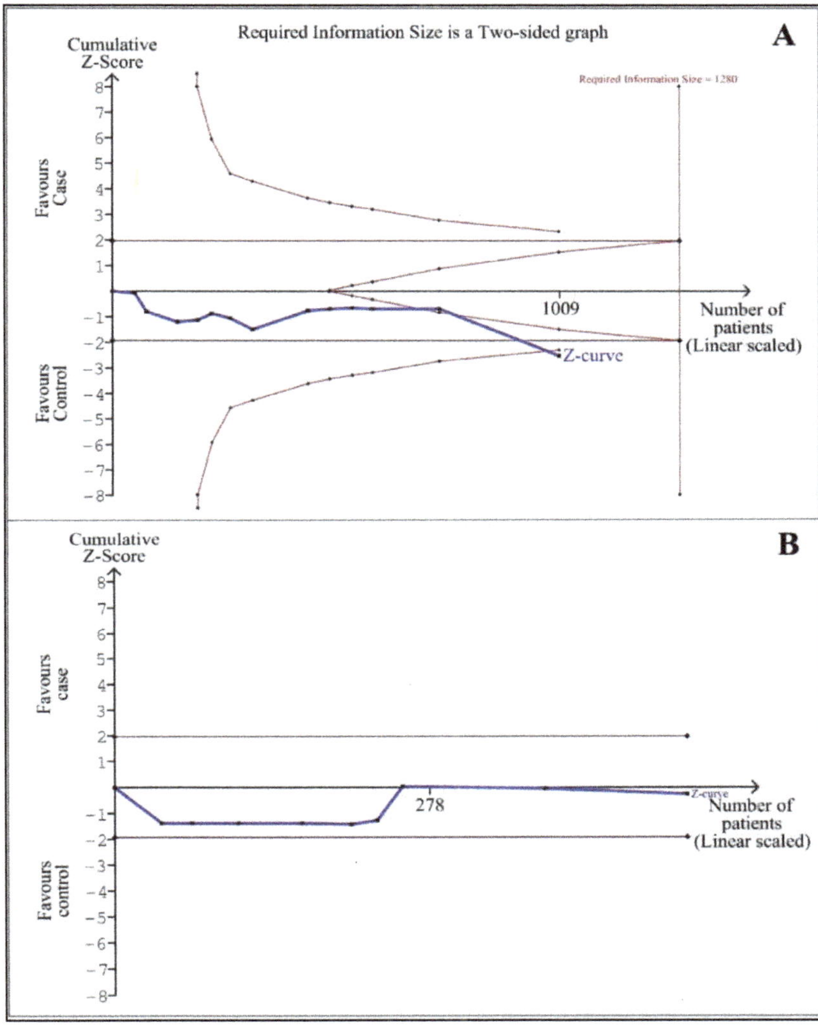

Figure 5. Trial sequential analysis of serum/plasma ghrelin levels. (**A**) Adults with obstructive sleep apnea compared to controls (D^2 = 98%). (**B**) Before and after continuous positive airways pressure therapy (D^2 = 81%) in adults with OSA. The red horizontal lines show monitoring boundaries for benefit (upper line), monitoring boundaries for harm (lower line), and futility boundaries (middle lines). The red vertical line is related to the required sample size.

3.10. Publication Bias

Figure 6 represents the funnel plots of serum/plasma ghrelin levels in both case–control and before–after studies. The results display no publication bias among case–control (*p*-values: Egger's = 0.357 and Begg's = 0.714) and before–after (*p*-values: Egger's = 0.891 and Begg's = 0.834) studies.

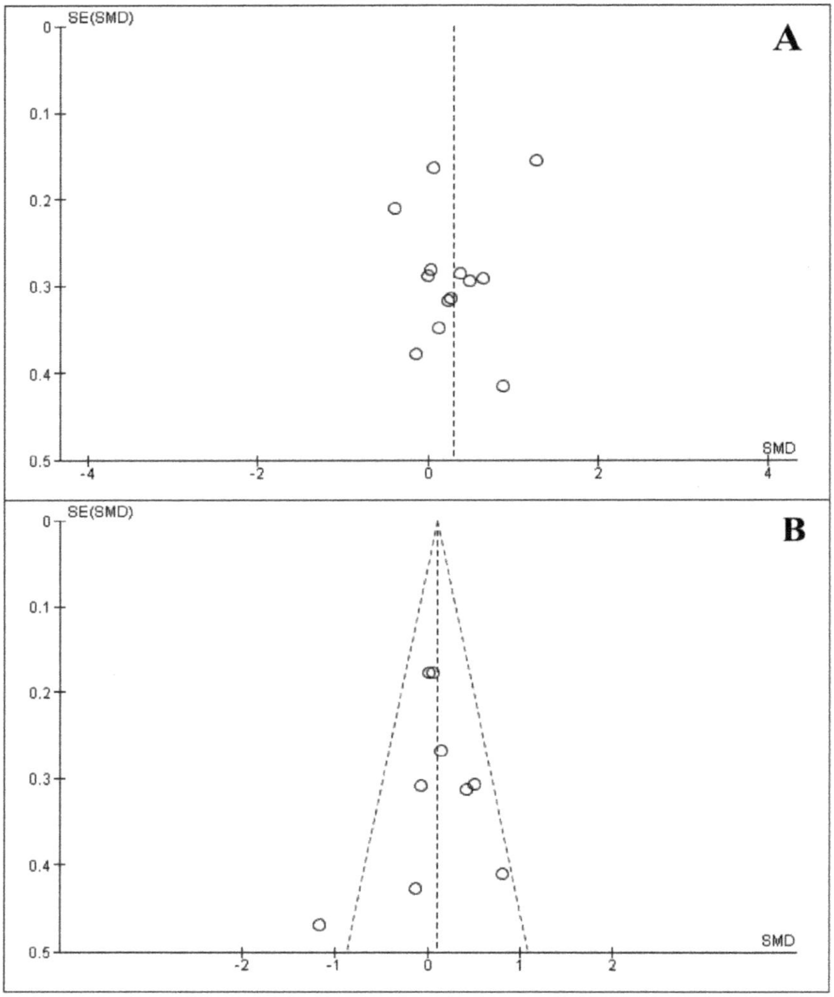

Figure 6. Funnel plots of serum/plasma ghrelin levels. (**A**) Adults with obstructive sleep apnea compared to controls. (**B**) Before and after continuous positive airways pressure therapy in adults with OSA. SMD: Standardized mean difference. SE: Standard error. Circles represent individual studies. The diagonal dashed lines represent the pseudo 95% confidence intervals around the pooled SMD for each standard error of the ordinate vertical axis values. The vertical dashed line represents the pooled SMD.

4. Discussion

The relationship between OSA and plasma/serum ghrelin levels and the effect of CPAP therapy on ghrelin levels have remained controversial [30]. The main results of the present meta-analysis recommended that the serum/plasma levels of ghrelin had no significant difference in the adults with OSA compared to the controls, moreover in adults with OSA before compared to after CPAP therapy. Removing outliers, the serum/plasma levels of ghrelin were significantly higher in the adults with OSA compared to the controls. Two analyses included low sample sizes based on TSA results. Blood sample, sample size, quality scores, means age, and AHI of adults with OSA were effective factors in case–control studies, and the mean AHI of adults with OSA before CPAP therapy in before–after studies.

Therefore, the present findings require to be confirmed in additional studies with more cases and higher qualities.

Among thirteen case–control studies, three studies [28,41,61] showed a significantly high level of ghrelin, whereas other studies did not find any significant difference between in adults with OSA versus controls. Among all before–after studies in the present meta-analysis, the CPAP therapy had a significant defect in increasing [40] and decreasing [41] the blood levels of ghrelin, but other studies did not find any effect of CPAP on the levels of ghrelin in adults with OSA.

A systematic review recommended the positive impact of older age, male gender, and higher BMI on OSA prevalence [4]. Research showed that plasma ghrelin decreased in obese people and increased in lean people [63]. Ciftci et al. [55] revealed that serum ghrelin level has a positive correlation with BMI and AHI. Other studies confirmed the positive correlation of serum ghrelin levels with BMI [40] and AHI [60]. However, a number of studies did not confirm the correlation of serum ghrelin level with BMI [56] and AHI [40,56]. Whatever the present meta-analysis showed the correlation of AHI and age with blood ghrelin levels in adults with OSA, but it did not find any significant correlation between blood levels of ghrelin and BMI.

One study [28] reported a significant association between serum ghrelin levels and the severity of OSA as serum level of ghrelin was significantly higher in adults with severe OSA vs. moderate OSA and moderate OSA compared to mild OSA. Unfortunately, most studies did not report the blood levels of ghrelin based on OSA severity and therefore we could not analyze the association between the blood ghrelin levels and the severity of OSA. The researchers need to perform this analysis among adults with OSA in their original articles in the future. In addition, results of this current meta-analysis were in line with the previously published meta-analysis [30].

There were three significant limitations during the meta-analysis design. (1) A low number of participants in the studies and low included studies in each analysis. (2) Less number of studies had high quality. (3) High heterogeneity among case–control studies. In contrast, there were two important strengths. (1) The stability of results. (2) A lack of publication bias across the studies.

5. Conclusions

The present meta-analysis recommended that the blood levels of ghrelin had no significant difference in the adults with OSA vs. the controls, moreover in adults with OSA before vs. after CPAP therapy. Notwithstanding the low number of individuals in the analyses, the study reported that blood sample, sample size, quality scores, mean age, and mean AHI of adults with OSA were effective factors in case–control studies, and mean AHI of adults with OSA before CPAP therapy in before–after studies. Therefore, the present findings require to be accepted in additional studies with more cases and higher qualities.

Supplementary Materials: The following supporting information can be downloaded at: https://www.mdpi.com/article/10.3390/life13010149/s1. I. The Joanna Briggs Institute (JBI) Critical Appraisal Checklist for case-control study (last amended in 2017). Q. The National Institutes of Health (NIH) quality as-sessment tool for before-after (Pre-Post) study with no control group.

Author Contributions: Conceptualization, A.G.; methodology, M.S.; software, M.S.; validation, M.S. and S.B.; formal analysis, M.S.; investigation, M.M.I.; resources, M.K.C.; data curation, M.S.; writing—original draft preparation, S.B.; writing—review and editing, and L.S.B. and A.B.B. visualization, M.S.; supervision, A.G.; project administration, A.G.; funding acquisition, M.M.I. All authors have read and agreed to the published version of the manuscript.

Funding: This research received no external funding.

Institutional Review Board Statement: Not applicable.

Informed Consent Statement: Not applicable.

Data Availability Statement: No new data were created or analyzed in this study. Data sharing is not applicable to this article.

Conflicts of Interest: The authors have declared that no competing interest exist.

References

1. Pretto, J.J.; Gyulay, S.G.; Hensley, M.J. Trends in anthropometry and severity of sleep-disordered breathing over two decades of diagnostic sleep studies in an Australian adult sleep laboratory. *Med. J. Aust.* **2010**, *193*, 213–216. [CrossRef] [PubMed]
2. Park, J.G.; Ramar, K.; Olson, E.J. Updates on definition, consequences, and management of obstructive sleep apnea. In *Mayo Clinic Proceedings*; Elsevier: Amsterdam, The Netherlands, 2011; pp. 549–555.
3. Quan, S.; Gillin, J.C.; Littner, M.; Shepard, J. Sleep-related breathing disorders in adults: Recommendations for syndrome definition and measurement techniques in clinical research. editorials. *Sleep* **1999**, *22*, 662–689. [CrossRef] [PubMed]
4. Senaratna, C.V.; Perret, J.L.; Lodge, C.J.; Lowe, A.J.; Campbell, B.E.; Matheson, M.C.; Hamilton, G.S.; Dharmage, S.C. Prevalence of obstructive sleep apnea in the general population: A systematic review. *Sleep Med. Rev.* **2017**, *34*, 70–81. [CrossRef] [PubMed]
5. Fallahi, A.; Jamil, D.I.; Karimi, E.B.; Baghi, V.; Gheshlagh, R.G. Prevalence of obstructive sleep apnea in patients with type 2 diabetes: A systematic review and meta-analysis. *Diabetes Metab. Syndr. Clin. Res. Rev.* **2019**, *13*, 2463–2468. [CrossRef]
6. Mirrakhimov, A.E.; Sooronbaev, T.; Mirrakhimov, E.M. Prevalence of obstructive sleep apnea in Asian adults: A systematic review of the literature. *BMC Pulm. Med.* **2013**, *13*, 10. [CrossRef]
7. Hnin, K.; Mukherjee, S.; Antic, N.A.; Catcheside, P.; Chai-Coetzer, C.L.; McEvoy, D.; Vakulin, A. The impact of ethnicity on the prevalence and severity of obstructive sleep apnea. *Sleep Med. Rev.* **2018**, *41*, 78–86. [CrossRef]
8. Xu, J.; Chen, J.; Li, Y.; Zhang, D.; Li, X. Association of angiotensin-converting enzyme gene insertion/deletion polymorphism and obstructive sleep apnoea in a Chinese population: A meta-analysis. *J. Renin-Angiotensin-Aldosterone Syst. JRAAS* **2020**, *21*, 1470320320934716. [CrossRef]
9. Zhong, A.; Xiong, X.; Shi, M.; Xu, H. Roles of interleukin (IL)-6 gene polymorphisms, serum IL-6 levels, and treatment in obstructive sleep apnea: A meta-analysis. *Sleep Breath.* **2016**, *20*, 719–731. [CrossRef]
10. Lv, D.; Tan, L.; Wu, Y.; Cao, C.; Deng, Z. Leptin and leptin receptor gene polymorphisms in obstructive sleep apnea: A HuGE review and meta-analysis. *Sleep Breath.* **2015**, *19*, 1073–1078. [CrossRef]
11. Veatch, O.J.; Bauer, C.R.; Keenan, B.T.; Josyula, N.S.; Mazzotti, D.R.; Bagai, K.; Malow, B.A.; Robishaw, J.D.; Pack, A.I.; Pendergrass, S.A. Characterization of genetic and phenotypic heterogeneity of obstructive sleep apnea using electronic health records. *BMC Med. Genom.* **2020**, *13*, 105. [CrossRef]
12. Imani, M.M.; Sadeghi, M.; Gholamipour, M.A.; Brühl, A.B.; Sadeghi-Bahmani, D.; Brand, S. Evaluation of Blood Intercellular Adhesion Molecule-1 (ICAM-1) Level in Obstructive Sleep Apnea: A Systematic Review and Meta-Analysis. *Medicina* **2022**, *58*, 1499. [CrossRef] [PubMed]
13. Imani, M.M.; Sadeghi, M.; Mohammadi, M.; Brühl, A.B.; Sadeghi-Bahmani, D.; Brand, S. Association of Blood MCP-1 Levels with Risk of Obstructive Sleep Apnea: A Systematic Review and Meta-Analysis. *Medicina* **2022**, *58*, 1266. [CrossRef] [PubMed]
14. Najafi, A.; Mohammadi, I.; Sadeghi, M.; Brühl, A.B.; Sadeghi-Bahmani, D.; Brand, S. Evaluation of Plasma/Serum Adiponectin (an Anti-Inflammatory Factor) Levels in Adult Patients with Obstructive Sleep Apnea Syndrome: A Systematic Review and Meta-Analysis. *Life* **2022**, *12*, 738. [CrossRef] [PubMed]
15. Imani, M.M.; Sadeghi, M.; Farokhzadeh, F.; Khazaie, H.; Brand, S.; Dürsteler, K.M.; Brühl, A.; Sadeghi-Bahmani, D. Evaluation of Blood Levels of C-Reactive Protein Marker in Obstructive Sleep Apnea: A Systematic Review, Meta-Analysis and Meta-Regression. *Life* **2021**, *11*, 362. [CrossRef]
16. Imani, M.M.; Sadeghi, M.; Khazaie, H.; Sanjabi, A.; Brand, S.; Brühl, A.; Sadeghi Bahmani, D. Associations Between Morning Salivary and Blood Cortisol Concentrations in Individuals With Obstructive Sleep Apnea Syndrome: A Meta-Analysis. *Front. Endocrinol.* **2020**, *11*, 568823. [CrossRef]
17. Imani, M.M.; Sadeghi, M.; Khazaie, H.; Emami, M.; Sadeghi Bahmani, D.; Brand, S. Evaluation of Serum and Plasma Interleukin-6 Levels in Obstructive Sleep Apnea Syndrome: A Meta-Analysis and Meta-Regression. *Front. Immunol.* **2020**, *11*, 1343. [CrossRef]
18. Imani, M.M.; Sadeghi, M.; Khazaie, H.; Emami, M.; Sadeghi Bahmani, D.; Brand, S. Serum and Plasma Tumor Necrosis Factor Alpha Levels in Individuals with Obstructive Sleep Apnea Syndrome: A Meta-Analysis and Meta-Regression. *Life* **2020**, *10*, 87. [CrossRef]
19. Rezaei, F.; Abbasi, H.; Sadeghi, M.; Imani, M.M. The effect of obstructive sleep apnea syndrome on serum S100B and NSE levels: A systematic review and meta-analysis of observational studies. *BMC Pulm. Med.* **2020**, *20*, 31. [CrossRef]
20. Kojima, M.; Hosoda, H.; Kangawa, K. Purification and distribution of ghrelin: The natural endogenous ligand for the growth hormone secretagogue receptor. *Horm. Res. Paediatr.* **2001**, *56*, 93–97. [CrossRef]
21. Date, Y.; Kojima, M.; Hosoda, H.; Sawaguchi, A.; Mondal, M.S.; Suganuma, T.; Matsukura, S.; Kangawa, K.; Nakazato, M. Ghrelin, a novel growth hormone-releasing acylated peptide, is synthesized in a distinct endocrine cell type in the gastrointestinal tracts of rats and humans. *Endocrinology* **2000**, *141*, 4255–4261. [CrossRef]
22. Kojima, M.; Kangawa, K. Structure and function of ghrelin. *Orphan G Protein-Coupled Recept. Nov. Neuropept.* **2008**, *46*, 89–115.
23. Agosti, E.; De Feudis, M.; Angelino, E.; Belli, R.; Teixeira, M.A.; Zaggia, I.; Tamiso, E.; Raiteri, T.; Scircoli, A.; Ronzoni, F.L. Both ghrelin deletion and unacylated ghrelin overexpression preserve muscles in aging mice. *Aging* **2020**, *12*, 13939. [CrossRef]

24. Wiedmer, P.; Nogueiras, R.; Broglio, F.; D'alessio, D.; Tschöp, M.H. Ghrelin, obesity and diabetes. *Nat. Clin. Pract. Endocrinol. Metab.* **2007**, *3*, 705–712. [CrossRef]
25. Van Der Lely, A.J.; Tschop, M.; Heiman, M.L.; Ghigo, E. Biological, physiological, pathophysiological, and pharmacological aspects of ghrelin. *Endocr. Rev.* **2004**, *25*, 426–457. [CrossRef] [PubMed]
26. Shechter, A. Obstructive sleep apnea and energy balance regulation: A systematic review. *Sleep Med. Rev.* **2017**, *34*, 59–69. [CrossRef] [PubMed]
27. Major, G.C.; Series, F.; Tremblay, A. Does the energy expenditure status in obstructive sleep apnea favour a positive energy balance? *Clin. Investig. Med.* **2007**, *30*, E262–E268. [CrossRef] [PubMed]
28. Öztürk, Ö.; Cebeci, D.; Şahin, U.; Tülüceoğlu, E.E.; Calapoğlu, N.Ş.; Gonca, T.; Calapoğlu, M. Circulating levels of ghrelin, galanin, and orexin-A orexigenic neuropeptides in obstructive sleep apnea syndrome. *Sleep Breath.* **2022**, *26*, 1209–1218. [CrossRef] [PubMed]
29. Zhang, D.M.; Pang, X.L.; Huang, R.; Gong, F.Y.; Zhong, X.; Xiao, Y. Adiponectin, omentin, ghrelin, and visfatin levels in obese patients with severe obstructive sleep apnea. *BioMed Res. Int.* **2018**, *2018*, 3410135. [CrossRef] [PubMed]
30. Sun, M.-L.; Niu, X.; Xiao, X.-Y.; Chen, X. The differences in plasma/serum ghrelin levels between obstructive sleep apnea-hypopnea patients and controls: A protocol for systematic review and meta-analysis. *Medicine* **2021**, *100*, e24368. [CrossRef]
31. Marzullo, P.; Verti, B.; Savia, G.; Walker, G.E.; Guzzaloni, G.; Tagliaferri, M.; Di Blasio, A.; Liuzzi, A. The relationship between active ghrelin levels and human obesity involves alterations in resting energy expenditure. *J. Clin. Endocrinol. Metab.* **2004**, *89*, 936–939. [CrossRef]
32. Yuan, M.-J.; Li, W.; Zhong, P. Research progress of ghrelin on cardiovascular disease. *Biosci. Rep.* **2021**, *41*, BSR20203387. [CrossRef] [PubMed]
33. Pulkkinen, L.; Ukkola, O.; Kolehmainen, M.; Uusitupa, M. Ghrelin in diabetes and metabolic syndrome. *Int. J. Pept.* **2010**, *2010*, 248948. [CrossRef] [PubMed]
34. Mao, Y.; Tokudome, T.; Kishimoto, I. Ghrelin and blood pressure regulation. *Curr. Hypertens. Rep.* **2016**, *18*, 1–6. [CrossRef] [PubMed]
35. Hamilton, G.S.; Naughton, M.T. Impact of obstructive sleep apnoea on diabetes and cardiovascular disease. *Med. J. Aust.* **2013**, *199*, S27–S30. [CrossRef]
36. Gonzaga, C.; Bertolami, A.; Bertolami, M.; Amodeo, C.; Calhoun, D. Obstructive sleep apnea, hypertension and cardiovascular diseases. *J. Hum. Hypertens.* **2015**, *29*, 705–712. [CrossRef]
37. Butt, M.; Dwivedi, G.; Khair, O.; Lip, G.Y. Obstructive sleep apnea and cardiovascular disease. *Int. J. Cardiol.* **2010**, *139*, 7–16. [CrossRef]
38. Mcardle, N.; Devereux, G.; Heidarnejad, H.; Engleman, H.M.; Mackay, T.W.; Douglas, N.J. Long-term use of CPAP therapy for sleep apnea/hypopnea syndrome. *Am. J. Respir. Crit. Care Med.* **1999**, *159*, 1108–1114. [CrossRef]
39. Tachikawa, R.; Ikeda, K.; Minami, T.; Matsumoto, T.; Hamada, S.; Murase, K.; Tanizawa, K.; Inouchi, M.; Oga, T.; Akamizu, T. Changes in energy metabolism after continuous positive airway pressure for obstructive sleep apnea. *Am. J. Respir. Crit. Care Med.* **2016**, *194*, 729–738. [CrossRef]
40. De Santis, S.; Cambi, J.; Tatti, P.; Bellussi, L.; Passali, D. Changes in ghrelin, leptin and pro-inflammatory cytokines after therapy in Obstructive Sleep Apnea Syndrome (OSAS) patients. *Otolaryngol. Pol.* **2015**, *69*, 1–8. [CrossRef]
41. Garbuio, S.; Salles, L.V.; D'Almeida, V.; Tufik, S.; Bittencourt, L.R.A. Study of metabolic changes in patients with obstructive sleep apnea syndrome before and after use of continuous positive airway pressure. *Sleep Sci.* **2009**, *2*, 76–81.
42. Takahashi, K.; Chin, K.; Akamizu, T.; Morita, S.; Sumi, K.; Oga, T.; Matsumoto, H.; Niimi, A.; Tsuboi, T.; Fukuhara, S.; et al. Acylated ghrelin level in patients with OSA before and after nasal CPAP treatment. *Respirology* **2008**, *13*, 810–816. [CrossRef]
43. Moher, D.; Shamseer, L.; Clarke, M.; Ghersi, D.; Liberati, A.; Petticrew, M.; Shekelle, P.; Stewart, L.A. Preferred reporting items for systematic review and meta-analysis protocols (PRISMA-P) 2015 statement. *Syst. Rev.* **2015**, *4*, 1–9. [CrossRef]
44. Morgan, R.L.; Thayer, K.A.; Bero, L.; Bruce, N.; Falck-Ytter, Y.; Ghersi, D.; Guyatt, G.; Hooijmans, C.; Langendam, M.; Mandrioli, D. GRADE: Assessing the quality of evidence in environmental and occupational health. *Environ. Int.* **2016**, *92*, 611–616. [CrossRef]
45. Morgan, R.L.; Thayer, K.A.; Santesso, N.; Holloway, A.C.; Blain, R.; Eftim, S.E.; Goldstone, A.E.; Ross, P.; Guyatt, G.; Schünemann, H.J. Evaluation of the risk of bias in non-randomized studies of interventions (ROBINS-I) and the 'target experiment'concept in studies of exposures: Rationale and preliminary instrument development. *Environ. Int.* **2018**, *120*, 382–387. [CrossRef] [PubMed]
46. Institute, J. The Joanna Briggs Institute Critical Appraisal Tools For use in JBI Systematic Reviews Checklist for Analytical Clinical Trial Studies. 2017. Available online: https://jbi.global/critical-appraisal-tools (accessed on 17 October 2022).
47. National Institutes of Health. Quality assessment tool for before-after (pre-post) studies with no control group. In *Systematic Evidence Reviews and Clinical Practice Guidelines*; National Institutes of Health: Washington, DC, USA, 2014. Available online: https://www.nhlbi.nih.gov/health-topics/study-quality-assessment-tools (accessed on 17 October 2022).
48. Der Simonian, R.; Laird, N. Meta-analysis in clinical trials revisited. *Contemp. Clin. Trials* **2015**, *45*, 139–145. [CrossRef] [PubMed]
49. Mantel, N.; Haenszel, W. Statistical aspects of the analysis of data from retrospective studies of disease. *J. Natl. Cancer Inst.* **1959**, *22*, 719–748.
50. Begg, C.B.; Mazumdar, M. Operating characteristics of a rank correlation test for publication bias. *Biometrics* **1994**, *50*, 1088–1101. [CrossRef]

51. Egger, M.; Smith, G.D.; Schneider, M.; Minder, C. Bias in meta-analysis detected by a simple, graphical test. *BMJ* **1997**, *315*, 629–634. [CrossRef] [PubMed]
52. Imberger, G.; Thorlund, K.; Gluud, C.; Wetterslev, J. False-positive findings in Cochrane meta-analyses with and without application of trial sequential analysis: An empirical review. *BMJ Open* **2016**, *6*, e011890. [CrossRef]
53. Wetterslev, J.; Jakobsen, J.C.; Gluud, C. Trial sequential analysis in systematic reviews with meta-analysis. *BMC Med. Res. Methodol.* **2017**, *17*, 39. [CrossRef]
54. Bİçer, Ü.; Kutlu, R.; Yosunkaya, S.; Kilinç, İ. Evaluation of Plasma Ghrelin, Omentin–1 Levels and Insulin Resistance in Patients With Obstructive Sleep Apnea Syndrome. *Konuralp Med. J.* **2021**, *13*, 114–121. [CrossRef]
55. Ciftci, T.U.; Kokturk, O.; Bukan, N.; Bilgihan, A. Leptin and ghrelin levels in patients with obstructive sleep apnea syndrome. *Respiration* **2005**, *72*, 395–401. [CrossRef] [PubMed]
56. Gharraf, H.S.; AlHadidy, A.; AlNehr, I. Study of leptin and ghrelin serum levels in patients with obstructive sleep apnea. *Egypt. J. Chest Dis. Tuberc.* **2019**, *68*, 567.
57. Liu, W.; Yue, H.; Zhang, J.; Pu, J.; Yu, Q. Effects of plasma ghrelin, obestatin, and ghrelin/obestatin ratio on blood pressure circadian rhythms in patients with obstructive sleep apnea syndrome. *Chin. Med. J.* **2014**, *127*, 850–855. [PubMed]
58. Papaioannou, I.; Patterson, M.; Twigg, G.L.; Vazir, A.; Ghatei, M.; Morrell, M.J.; Polkey, M.I. Lack of association between impaired glucose tolerance and appetite regulating hormones in patients with obstructive sleep apnea. *J. Clin. Sleep Med.* **2011**, *7*, 486–492. [CrossRef]
59. Sánchez-de-la-Torre, M.; Mediano, O.; Barceló, A.; Piérola, J.; de la Peña, M.; Esquinas, C.; Miro, A.; Durán-Cantolla, J.; Agustí, A.G.; Capote, F. The influence of obesity and obstructive sleep apnea on metabolic hormones. *Sleep Breath.* **2012**, *16*, 649–656. [CrossRef]
60. Ursavas, A.; Ilcol, Y.O.; Nalci, N.; Karadag, M.; Ege, E. Ghrelin, leptin, adiponectin, and resistin levels in sleep apnea syndrome: Role of obesity. *Ann. Thorac. Med.* **2010**, *5*, 161. [CrossRef]
61. Yang, D.; Liu, Z.; Luo, Q. Plasma ghrelin and pro-inflammatory markers in patients with obstructive sleep apnea and stable coronary heart disease. *Med. Sci. Monit. Int. Med. J. Exp. Clin. Res.* **2013**, *19*, 251.
62. Yang, D.; Liu, Z.-H.; Zhao, Q.; Luo, Q. Effects of nasal continuous positive airway pressure treatment on insulin resistance and ghrelin levels in non-diabetic apnoeic patients with coronary heart disease. *Chin. Med. J.* **2013**, *126*, 3316–3320.
63. Tschop, M.; Weyer, C.; Tataranni, P.A.; Devanarayan, V.; Ravussin, E.; Heiman, M.L. Circulating ghrelin levels are decreased in human obesity. *Diabetes* **2001**, *50*, 707–709. [CrossRef] [PubMed]

life

MDPI

Article

Sleep Endoscopy with Positive Airway Pressure: A Method for Better Compliance and Individualized Treatment of Patients with Obstructive Sleep Apnea

Michaela Masárová [1,2], Petr Matoušek [1,2], Ondřej Jor [3], Vilém Novák [4], Adéla Vrtková [5], Vojtěch Kubec [6], Karol Zeleník [1,2], Pavel Komínek [1,2] and Martin Formánek [1,2,*]

1 Department of Otorhinolaryngology and Head and Neck Surgery, University Hospital Ostrava, 708 52 Ostrava, Czech Republic
2 Faculty of Medicine, University of Ostrava, 703 00 Ostrava, Czech Republic
3 Department of Anesthesiology, Resuscitation and Intensive Medicine, University Hospital Ostrava, 708 52 Ostrava, Czech Republic
4 Department of Pediatric Neurology, University Hospital Ostrava, 708 52 Ostrava, Czech Republic
5 Department of Applied Mathematics, Faculty of Electrical Engineering and Computer Science, VSB-Technical University of Ostrava, 708 00 Ostrava, Czech Republic
6 Department of Otorhinolaryngology, Faculty of Medicine in Pilsen, University Hospital in Pilsen, Charles University, 305 99 Pilsen, Czech Republic
* Correspondence: martin.formanek@fno.cz; Tel.: +420-597-375-812

Abstract: In this study, we aimed to observe the effects of positive airway pressure (PAP) on individual levels of obstruction during drug-induced sleep endoscopy (DISE) of the upper airways (UA), to evaluate at which pressures the obstruction disappeared or worsened, and to identify cases in which PAP was ineffective. This prospective study was conducted from June 2018 to June 2022. PAP testing was performed during DISE in patients with moderate and severe OSA. The pressure was gradually increased over the range from 6.0 to 18.0 hPa. Our findings were evaluated using the VOTE classification. The examination was performed in 56 patients, with a median apnea–hypopnea index (AHI) of 26.4. Complete obstruction of the soft palate was observed in 51/56 patients (91%), oropharyngeal obstruction in 15/56 patients (27%), tongue base obstruction in 23/56 patients (41%), and epiglottic collapse in 16/56 patients (29%). PAP was most effective in cases of complete oropharyngeal obstruction, and least effective in cases of epiglottic collapse, where it was ineffective in 11/16 patients. DISE with PAP is a simple diagnostic method that can be helpful for identifying anatomic and dynamic reasons for PAP intolerance. The main indication is ineffective PAP treatment.

Keywords: obstructive sleep apnea; drug-induced sleep endoscopy; positive airway pressure

Citation: Masárová, M.; Matoušek, P.; Jor, O.; Novák, V.; Vrtková, A.; Kubec, V.; Zeleník, K.; Komínek, P.; Formánek, M. Sleep Endoscopy with Positive Airway Pressure: A Method for Better Compliance and Individualized Treatment of Patients with Obstructive Sleep Apnea. *Life* **2022**, *12*, 2108. https://doi.org/10.3390/life12122108

Academic Editor: Luca Steardo

Received: 21 November 2022
Accepted: 12 December 2022
Published: 15 December 2022

1. Introduction

For adults with moderate and severe obstructive sleep apnea (OSA), treatment with positive airway pressure (PAP) is currently considered the "gold standard" [1,2]. Despite this consensus, some of the patients show poor compliance for a variety of reasons—including psychological reasons, allergic contact dermatitis to the mask material, latex allergy, anatomical abnormalities, rhinitis, etc. Additionally, PAP is ineffective in some patients, or works only at intolerably high pressures [3,4]. These patients are then exposed to the risks associated with untreated OSA, which can be life-threatening. In addition, it is assumed that the application of PAP affects all areas of the upper respiratory tract equally, regardless of the individual characteristics of the patient. Therefore, factors such as the location and range of obstruction may often be ignored [4,5].

Drug-induced sleep endoscopy (DISE) is currently the main diagnostic method for identifying sites of obstruction in patients with OSA [6,7]. Since it is mainly performed before planned surgical treatment [7,8], the majority of PAP-treated OSA patients do not undergo DISE [4,5].

PAP performed during DISE is a new combination of diagnostic methods that allows the direct visualization and evaluation of how PAP affects individual collapsing areas of the airway. This method can provide a better understanding how PAP actually works, and possibly predict a good response to and compliance with future PAP treatment [4,5,9–11].

A limited number of studies have dealt with this issue. According to the preliminary results of these studies, it appears that overpressure ventilation is more effective in cases of obstruction of the soft palate and oropharynx compared to obstructions of the base of the tongue or cases of epiglottic collapse. These results logically suggest that for certain sites of obstruction, surgical treatment might be more appropriate compared to PAP, regardless of factors such as AHI, BMI, etc. [5,10,11].

In the present study, we aimed to observe the effects of PAP on individual levels of airways obstruction during DISE, to evaluate the reaction of obstruction upon overpressure, and to identify cases in which PAP is ineffective.

2. Materials and Methods

This prospective study was performed in accordance with the Declaration of Helsinki and the requirements of Good Clinical Practice, and was approved by the Ethics Committee of the University Hospital Ostrava (identifier: 360/2021). The study was registered in ClinicalTrials under the number NCT02855515. Written informed consent was obtained from each patient before any procedure was initiated.

2.1. Study Design and Patients

This prospective study was performed at the tertiary referral center University Hospital Ostrava from June 2018 to June 2022. We consecutively enrolled adult patients with moderate or severe OSA with an apnea–hypopnea index (AHI) of $\geq$15 episodes/hour. Patients were excluded if they had major comorbidities representing excessive risk for general anesthesia (decompensation phase)—cardiac, liver, or kidney disease, cancer; craniofacial malformations; neurological pathologies; if they were pregnant; or if they did not agree to be included in the study.

Sixty-four consecutive patients were included in the study; eight patients were excluded and examination was performed in fifty-six patients.

2.2. Clinical Evaluation

Patients were evaluated by collection of a comprehensive history that covered sleep habits and disturbances. As a subjective measure of a patient's sleepiness, the Epworth sleepiness scale (ESS) was used. Clinical evaluation included a complete head and neck examination. The upper airways and digestive tract were examined using a flexible videoendoscope with a 3.5 mm diameter (Olympus, Tokyo, Japan).

2.3. Drug-Induced Sleep Endoscopy

Sleep endoscopy was performed in the operating room. Intramuscular administration of Dormicum (midazolam) 5 mg and atropine 0.5 mg was performed 30 min before the sleep endoscopy. Subsequently, the patient was induced to sleep with intravenous propofol (an initial 1 mg/kg bolus, followed by 20–30 mg every 3–5 min). The depth of anesthesia was measured using the bispectral index. Vital signs were monitored. Sleep endoscopy was performed using a flexible videoendoscope with a diameter of 3.5 mm (Olympus, Tokyo, Japan). The examination length was 15–20 min. The results were evaluated using the Kezirian VOTE classification, for which obstruction is evaluated in the four localities of the upper airways [12].

2.4. Positive Airway Pressure Titration during the Sleep Endoscopy

Titration was performed using the BiPAP A40 (Philips Respironics, Florida, USA) in PAP mode. PAP titration was performed immediately after DISE. An appropriately sized overpressure ventilation mask (Respironics PerforMax Full-face mask; Philips Respironics, Florida, USA) was applied to the patient's face. A special connecting valve (Philips Respironics, Florida, USA) was inserted between the mask and the device hose, through which a flexible endoscope was inserted into the nasopharynx and upper airways (Figure 1).

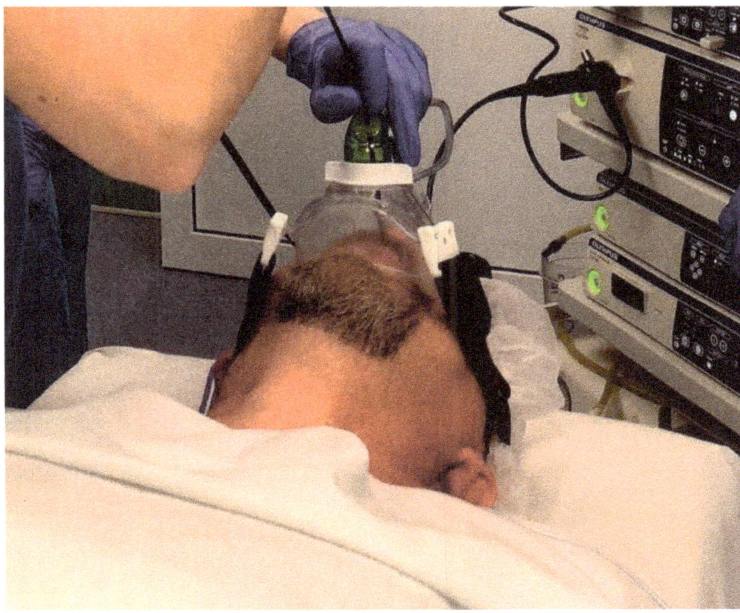

Figure 1. Drug-induced sleep endoscopy (DISE) with simultaneous positive airway pressure (PAP); external view.

Sleep endoscopy was then performed under overpressure ventilation. The examination started at a pressure of 6.0 hPa. The pressure on the PAP was gradually elevated (in the range of 6.0, 8.0, 10.0, 12.0, 14.0, and 18.0 hPa (Figure 2). The efficiency of treatment was visually assessed, with simultaneous monitoring of the blood oxygen saturation. At each tested PAP pressure, the VOTE classification was used for evaluation [12].

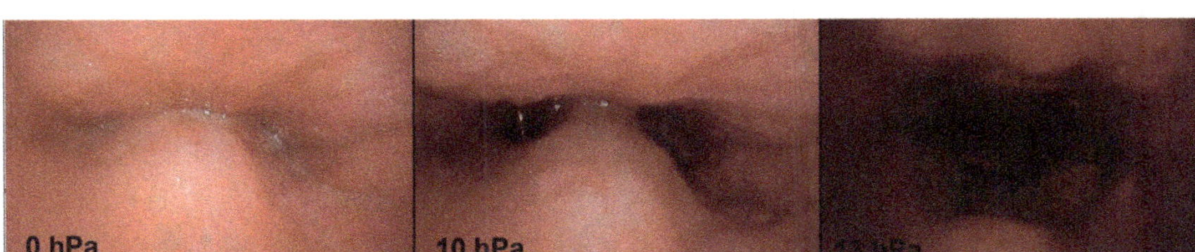

Figure 2. The effect of positive airway pressure (PAP) during drug-induced sleep endoscopy (DISE) at different pressures; endoscopic view.

We compared the findings among the examined patients. Notably, we observed which areas of the upper airways responded better to PAP treatment and in which areas PAP had worse effects.

2.5. Statistical Analysis

Numerical variables were presented as the median and interquartile range (IQR). Categorical variables were presented as absolute and relative frequencies (%). The chi-square test for equality of proportions was used for the comparison of the effect of PAP. The significance level was set to 0.05 and all statistical analyses were performed using R software (version 4.2.1).

3. Results

3.1. Patients Characteristics

This study included a total of 56 patients (9 women, 47 men) aged 22–59 years. The median AHI was 26.4, and median BMI was 29.2 kg/m^2 (Table 1). During DISE, complete obstruction was observed in the soft palate region in 51/56 patients (91%), in the oropharynx in 15/56 patients (27%), in the tongue base in 23/56 patients (41%), and due to the epiglottis (epiglottic collapse) in 16/56 patients (29%) (Table 2).

Table 1. Demographic data, entrance limited polygraphy, and ENT examination while conscious.

	Median (IQR) or n (%)
Age, years	46.0 (39; 55)
BMI, kg/m^2	29.2 (27.4; 31.3)
AHI	26.4 (18.9; 31.8)
T90, %	2.3 (0.6; 9.1)
Soft palate obstruction	53 (94.6)
Oropharynx obstruction	13 (23.2)
Tongue base obstruction	39 (69.6)
Epiglottic pathology	—
Mallampati	
I	2 (3.6)
II	10 (17.9)
III	30 (53.6)
IV	14 (25.0)
Friedman	
0	4 (7.1)
1	28 (50.0)
2	22 (39.3)
3	2 (3.6)

The values represent the median and interquartile range (IQR) or absolute and relative frequencies (%). AHI = apnea–hypopnea index; T90 = time under $SaO_2 < 90\%$.

Table 2. Degree and type of obstruction and analysis of the overall improvement with PAP.

	Soft Palate	**Oropharynx**	**Tongue Base**	**Epiglottis**	***p***
Obstruction—degree					
Complete	51 (91.1)	15 (26.8)	23 (41.1)	16 (28.6)	
Partial	4 (7.1)	27 (48.2)	17 (30.4)	5 (8.9)	
No	1 (1.8)	14 (25.0)	16 (28.6)	35 (62.5)	
Obstruction—type					
Concentric	33 (58.9)	—	—	—	
Laterolateral	2 (3.6)	42 (75.0)	—	2 (3.6)	
Anteroposterior	20 (35.7)	—	40 (71.4)	19 (33.9)	
No	1 (1.8)	14 (25.0)	16 (28.6)	35 (62.5)	
Improvement	38/51 (74.5)	15/15 (100.0)	20/23 (87.0)	5/16 (31.3)	<0.001

The numbers represent absolute frequencies and relative frequencies (%). The *p* value was obtained using the chi-square test for equality of proportions.

3.2. Effect of PAP

PAP was most effective in complete oropharyngeal obstruction, where we observed improvement in all 15 (100%) obstructions. PAP was least effective in epiglottic collapse, with improvement observed in only 5/16 cases with obstruction due to the epiglottis. PAP was ineffective in the remaining 11/16 cases with epiglottic collapse (Table 2).

Single-level obstruction was detected in 3/56 patients (5%), and multilevel obstruction (two or more regions) in 53/56 patients (95%). Obstruction was observed at three sites in 25/56 patients (45%), and at all four locations in 11/56 patients (20%).

Within the area of the soft palate, we observed all three types of obstruction (Table 2). Among the 20 patients with anteroposterior obstruction, 16 (80%) exhibited a median opening pressure of 10 hPa, while PAP had no effect on the remaining 4 (20%). Among the 33 patients with concentric obstruction, 21 (64%) exhibited a median opening pressure of 10 hPa, while PAP was ineffective in the remaining 12 (36%) ($p = 0.343$) (Figure 3, Table 3). In two patients with concentric obstruction, the application of higher pressure (10.0 hPa) resulted in a change of the obstruction to the anteroposterior configuration. Moreover, at a pressure of 12.0 hPa, these patients no longer had obstruction in this area.

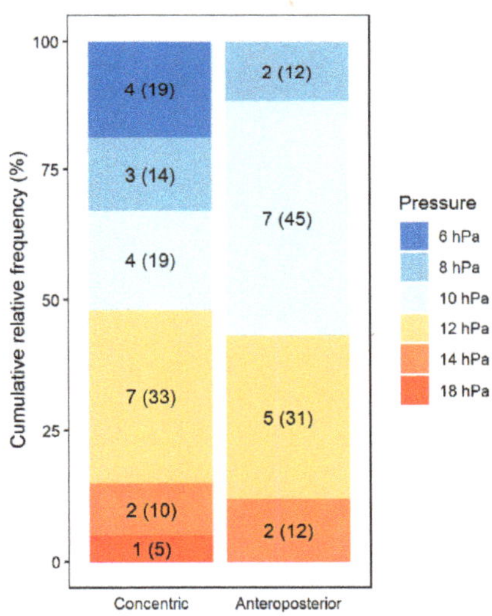

Figure 3. Analysis of the association between concentric and anteroposterior obstruction of the soft palate and the pressure at which improvement occurred.

Table 3. Median opening pressure in individual areas, and analysis of the inefficiency of the PAP.

Level of Obstruction	Type of Obstruction	Median Opening Pressure (hPa)	Ineffectiveness of PAP (% Patients)
Soft palate	Anteroposterior	10	20.0
	Concentric	10	36.4
Oropharynx	Laterolateral	12	0.0
Tongue base	Anteroposterior	12	13.0
Epiglottis	Anteroposterior (collapse)	14	68.8

PAP yielded improvement in all 15 patients (100%) with complete oropharyngeal obstruction; the median opening pressure was 12 hPa. Among the 23 patients with complete obstruction in the region of the base of the tongue, PAP was effective in 20 patients, and the median opening pressure was 12 hPa; in the remaining 3 patients, PAP had no effect at all, even with the highest tested pressure of 18.0 hPa. Among the 16 patients with epiglottic collapse, simultaneous PAP yielded improvement in 5 (31%), with a median opening pressure of 14 hPa. In the remaining 11 patients, PAP was ineffective, and resulted in the epiglottis being pushed even more strongly against the back wall of the pharynx, which was clinically correlated with persistence of apnea and simultaneous drop in SpO_2. Figure 4 and Table 3 summarize the results regarding the mean opening pressure required to overcome obstructions in the individual upper airway regions.

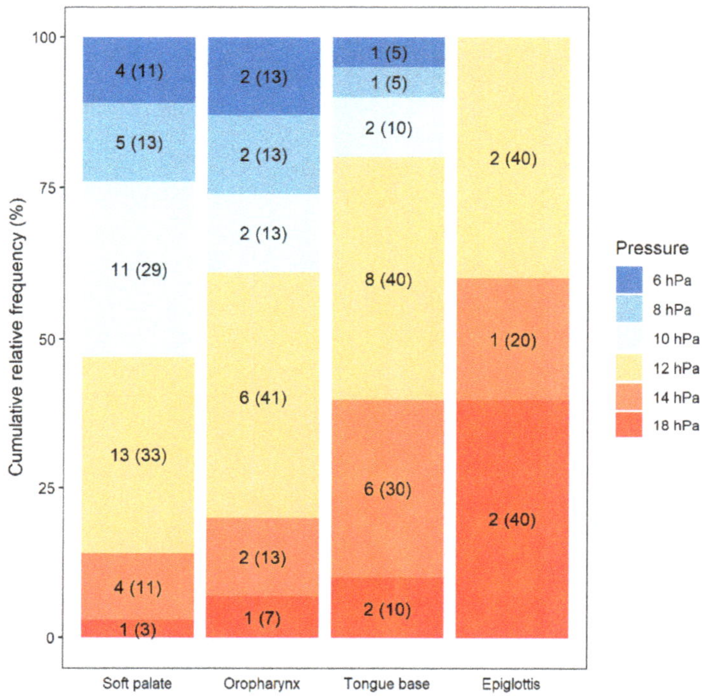

Figure 4. Analysis of the association between the level of the obstruction and the pressure at which improvement occurred.

3.3. *Ineffectiveness of PAP*

PAP was ineffective in 45% cases of complete soft palate obstruction, in 13% cases of complete tongue base obstruction, and in 69% cases of complete epiglottic collapse.

4. Discussion

DISE was first described in 1991 [13,14]. Since then, it has gradually become the most widely used method for upper airway examination while sleeping [8,15]. DISE is currently the main diagnostic method for determining the location of upper airways obstruction, and these findings can help with treatment optimization [6–8,16–19]. DISE is based on the assumption that the upper airway configuration differs when a person is awake versus sleeping, which has been proven by many authors [7,8,13,15]. DISE is mainly indicated in patients under consideration for surgical treatment of OSA [7,8]; it is also performed in patients who have either failed or are noncompliant with PAP therapy [20]. Over recent years, it has been demonstrated that DISE can be effectively used simultaneously with a PAP device, where optical control enables the direct monitoring of the effect of PAP [4,5]. However, only a limited number of studies have examined the importance of DISE in the current use of PAP [5,10,11].

PAP, which is considered a treatment of choice for patients suffering from moderate to severe OSA, is not easily accepted. Many diverse factors may cumulate leading to patients' failure to initiate PAP or non-compliance [21]. There are numerous factors that may affect adherence to PAP treatment. For instance, several demographic and clinical variables, e.g., age, sex, BMI (body mass index), race, AHI, ESS score, and the presence of comorbidities, were shown to have an effect on compliance [22]. However, there are few studies that deal with the effect of PAP on individual locations of the upper airways. Too high an excess pressure, which in some cases is necessary to overcome an obstruction in a certain location of the upper airways, can be a particularly significant factor affecting PAP compliance [4,5].

According to current knowledge, overpressure ventilation is more effective in cases of obstructions of the soft palate and oropharynx, compared to obstructions of the base of the tongue or cases of epiglottic collapse [5,10,11]. In a study of 30 patients, Jung et al. confirmed that the oropharynx has a major influence on the effectiveness of PAP [10]. Accordingly, Schwab et al. assessed PAP effectiveness by conducting MRI examinations at different pressure levels (0.0, 5.0, 10.0, and 15.0 hPa), and observed that PAP had the greatest effect on lateral oropharyngeal walls [23].

Torre et al. also reported that PAP treatment is most effective for laterolateral oropharyngeal obstruction [5]. They found that in cases with complete concentric obstruction of the soft palate, the application of higher pressure led to a change in the anteroposterior configuration. With this type of obstruction, channels are laterally created, which enables airflow, such that PAP is effective. Their study also revealed that a mean pressure of 10.0 hPa was required for PAP to be effective in two-level obstruction of the soft palate and oropharynx. However, their work also describes the case of a patient with multilevel obstruction with hypertrophic palatine tonsils that completely blocked the upper airways. In that case, a pressure of up to 15.0 hPa was needed to open the upper airways. As this pressure was uncomfortable for the patient, a bilateral tonsillectomy was performed, which enabled the patient to tolerate the PAP pressure [5]. This case further illustrates the importance of PAP examination under the control of DISE. Direct visualization of the upper airways, with the simultaneous use of PAP, enables the detection of any anatomical abnormalities that could be surgically or orthodontically resolved to enable the reduction of PAP pressure with subsequent better tolerance of treatment. Torre et al. also reported that obstructions of the base of the tongue and epiglottic collapse require higher pressure to open the upper airways, with an average pressure of 15.0 hPa required to eliminate obstruction of the base of the tongue. PAP did not resolve primary epiglottic collapse, and was particularly ineffective in cases of laterolateral epiglottic obstruction, where the problem persists even with pressures higher than 15.0 hPa [5].

Lai et al. also reported that higher pressure is needed in cases where the obstruction is in the area of the base of the tongue, and that the area of the base of the tongue is essential for correct PAP setting [24]. In cases of primary epiglottic collapse, several studies have confirmed that increased PAP pressure leads to exacerbation of the obstruction, with the epiglottis being pushed onto the back wall of the pharynx [9,25–27]. Epiglottic collapse has been found in 15.0–31.4% of adult patients with OSA in whom PAP treatment was ineffective [2,26,28,29]. Unfortunately, epiglottic collapse is not observed during the examination of an awake patient. Our present results confirmed that PAP was most effective in the oropharynx, particularly in cases of its laterolateral obstruction, while PAP was least effective against obstructions caused by the epiglottis and the base of the tongue.

Overpressure must be correctly set to achieve a therapeutic effect, sufficient adherence to treatment, and a low level of side effects. Insufficient pressure reduces the number of apneas, but leaves hypopnea and awakening reactions, with the associated high cardiovascular risk. Insufficient positive pressure also fails to eliminate the subjective symptoms of OSA, resulting in low willingness to regularly use PAP. In contrast, excessively high overpressure can burden the patient with treatment-induced central sleep apnea, middle ear ventilation disorders, and other problems that ultimately contribute to inadequate treatment or low compliance [30].

Although PAP is considered the gold standard of OSA treatment, some patients experience no improvement and a persisting high residual AHI. Subjective problems can lead to a patient refusing to use overpressure and treatment failure. DISE with PAP is a very simple method that can be used to predict the success of PAP in individual patients. It can also reveal the anatomical and dynamic causes of PAP dysfunction, which may be resolved surgically to improve subsequent adherence to overpressure treatment [30].

This especially applies to cases involving the obstruction in the area of the epiglottis. Based on both our present results and previously published studies, we are inclined to recommend that when OSA is caused by epiglottis collapse, surgical treatment should be indicated. Possible surgical options include epiglottidectomy (total, partial, or V-shaped) or the more elegant and gentle method of transoral epiglottopexy [28,29,31–34]. Obstruction on the level of the base of the tongue also seems to be less suitable for overpressure treatment; therefore, additional care should be taken in setting up and following up treatment of these patients.

Our present results are limited by the small number of patients, the effect of the presence of optical fibers on PAP, and by the use of the VOTE classification, which is a subjective evaluation system. However, it should be noted that this classification is recognized worldwide, and we tried to account for its limitations by having two experienced physicians independently evaluate our findings. Subjective video closure analysis could be objectified by implementing DISE with type 3 PSG monitoring [35]. Another limitation of the study is that the patient position was only supine and only an oronasal mask was used. Some studies have suggested that higher PAP pressures may be required when this type of mask is used [36]. For technical reasons, it was only possible to use an oronasal mask. Only this type of mask is able to connect to a "connecting elbow" with a perforation for the entry of the endoscope. The short period of DISE and the lack of different sleep phases (REM, NREM) were other limitations, but the depth of anesthesia was monitored using the BIS (bispectral index) within the specified range.

5. Conclusions

PAP performed in combination with DISE is a simple, valid, safe, effective, and easy-to-use diagnostic method, which directly allows visualization of the effects of PAP ventilation on different types of upper airway obstructions at specific sites. It can be helpful for identifying anatomic and dynamic reasons for PAP intolerance or ineffectiveness. Our present results indicated that PAP was most effective in cases of laterolateral oropharyngeal obstruction, while PAP was least effective in cases of obstruction on the level of the epiglottis and the base of the tongue. DISE and PAP could be used to improve PAP treatment compliance and effectiveness.

Author Contributions: Conceptualization, K.Z., P.M. and P.K.; Data curation, M.M. and A.V.; Formal analysis, M.M. and A.V.; Funding acquisition, M.M., M.F., P.M. and P.K.; Investigation, M.M., O.J., V.N. and V.K.; Methodology, M.F., K.Z. and P.M.; Supervision, M.F. and P.K.; Writing—original draft, M.M.; Writing—review & editing, M.F. and P.K. All authors have read and agreed to the published version of the manuscript.

Funding: This work was supported by the Ministry of Health, Czech Republic, Conceptual Development of Research Organization, MH CZ–17/DRO–FNOs/2021.

Institutional Review Board Statement: Approved by the Ethics Committee of the University Hospital Ostrava (identifier: 360/2021). The study was registered in ClinicalTrials under the number NCT02855515.

Informed Consent Statement: Informed consent was obtained from the patients in this study.

Data Availability Statement: All available data presented.

Conflicts of Interest: The authors declare that they have no conflict of interest.

References

1. Pavwoski, P.; Shelgikar, A.V. Treatment options for obstructive sleep apnea. *Neurol. Clin. Pract.* **2016**, *7*, 77–85. [CrossRef] [PubMed]
2. Waters, T. Alternative interventions for obstructive sleep apnea. *Clevel. Clin. J. Med.* **2019**, *86*, 34–41. [CrossRef] [PubMed]
3. Weaver, T.E.; Grunstein, R.R. Adherence to continuous positive airway pressure therapy: The challenge to effective treatment. *Proc. Am. Thorac. Soc.* **2008**, *5*, 173–178. [CrossRef]
4. Freiser, M.E.; Schell, A.E.; Soose, R.J. DISE-PAP: A method for troubleshooting residual AHI elevation despite positive pressure therapy. *J. Clin. Sleep Med.* **2020**, *16*, 631–633. [CrossRef] [PubMed]
5. Torre, C.; Liu, S.Y.; Kushida, C.A.; Nekhendzy, V.; Huon, L.K.; Capasso, R. Impact of continuous positive airway pressure in patients with obstructive sleep apnea during drug-induced sleep endoscopy. *Clin. Otolaryngol.* **2017**, *42*, 1218–1223. [CrossRef] [PubMed]
6. Cavaliere, M.; Russo, F.; Iemma, M. Awake versus druginduced sleep endoscopy: Evaluation of airway obstruction in obstructive sleep apnea/hypopnoea syndrome. *Laryngoscope* **2013**, *123*, 2315–2318. [CrossRef] [PubMed]
7. Hybášková, J.; Jor, O.; Novák, V.; Matoušek, P.; Komínek, P. Možné využití spánkové endoskopie pro zvýšení efektivity léčby (operační i neoperační) u pacientů s obstrukční spánkovou apnoí. *Cesk Slov Neurol N* **2017**, *80/113*, 428–433. [CrossRef]
8. Certal, V.F.; Pratas, R.; Guimaraes, L.; Lugo, R.; Tsou, Y.; Camacho, M.; Capasso, R. Awake examination versus DISE for surgical decision making in patients with OSA: A systematic rewiew. *Laryngoscope* **2016**, *126*, 768–774. [CrossRef]
9. Civelek, S.; Emre, I.E.; Dizdar, D.; Cuhadaroglu, C.; Eksioglu, B.K.; Eraslan, A.K.; Turgut, S. Comparison of conventional continuous positive airway pressure to continuous positive airway pressure titration performed with sleep endoscopy. *Laryngoscope* **2012**, *122*, 691–695. [CrossRef]
10. Jung, S.H.; Koo, S.K.; Choi, J.W.; Moon, J.S.; Lee, S.H. Upper airway structural changes induced by CPAP in OSAS patients: A study using drug-induced sleep endoscopy. *Eur. Arch. Otorhinolaryngol.* **2017**, *274*, 247–252. [CrossRef]
11. Lan, M.C.; Hsu, Y.B.; Lan, M.Y.; Huang, Y.C.; Kao, M.C.; Huang, T.T.; Chiu, T.J.; Yang, M.C. The predictive value of drug-induced sleep endoscopy for CPAP titration in OSA patients. *Sleep Breath.* **2018**, *22*, 949–954. [CrossRef] [PubMed]
12. Kezirian, E.J.; Hussey, H.M.; Brietzke, S.E.; Cohen, S.M.; Davis, G.E.; Shin, J.J.; Weinberger, D.G.; Cabana, M.D. Hypopharyngeal surgery in obstructive sleep apnea: Practice patterns, perceptions, and attitudes. *Otolaryngol. Head Neck Surg.* **2012**, *147*, 964–971. [CrossRef] [PubMed]
13. De Vito, A.; Carrasco, L.M.; Vanni, A.; Bosi, M.; Braghiroli, A.; Campanini, A.; De Vries, N.; Hamans, E.; Hohenhorst, W.; Kotecha, B.T.; et al. European position paper on drug-induced sedation endoscopy (DISE). *Sleep Breath.* **2014**, *18*, 453–465. [CrossRef] [PubMed]
14. Croft, C.B.; Pringle, M. Sleep nasendoscopy: A technique of assessment in snoring and obstructive sleep apnea. *Clin. Otolaryngol. Allied Sci.* **1991**, *16*, 504509. [CrossRef] [PubMed]
15. Veer, V.; Zhang, H.; Mandavia, R.; Mehta, N. Introducing a new classification for drug-induced sleep endoscopy (DISE): The PTLTbE system. *Sleep Breath.* **2020**, *24*, 1685–1693. [CrossRef] [PubMed]
16. De Corso, E.; Fiorita, A.; Rizzotto, G.; Mennuni, G.F.; Meucci, D.; Giuliani, M.; Marchese, M.R.; Levantesi, L.; Marca, G.D.; Paludetti, G.; et al. The role of drug-induced sleep endoscopy in the diagnosis and management of obstructive sleep apnoea syndrome: Our personal experience. *Acta Otorhinolaryngol. Ital.* **2013**, *33*, 405–413. [CrossRef]
17. Bachar, G.; Nageris, B.; Feinmesser, R.; Hadar, T.; Yaniv, E.; Shpitzer, T.; Eidelman, L. Novel grading system for quantifying upper-airway obstruction on sleep endoscopy. *Lung* **2012**, *190*, 313–318. [CrossRef]
18. Cho, J.S.; Soh, S.; Kim, E.J.; Cho, H.J.; Shin, S.; Kim, H.J.; Koo, B.N. Comparison of three sedation regimens for drug-induced sleep endoscopy. *Sleep Breath.* **2015**, *19*, 711–717. [CrossRef]
19. Hybášková, J.; Babiarová, V.; Jor, O.; Novák, V.; Matoušek, P.; Komínek, P. Flexibilní endoskopie horních dýchacích cest v léky navozeném spánku. *Otorinolaryngol. Foniatr.* **2016**, *65*, 30–34.
20. Atkins, J.H.; Mandel, J.E. Drug-induced sleep endoscopy: From obscure technique to diagnostic tool for assessment of obstructive sleep apnea for surgical interventions. *Curr. Opin. Anaesthesiol.* **2018**, *31*, 120–126. [CrossRef]
21. Pépin, J.L.; Krieger, J.; Rodenstein, D.; Cornette, A.; Sforza, E.; Delguste, P.; Deschaux, C.; Grillier, V.; Lévy, P. Effective Compliance during the First 3 Months of Continuous Positive Airway Pressure A European Prospective Study of 121 Patients. *Am. J. Respir. Crit. Care Med.* **1999**, *160*, 1124–1129. [CrossRef] [PubMed]
22. Sabil, A.; le Vaillant, M.; Stitt, C.; Goupil, F.; Pigeanne, T.; Leclair-Visonneau, L.; Masson, P.; Bizieux-Thaminy, A.; Humeau, M.-P.; Meslier, N.; et al. A CPAP Data–Based Algorithm for Automatic Early Prediction of Therapy Adherence. *Sleep Breath.* **2021**, *25*, 957–962. [CrossRef] [PubMed]
23. Schwab, R.J.; Pack, A.I.; Gupta, K.B.; Metzger, L.J.; Oh, E.; Getsy, J.E.; Hoffman, E.A.; Gefter, W.B. Upper airway and soft tissue structural changes incuded by CPAP in normal subjects. *Am. J. Respir. Crit. Care Med.* **1996**, *154*, 1106–1116. [CrossRef] [PubMed]
24. Lai, C.C.; Friedman, M.; Lin, H.C.; Wang, P.C.; Hwang, M.S.; Hsu, C.M.; Lin, M.C.; Chin, C.H. Clinical predictors of effective-continuous positive airway pressure in patients with obstructive sleep apnea/hypopnea syndrome. *Laryngoscope* **2015**, *125*, 1983–1987. [CrossRef] [PubMed]
25. Torre, C.; Camacho, M.; Liu, S.Y.; Huon, L.K.; Capasso, R. Epiglottis collapse in adult obstructive sleep apnea: A systematic review. *Laryngoscope* **2016**, *126*, 515–523. [CrossRef]

26. Dedhia, E.C.; Rosen, C.A.; Soose, R.J. What is the role of the larynx in adult obstructive sleep apnea? *Laryngoscope* **2014**, *124*, 1029–1034. [CrossRef]
27. Verse, T.; Pirsig, W. Age-related changes in the epiglottis causing failure of nasal continuous positive airway pressure therapy. *J. Laryngol. Otol.* **1999**, *113*, 1022–1025. [CrossRef]
28. Kwon, E.O.; Jung, S.Y.; Al-Dilaijan, K.; Min, J.Y.; Lee, K.H.; Kim, S.W. Is Epiglottis Surgery Necessary for Obstructive Sleep Apnea Patients With Epiglottis Obstruction? *Laryngoscope* **2019**, *129*, 2658–2662. [CrossRef]
29. Roustan, V.; Barbieri, M.; Incandela, F.; Missale, F.; Camera, H.; Braido, F.; Mora, R.; Peretti, G. Transoral glossoepiglottopexy in the treatment of adult obstructive sleep apnoea: A surgical approach. *Acta Otorhinolaryngol. Ital.* **2018**, *38*, 38–44. [CrossRef]
30. Betka, J.; Kubíčková, J.; Klozar, J.; Šonka, K. *Poruchy Dýchání ve Spánku*, 1st ed.; Tobiáš: Hradec Králové, Czech Republic, 2019; pp. 106–113.
31. Mickelson, S.A.; Rosenthal, L. Midline glossectomy and epiglottidectomy for obstructive sleep apnea syndrome. *Laryngoscope* **1997**, *107*, 614–619. [CrossRef]
32. Cassano, M. Endoscopic coblator-assisted epiglottoplasty in "obstructive sleep apnoea syndrome" patients. *Clin. Otolaryngol.* **2017**, *42*, 1112–1114. [CrossRef] [PubMed]
33. Oluwasanmi, A.F.; Mal, R.K. Diathermy epiglottectomy: Endoscopic technique. *J. Laryngol. Otol.* **2001**, *115*, 289–292. [CrossRef] [PubMed]
34. Catalfumo, F.J.; Golz, A.; Westerman, S.T.; Gilbert, L.M.; Joachims, H.Z.; Goldenberg, D. The epiglottis and obstructive sleep apnoea syndrome. *J. Laryngol. Otol.* **1998**, *112*, 940–943. [CrossRef] [PubMed]
35. Gobbi, R.; Baiardi, S.; Mondini, S.; Cerritelli, L.; Piccin, O.; Scaramuzzino, G.; Milano, F.; Melotti, M.R.; Mordini, F.; Pirodda, A.; et al. Technique and Preliminary Analysis of Drug-Induced Sleep Endoscopy With Online Polygraphic Cardiorespiratory Monitoring in Patients With Obstructive Sleep Apnea Syndrome. *JAMA Otolaryngol. Head Neck Surg.* **2017**, *143*, 459–465. [CrossRef]
36. Ebben, M.R.; Narizhnaya, M.; Segal, A.Z.; Barone, D.; Krieger, A.C. A randomised controlled trial on the effect of mask choice on residual respiratory events with continuous positive airway pressure treatment. *Sleep Med.* **2014**, *15*, 619–624. [CrossRef]

life

Article

The Predictive Role of the Upper-Airway Adipose Tissue in the Pathogenesis of Obstructive Sleep Apnoea

Viktória Molnár [1,*], Zoltán Lakner [2], András Molnár [1], Dávid László Tárnoki [3], Ádám Domonkos Tárnoki [3], László Kunos [4], Zsófia Jokkel [3] and László Tamás [1,5]

1 Department of Otolaryngology and Head and Neck Surgery, Semmelweis University, 1083 Budapest, Hungary
2 Szent István Campus, Hungarian University of Agriculture and Life Sciences, 2100 Gödöllő, Hungary
3 Medical Imaging Centre, Semmelweis University, 1082 Budapest, Hungary
4 Institute of Pulmonology, Törökbálint, 2045 Törökbálint, Hungary
5 Department of Voice, Speech and Swallowing Therapy, Faculty of Health Sciences, Semmelweis University, 1088 Budapest, Hungary
* Correspondence: viktoria.molnar.se@gmail.com; Tel.: +36-20-663-2402

Simple Summary: Obstructive sleep apnoea (OSA) is an underdiagnosed disorder from which many patients are suffering, and may lead to severe complications. The adipose tissue near the upper airways is essential in upper-airway collapses and OSA severity. The present investigation aimed to determine the correlations between upper-airway adipose tissue MRI parameters and OSA, using artificial intelligence to analyse the pathophysiology of OSA and predict obstruction location. Including anthropometric and MRI adipose tissue parameters, OSA and upper-airway obstruction can be predicted with high precision. Artificial intelligence can effectively be used in OSA diagnostics as it can analyse non-linear correlations; thus, it can be helpful for undiagnosed OSA cases.

Citation: Molnár, V.; Lakner, Z.; Molnár, A.; Tárnoki, D.L.; Tárnoki, Á.D.; Kunos, L.; Jokkel, Z.; Tamás, L. The Predictive Role of the Upper-Airway Adipose Tissue in the Pathogenesis of Obstructive Sleep Apnoea. *Life* **2022**, *12*, 1543. https://doi.org/10.3390/life12101543

Academic Editors: David Slouka and Milan Štengl

Received: 3 September 2022
Accepted: 29 September 2022
Published: 4 October 2022

Publisher's Note: MDPI stays neutral with regard to jurisdictional claims in published maps and institutional affiliations.

Abstract: This study aimed to analyse the thickness of the adipose tissue (AT) around the upper airways with anthropometric parameters in the prediction and pathogenesis of OSA and obstruction of the upper airways using artificial intelligence. One hundred patients were enrolled in this prospective investigation, who were divided into control (non-OSA) and mild, moderately severe, and severe OSA according to polysomnography. All participants underwent drug-induced sleep endoscopy, anthropometric measurements, and neck MRI. The statistical analyses were based on artificial intelligence. The midsagittal SAT, the parapharyngeal fat, and the midsagittal tongue fat were significantly correlated with BMI; however, no correlation with AHI was observed. Upper-airway obstruction was correctly categorised in 80% in the case of the soft palate, including parapharyngeal AT, sex, and neck circumference parameters. Oropharyngeal obstruction was correctly predicted in 77% using BMI, parapharyngeal AT, and abdominal circumferences, while tongue-based obstruction was correctly predicted in 79% using BMI. OSA could be predicted with 99% precision using anthropometric parameters and AT values from the MRI. Age, neck circumference, midsagittal and parapharyngeal tongue fat values, and BMI were the most vital parameters in the prediction. Basic anthropometric parameters and AT values based on MRI are helpful in predicting OSA and obstruction location using artificial intelligence.

Keywords: obstructive sleep apnoea; MRI; obesity; parapharyngeal adipose tissue; artificial intelligence; drug-induced sleep endoscopy

1. Introduction

Obstructive sleep apnoea is the most common sleep-related breathing disorder and, in unattended cases, is a major public health problem due to the background comorbidities [1]. Its increasing prevalence can be explained by the dynamic increase in obesity, the most crucial risk factor for OSA [2]. The prevalence of obesity has tripled since 1975. In 2016, 39%

of adults were overweight and 13% obese, representing 1.9 billion and 650 million people, respectively [3]. Obesity is typical for developed countries, and explained by increased calorie intake, physical inactivity, and changes in the gut microbiome [4]. In addition to OSA, obesity is also a risk factor for other conditions, such as insulin resistance, diabetes mellitus, hypertension, atherosclerosis, stroke, or myocardial infarction [5]. Obesity can be classified into visceral and general types, of which visceral obesity is critical due to its decreasing effects on lung volumes and pharyngeal wall tension [6]. Although the pathophysiology of OSA is complex and multifactorial, impaired dilator muscle functions, ineffective loop gain, and low arousal threshold are its essential background [7]. Upper-airway obstruction can be the result of deposits of adipose tissue near the upper airways, which can be examined by CT or MRI. The significance of the parapharyngeal adipose corpus was first mentioned by Wlofram-Gabel et al., in 1996 [8]. The parapharyngeal region, the tongue, and subcutaneous adipose tissue of the neck lead to upper-airway obstruction in different ways. The correlations between OSA, obesity, and anthropometric parameters have been particularly investigated in several studies to analyse the pathophysiology of OSA in a more detailed manner and to predict OSA. Of the anthropometric parameters, BMI, neck, abdominal, and hip circumferences and waist–hip ratio are mainly investigated. In the recent ELSA-Brasil study, which included 2059 patients, all parameters mentioned above were found to be significantly higher in the OSA group than in the non-OSA group [9]. The Sleep Heart study, which included 6167 patients, observed significantly higher BMI and neck and hip circumferences in the case of severe OSA. However, it was also concluded that the BMI cut-off does not precisely represent the severity of obesity in different races and sexes [10]. The predictive role of anthropometric parameters in OSA depends on the sexes. In women, waist circumference and waist-to-height ratio were the most crucial parameters in predicting OSA, while in men, neck circumference and waist-to-height ratio were crucial in predicting mild OSA, and BMI in severe OSA [11]. The Wisconsin Sleep Cohort study, conducted in the USA, including 1520 participants between 30 and 70 years of age, observed a higher prevalence of sleep-related breathing disorders in older men with higher BMI values. Moreover, BMI was also the most strongly correlated with sleep-disordered breathing in younger participants [12].

The anthropometric parameters and the parameters of the adipose tissue near the upper airways can be analysed using medical imaging methods, resulting in large databases. The complex pathophysiology behind OSA cannot be described using simple statistical methods in all cases. Given the fast improvement in sciences, using artificial intelligence is advantageous in diagnostics, prediction, and therapy. Although many possibilities regarding OSA diagnostics are accessible (e.g., self-administered questionnaires, home sleep tests, or polysomnography), the ratio of undiagnosed cases is still high.

In the last two decades, the improvement in bioinformatics and artificial intelligence has allowed easy and rapid detection of OSA. At first, machine learning-based models included essential risk factors for OSA (i.e., age, sex, BMI, or neck circumference) [13,14], while others performed a prediction using anthropometric and faciocervical measurements [15]. Regarding the methods, the most vital expectations were simplicity and rapidity; therefore, ECG [16] and oxyhaemoglobin saturation [17], despite their effectiveness in prediction, cannot be integrated into daily practice. Other studies investigated the use of artificial intelligence to predict OSA using questionnaires [18]. The prediction of OSA was also successful by 2D imaging [19] and 3D face reconstruction using artificial intelligence [20].

In the present study, in addition to anthropometric parameters, the upper-airway adipose tissue was examined using MRI in an OSA population to analyse its effects on OSA pathogenesis. Furthermore, the prediction of OSA and upper-airway obstruction, including the parameters mentioned above, was also investigated using artificial intelligence (the Flexible Discriminance analysis and the Multivariate Adaptive Regression Splines).

2. Materials and Methods

2.1. Participants

This prospective investigation was conducted at the Department of Otolaryngology and Head and Neck Surgery of Semmelweis University, and included one hundred participants (74 men and 26 women, mean age ± SD, 42.15 ± 11.7 years). Those over 18 years of age with snoring or suspected OSA, who gave their consent to participate in the investigation, were enrolled. Those who previously had oral or otorhinolaryngological surgeries, those who had craniofacial malformations (e.g., Down syndrome), had claustrophobia, soft tissue or thyroid gland disorders, neurological or psychiatric diseases, and those with alcohol or drug abuse or pregnancy were excluded. All participants were examined using a general otorhinolaryngological examination, a sleep test (i.e., polysomnography), drug-induced sleep endoscopy, and MRI of the neck region. The flow chart is presented in Figure 1.

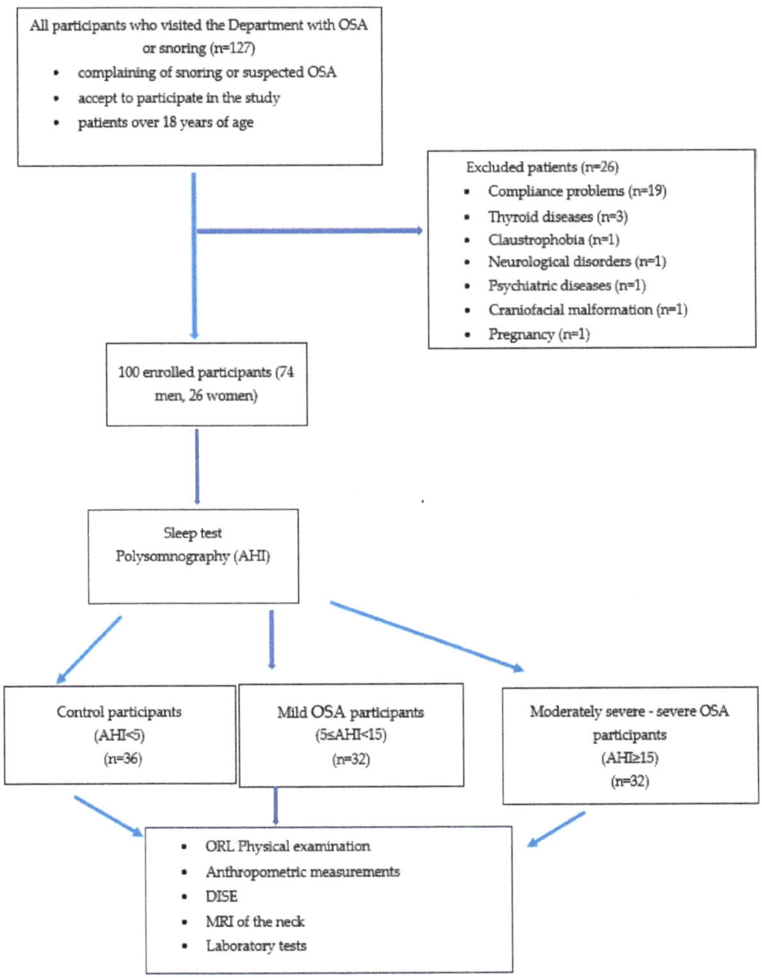

Figure 1. Study participants' flow chart. AHI = apnoea-hypopnoea index, DISE = drug-induced sleep endoscopy, MRI = magnetic resonance imaging, OSA = obstructive sleep apnoea, ORL = otorhinolaryngological.

The study was approved by the Hungarian Research Ethics Authority (National Institute of Pharmacy and Nutrition, approval reference number: 2788/2019). All patients gave their informed consent in writing.

Figure 1 shows the study population's flow chart.

2.2. Anthropometric Measurements

Participants' general anthropometric parameters, such as gender, age, body height, weight, and BMI, were calculated. Neck circumference was measured in the cricothyroid membrane, hip circumference in the anterior superior iliac spine, and abdominal circumference in the umbilicus, using a tape measure in each case.

2.3. Sleep Test

A SOMNOscreen Plus PSG device (SOMNOmedics GMBH Germany) was applied for overnight polysomnography, at the Institute of Pulmonology Törökbálint, under medical supervision. The examination results were adapted according to the American Academy of Sleep Medicine. Apnoea is determined as a reduction of 90% or more airflow through the oronasal thermistor for 10 s or more and hypopnoea as a reduction of 30% or more airflow, accompanied by a desaturation or arousal of 3% or more oxyhaemoglobin. The severity of OSA can be classified according to the apnoea-hypopnoea index (AHI) [21]. Due to the relatively low number of participants, they were classified into control (AHI < 5), mild OSA ($5 \leq$ AHI < 15), and moderately severe–severe OSA (AHI ≥ 15) groups.

2.4. MRI

The MRI examinations were performed at the Medical Imaging Centre of Semmelweis University, using a Philips Ingenia 1.5 T MRI device. Neck MRI was conducted using coronal T1 TSE (with 3.5 mm slice thicknesses without a gap), axial T2 SPIR, T1 TSE, and DWI measurements (with 3 mm slice thicknesses with a 1.5 mm gap), and sagittal T2 TSE, STIR, and T1 TSE analyses. Examinations were performed from the posterior nasal spine to the hyoid bone. Participants were instructed to breathe normally through their nose and avoid movements and swallowing. The images were analysed by an experienced radiologist using a Philips IntelliSpace Portal (Philips Healthcare, Best, The Netherlands).

Parapharyngeal adipose tissue was defined as the largest extent of parapharyngeal fat tissue in the axial plane on both sides of the parapharyngeal wall and its areas were calculated using the region of interest (ROI) tool in the DICOM viewer. The thicknesses were also determined in the axial plane, using T1-weighed measurements using the ruler tool of the software. To calculate the estimated percentage of tongue fat, the area of the tongue in the midsagittal axis of the T1-weighted images with the ROI tool was also measured. In the next analysis step, the well-defined contiguous areas of the tongue adipose tissue in the same plane were differentiated using the same tool. Then, the tongue area and the areas of the tongue fat tissue were compared and a rough percentage of the ratio of the adipose tissue to the tongue was calculated from the measured data. The neck SAT was determined in the midsagittal region and was of the parapharyngeal AT in the axial plane, using T1-weighed measurements and the contour of the analysed region [22–24]. The MRI parameters are presented in Figure 2.

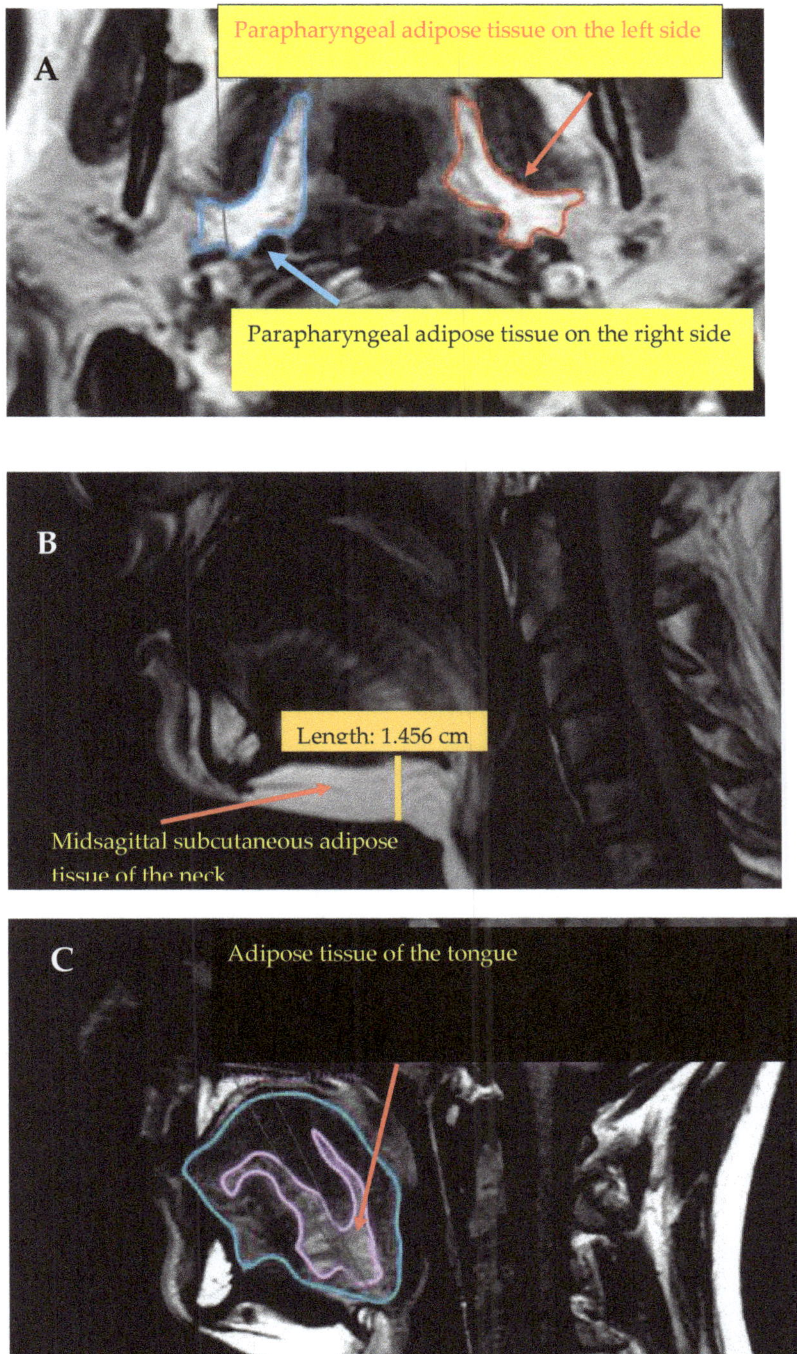

Figure 2. (**A**) T1-weighed measurements in the axial plane showing parapharyngeal adipose tissue on the left and right sides; (**B**,**C**) midsagittal axis of the T1-weighted MRI scans showing the midsagittal subcutaneous adipose tissue of the neck and adipose tissue of the tongue (taken from our data).

2.5. Drug-Induced Sleep Endoscopy

Drug-induced sleep endoscopy was performed in an operating room. A quantity of 1.5 mg per kilogram of propofol was applied for sedation and an Olympus flexible endoscope was inserted through the nose to the larynx. The results were adapted according to the VOTE classification, making it possible to determine precisely the location, severity, and configuration of the obstruction. Consequently, location could be determined as 'V', velum; 'O', oropharynx; 'T', tongue base; and 'E', epiglottis. The severity of obstruction could be 0, indicating that there is no obstruction; '1', indicating a partial obstruction; or '2', a total obstruction. 'X' means that the obstruction cannot be visualised. Configuration 'L' means a lateral, 'AP' an anteroposterior, and 'C' a concentric type of obstruction [25].

2.6. Data Processing

The correlations were examined using linear regression analysis. The differences between the grouping criteria were examined using the one-way analysis of variance (ANOVA). A critical condition of ANOVA is the homogeneity of variances, which was tested by Levene's test. The test values were above the critical level of 0.05; therefore, the homogeneity of variance criterion was satisfied. Based on this, the differences between groups were analysed by the Bonferroni test. This is the most widely used test for multiple comparisons, and is capable of detecting differences similar to the relatively conservative tests (e.g., Schaffé's S test) [26].

Our original idea was to determine the relationships between AHI and different parameters of the anthropometric and adipose tissue by a multivariable regression equation. However, the correlations between the parameters of the adipose tissue (i.e., independent variable) and the AHI values (i.e., dependent variable) were seen to be weak. This can be explained by the relatively high variance of the independent variables. Similar results were also obtained using the linear quadratic discriminant analysis. Consequently, more robust methods were selected which are less sensitive to the variance of the input variables. Therefore, each input variable was classified into three equal groups. In this case, the traditional logistic regression method was unable to be applied, as the number of empty cells was more than two-thirds of the total cells. At the same time, the unpredictable development of artificial intelligence, generally, and specifically machine learning, offers new tools for automatic, supervised patients' classification in the case of AHI and obstructions based on their demographic, anthropometric, and adipose tissue parameters. Different classification algorithms have been tested, indicating an abundance of methods [27]. The efficiency of the 'classic' machine learning algorithms (e.g., random forest method) was found to be relatively poor. Therefore, the methods and algorithms developed for classification problems in chemometrics were selected because, in this field, the number of independent variables (inputs) is generally relatively high, compared to the number of samples (records) [28]. Consequently, it was possible to find algorithms to classify patients according to input parameters with unexpected efficiency. The most favourable results were produced using the Flexible Discriminance analysis and the Multivariate Adaptive Regression Splines. The algorithms can be found in the freely downloadable 'mda' R-package [29]. Garson's method was used to test the relative significance of each parameter [30]. The cross-validation index was applied to detect the location of the obstruction [31].

The present investigation, such as all academic endeavours, had to combine the ambition to achieve well-founded results with the restriction of limited resources. Based on our preliminary calculations, following the recommendations of Shuster, an ideal sample should be three–five times larger [32]. This is a considerable difference, but we did not achieve one order of magnitude. The relatively low but not extraordinarily small sample size is an inherent limitation of the generalisability of results, although modern statistical methods, mainly bootstrapping, offer a favourable possibility to evaluate the robustness of results [33]. However, the application of cross-validation methods in sampling in the case of classifications considerably contributes to increasing reliability. Notwithstanding this, it should be noted that our results can only be considered to be preliminary. Validity must be further analysed and improved by increasing patients' numbers and involving other races with different craniofacial and obesity characteristics.

3. Results

Of the 100 participants, 36 belonged to the control, and 32 to the mild and 32 to the moderately severe–severe OSA group. A male predominance was observed in all groups. Additionally, a higher ratio of participants under 40 years of age and a lower BMI was detected in the control group. Patients over 40 years of age and with higher BMI values were found in the OSA groups.

3.1. Basic Demographic Values, Laboratory Test, and AT MRI Parameters

The groups' basic demographic values, laboratory test parameters, and AT MRI parameters are summarised in Table 1.

Table 1. Patients' basic demographic, MRI, and laboratory test results. The parameters show the mean ± SD values. *** indicates the significant difference at $p < 0.01$ level, while ** the significant difference at $p < 0.05$ level and * the significant differences at $p < 0.1$.

Indicators	Control Group $n = 36$ (A)	Mild OSA $n = 32$ (B)	Moderately Severe + Severe OSA $n = 32$ (C)	p-Value	Differences
Age (years)	38.42 ± 12.13	45.34 ± 11.17	43.16 ± 10.9	0.042 **	A-B
Weight (kg)	78.94 ± 13.15	93.03 ± 14.58	101.97 ± 17.21	0.000 ***	A-B; A-C; B-C
Hip circumference (cm)	100.49 ± 11.97	106.28 ± 10.46	111.15 ± 10.96	0.001 ***	A-C
Abdominal circumference (cm)	94.73 ± 12.7	104.97 ± 11.68	111.01 ± 12.51	0.000 ***	A-B; A-C
Neck circumference (cm)	37.95 ± 4.12	40.69 ± 3.42	42.73 ± 3.33	0.000 ***	A-B; A-C; B-C
Tongue fat midsagittal (cm^2)	824.82 ± 159.43	928.64 ± 154.39	933.66 ± 176.27	0.01**	A-B; A-C
Tongue fat (%)	0.33 ± 0.05	0.33 ± 0.05	0.33 ± 0.06	0.957	No significant difference
Midsagittal SAT of the neck (mm)	6.1 ± 1.69	6.62 ± 1.75	7.26 ± 1.57	0.019 **	A-C
Parapharyngeal AT on the right side (cm^2)	253.47 ± 62.88	269.89 ± 64.37	304.58 ± 61.64	0.004 **	A-C; B-C
Parapharyngeal AT on the left side (cm^2)	256.19 ± 63.67	285.45 ± 83.97	311.63 ± 60.54	0.006 **	A-C
Sum of the parapharyngeal AT (cm^2)	509.66 ± 121.96	555.34 ± 141.88	616.21 ± 110.07	0.003 **	A-C
Total cholesterol (mmol/L)	5.59 ± 1.15	5.9 ± 1.17	5.47 ± 1.01	0.279	No significant difference
HDL-cholesterol (mmol/L)	1.31 ± 0.29	1.22 ± 0.32	1.13 ± 0.17	0.024 **	A-C
LDL-cholesterol (mmol/L)	3.58 ± 0.81	3.89 ± 0.82	3.7 ± 0.8	0.282	No significant difference
Triglycerides (mmol/L)	1.82 ± 1.33	2.58 ± 1.8	2.04 ± 1.17	0.097 *	A-B

As Table 1 reveals, in the case of the anthropometric and most MRI parameters, a significant difference was observed between the OSA and control groups. Of the parameters examined, only the values of tongue fat%, total cholesterol, and LDL-cholesterol did not differ between the groups.

3.2. Correspondence between Demographic and MRI Parameters along with AHI and BMI

The correspondence between demographic and MRI parameters, along with AHI and BMI, is presented in Table 2.

Table 2. Pearson's correlation coefficients (r^2) between basic demographic values and MRI, and AHI and BMI. The parameters show the correlation coefficients. ** indicates the significant difference at $p < 0.05$ level.

	Correlation with	
Indicators	**BMI (kg/m²)**	**AHI (events/hour)**
	***p*-Value**	***p*-Value**
Age (years)	0.015	0.066
Hip circumference (cm)	0.793 **	0.011
Abdominal circumference (cm)	0.872 **	−0.014
Neck circumference (cm)	0.357 **	−0.035
Tongue fat midsagittal (cm²)	0.358 **	−0.035
Tongue fat (%)	0.146	0.022
Midsagittal SAT of the neck (mm)	0.509 **	−0.167
Parapharyngeal AT on the right side (cm²)	0.311 **	−0.067
Parapharyngeal AT on the left side (cm²)	0.299 **	−0.125
Sum of the parapharyngeal AT (cm²)	0.322 **	−0.103

As Table 2 reveals, most anthropometric and BMI parameters were correlated with BMI, of which the correlations with abdominal and hip circumferences and midsagittal neck SAT were the strongest. However, no significant correlations with AHI were observed.

3.3. Prediction of Velopharyngeal Obstruction

The prediction of velopharyngeal obstruction using anthropometric and MRI parameters is summarised in Tables 3 and 4.

Table 3. Prediction of velopharyngeal obstruction using anthropometric and MRI parameters. The table presents the categorisation of real and predicted velopharyngeal obstruction, along with number of patients in each group.

		Reference	
		Non-Velopharyngeal Obstruction	**Velopharyngeal Obstruction**
Prediction	Non-Velopharyngeal Obstruction	16	5
	Velopharyngeal Obstruction	15	64

Table 4. Relative significances of the parameters in the prediction of velopharyngeal obstruction, applying two statistical approaches (i.e., general cross-validation and residual sum-square methods). The table shows the role of different factors in predicting velopharyngeal obstruction, indicating the relative importance of each parameter.

Number of Subsets	General Cross Validation	Residual Sum Squares
Sum of the parapharyngeal AT (cm²)	100.0	100.0
Neck circumference (cm)	53.8	65.9
Age (years)	30.7	47.1

As Tables 3 and 4. present, in the case of both of the above-mentioned statistical methods, the most vital parameters were the parapharyngeal AT, followed by the circumference and age. Velopharyngeal obstruction could be predicted in 80% of the cases using these parameters. In the other 16% of the cases, the algorithm incorrectly predicted velopharyngeal obstruction, and, in 5%, the algorithm did not detect the presence of obstruction.

3.4. Prediction of Oropharyngeal Obstruction

The correlation between oropharyngeal obstruction, and anthropometric and MRI parameters and their predictive values, is presented in Tables 5 and 6.

Table 5. Prediction of oropharyngeal obstruction using anthropometric and MRI parameters. The table presents the categorisation of real and predicted oropharyngeal obstruction, together with patient numbers in each group.

		Reference	
		Non-Oropharyngeal Obstruction	Oropharyngeal Obstruction
Prediction	Non-Oropharyngeal obstruction	20	9
	Oropharyngeal obstruction	14	57

Table 6. Prediction of oropharyngeal obstruction using anthropometric and MRI parameters, by general cross-validation index. The table shows the role of the different parameters in predicting oropharyngeal obstruction in%, indicating the relative importance of each parameter.

Indicators	General Cross-Validation Index Value	Residual Sum of Squares
BMI (kg/m^2)	100.0	100.0
Sum of the parapharyngeal AT (cm^2)	44.1	68.4
Abdominal circumference (cm)	44.7	57.2

As can be seen from Table 5, the prediction of oropharyngeal obstruction was efficient in 77% of cases using anthropometric and MRI parameters. In the other 14% of cases, the algorithm indicated false obstruction, and in 9%, it was unable to predict the presence of obstruction.

The relative importance of different parameters in predicting obstruction is summarised in Table 6.

Table 6 shows that BMI played the most crucial role in the prediction, although abdominal circumference and the sum of parapharyngeal AT also contained important information. The relatively low confidence interval indicates that the possibility of an obstruction in the case of higher BMI values is relatively high. In the case of lower BMI values, the sum of the parapharyngeal AT parameter is essential to predict obstruction.

3.5. Prediction of Tongue-Based Obstruction

The prediction of tongue-based obstruction applying anthropometric and MRI parameters is summarised in Table 7.

Table 7. Prediction of tongue-based obstruction, including anthropometric and MRI parameters. The table presents the 100 patients' real and predicted categorisation of obstruction along with patient numbers in each group.

		Reference	
		Non-Tongue-Based Obstruction	Tongue-Based Obstruction
Prediction	Non-Tongue-based obstruction	12	4
	Tongue-based obstruction	17	67

As shown in Table 7, the tongue-based obstruction could be predicted in 79% of cases, using anthropometric and MRI parameters. However, the algorithm indicated a false obstruction in 17% of cases and a false negative obstruction in 4% of cases.

3.6. Prediction of OSA

The prediction of OSA, applying anthropometric, laboratory test, and MRI parameters, and the relative significance of each parameter, were analysed using artificial intelligence (i.e., the Garson test). The efficiency of OSA categorisation is presented in Table 8, while the relative significance of each parameter is presented in Table 9.

Table 8. Prediction of OSA categories, including anthropometric, laboratory test, and MRI parameters. The table presents 100 patients' real and predicted OSA categorisation, along with patient numbers in each group.

Estimated	OSA Categories		
	Control	**Mild OSA**	**Moderately Severe + Severe OSA**
Non-OSA	34		
Mild OSA	1	33	
Moderately severe + Severe OSA			32

Table 9. Relative importance of different factors (%) by multivariate discriminant analysis in the prediction of OSA subcategories, including anthropometric, laboratory test, and MRI parameters. The table shows the role of the different parameters in predicting OSA in%, indicating the relative importance of each parameter.

Indicators	Importance (%)
Age (years)	10.8
Tongue fat midsagittal (cm^2)	7.8
Neck circumference (cm)	7.7
Triglycerides (mmol/L)	7.5
HDL-cholesterol (mmol/L)	7.3
Parapharyngeal AT on the left side (cm^2)	7.1
Total cholesterol (mmol/L)	6.8
Hip circumference (cm)	6.1
BMI (kg/m^2)	5.65
Weight (kg)	4.9
Height (cm)	4.75
Abdominal circumference (cm)	4.5
LDL-cholesterol (mmol/L)	4.4
Tongue fat %	4.1
Parapharyngeal AT on the right side (cm^2)	3.8
Gender	3.6
Midsagittal SAT of the neck (mm)	3.2

Including anthropometric, laboratory test, and MRI parameters, using artificial intelligence, the presence of OSA could be predicted in 99% of cases. This means that the algorithm performed a false calculation in only one non-OSA case. To validate the results, the data were randomly divided into 'teaching' and 'test' parts in a 75:25 ratio. After hundreds of analyses, the average OSA prediction was over 90%. Age, tongue fat%, and neck circumference were determined as the essential parameters in the prediction, followed by laboratory test parameters (i.e., triglyceride and HDL-cholesterol). Left-sided parapharyngeal AT was also essential in the prediction, preceding other parameters, such as BMI and hip circumference.

4. Discussion

OSA affects a significant proportion of society and the ratio of undiagnosed cases is high; therefore, its diagnosis must be improved. Although many possibilities regarding the diagnosis of OSA are accessible, the earliest diagnosis is essential, due to the appearance of comorbidities. The primary purpose of diagnostic methods is to easily and quickly screen for OSA or diagnose the disorder with high specificity and sensitivity. Efficient screening is possible using self-administered questionnaires, of which the STOP-BANG (i.e., snoring, tiredness, observed apnoea, high blood pressure, BMI, age, neck circumference, and male gender) is generally used with reliable results. This questionnaire contains eight questions, and patients can answer with 'yes' or 'no' [34]. In addition, screening is also possible using the Berlin, Epworth, or STOP questionnaires. A meta-analysis that included 108 investigations with 47,989 participants determined a significantly higher sensitivity of the STOP-BANG questionnaire, although its specificity was lower than that of the Epworth questionnaire [35]. An alternative diagnostic approach is a home-sleep test (HST, Types III or IV), which can be effectively used when there is a high risk of moderate or severe OSA. The one-night polysomnography, in which sleep specialists interpret the results, is essential in the follow-up of the effectiveness of therapy and the diagnosis of OSA [36]. Notwithstanding the relatively low specificity of the questionnaires and their time requirement, and the necessity for qualified staff for home sleep tests, alternative methods, e.g., using artificial intelligence, are necessary. Previous results indicated that anthropometric parameters, the Epworth questionnaire, and expired gas analysis using machine learning could effectively predict OSA; only in 5.7% was a false mild instead of severe classification found [13]. The prediction of OSA based on machine learning was improved when the model was completed with physical examination parameters [14]. The prediction based on anthropometric and craniofacial parameters and the STOP-BANG questionnaire was more efficient in cases of moderate to severe OSA with no daytime symptoms [15]. Based on the correspondences mentioned above, the examination of vital OSA risk factors is not only essential regarding OSA pathophysiology, but using modern statistical methods (e.g., artificial intelligence), their role in OSA prediction can also be analysed.

The correlation between OSA and obesity is highly complex and has been particularly investigated; however, there are still some questions remaining. The present study aimed to investigate the role of anthropometric and AT MRI parameters of the neck, tongue, and parapharyngeal regions in the pathogenesis and prediction of OSA, plus the obstruction and location of the upper airways. Determining the correlation between BMI and AT is relatively easy; however, the correspondence between OSA and AT is more complex, as these correlations are not intuitive and cannot be described using simple functions. Therefore, other methods must be applied to analyse the correspondence between AT and anthropometric parameters (i.e., independent variables) and OSA (i.e., dependent variable) and predictive values. Consequently, artificial intelligence (i.e., Flexible Discriminance analysis and Multivariate Adaptive Regression Splines) was applied in our analyses.

The significance of the present investigation is that the use of artificial intelligence in OSA diagnostics on a relatively large sample was analysed.

Based on the fact that obesity is one of the most critical risk factors for OSA and is also correlated depending on age and sex with the severity of OSA, including anthropometric and AT MRI parameters of the upper airways, the severity categories of OSA could be correctly determined in 99% of cases. In the prediction, gender and hip circumference showed the most vital role. Carlisle et al. also observed the effect of age on pharyngeal morphology. In the case of older males, a higher retropharyngeal and retroglossal length was observed, along with the cross-sectional area of the soft palate and the diameter and cross-sectional area of the parapharyngeal fat pad, in that study [37,38]. The effect of age is also presented in increased genioglossus muscle activity in older awake males [39], which decreases during sleep, leading to vulnerability and collapsibility of the upper airways [40]. In the OSA prediction, age was a key factor based on the results of the current investigation, although no significant correlation with AHI and BMI was observed. Neck circumference was defined as the second essential parameter in the OSA prediction of anthropometric parameters. The literature contains conflicting data on neck circumference in OSA; some investigations have concluded a strong correlation between neck circumference and OSA severity [41,42], while others have not [43,44]. A meta-analysis regarding the correlation between neck circumference and obesity stated a sensitivity of neck circumference in the prediction of obesity of 80% and a specificity of 85% [45]. Neck circumference was defined to correlate with age, BMI, and hip and waist circumference, in both men and women [46].

In the OSA prediction, the algorithm indicated the anthropometric and left-sided parapharyngeal and tongue fat midsagittal MRI parameters as being the most crucial, showing a strong correlation with obesity. Accumulation of AT near the upper airways (i.e., tongue, parapharyngeal space, and central region) in obesity leads to increased collapsibility of the pharynx by mechanical effects and based on neuromuscular regulations in the central nervous system [47]. Compared to other somatic muscles, AT accumulation in the tongue showed a higher correlation with BMI and therefore, with obesity severity, which is strongly correlated with OSA severity [48]. Jugé et al. found similar results, and observed a significant positive correlation between tongue AT and BMI and older ages [49]. Our results showed that tongue fat% did not significantly differ between OSA categories and the control group, but the midsagittal region fat% parameter did. The tongue fat% neither correlated with AHI nor BMI; however, tongue midsagittal region fat% significantly correlated with BMI. Kim et al. observed a significant positive correlation between tongue fat volumes and AHI and BMI. Furthermore, a higher percentage of tongue fat% in the OSA group was detected; however, there was no significant difference compared to the control group [50]. Parapharyngeal AT is strongly correlated with obesity, highlighted by the correlation between parapharyngeal AT and BMI. However, no correlation with AHI was detected. Consequently, parapharyngeal AT parameters contained essential information in the algorithm; however, they did not significantly correlate with OSA severity. Chen et al. detected a significant correlation between AHI and the subglosso-supraglottic-level parapharyngeal fat pad, independently of BMI and neck circumference parameters [51]. According to Gao et al., in patients with a BMI over 28 kg/m^2, a significant positive effect of age on parapharyngeal AT volumes was detected [52].

In predicting velopharyngeal obstruction, the algorithm determined parapharyngeal AT as the most vital parameter, followed by neck circumference and age. Using these parameters, by artificial intelligence, the velopharyngeal obstruction could be correctly detected in 80% of cases. Jang et al. detected a higher percentage of retropalatal concentric obstruction in patients with OSA with higher parapharyngeal AT volumes [53]. This is in agreement with our results, referring to the significant role of parapharyngeal AT in velopharyngeal obstruction.

The importance of parapharyngeal AT in predicting oropharyngeal obstruction was not found, in contrast to the prediction of velopharyngeal obstruction, since BMI was indicated as the most crucial parameter, followed by abdominal circumference and parapharyngeal AT, with the latter two showing the same importance. Applying these three parameters, using artificial intelligence, the oropharyngeal obstruction could be predicted in 77% of cases. However, interestingly, the algorithm did not determine the other anthropometric and MRI parameters that are essential for prediction. Pahkala et al. highlighted the importance of increased lateral pharyngeal collapsibility associated with accumulation of parapharyngeal adipose tissue in obese patients, explained by impaired mechanisms controlling passive collapse of the pharyngeal wall [54]. However, Li et al. determined the increased mechanical loading of parapharyngeal AT on the lateral pharyngeal wall as a possible background [22]. Chen et al. observed a strong correlation between subglosso-supraglottic-level AT and lateral pharyngeal obstruction at the same level [51].

Regarding tongue-based obstruction, the higher tongue volume, the adipose deposits accumulated in the tongue, and the decreased muscle activity during sleep can be defined, and are also negatively influenced by the accumulated intramuscular AT [55]. To predict tongue-based obstruction, BMI was defined as the most vital parameter; thus, tongue-based obstruction could be predicted in 79% of cases. The correspondence between BMI and tongue volumes is highly complex; some researchers have indicated a strong correlation between them [56], while others have not [50]. According to our investigation results, tongue volumes significantly correlated with AHI and BMI in both sexes, while tongue fat significantly correlated with BMI.

Finally, it can be concluded that an MRI of the adipose tissue surrounding the upper airways can be an alternative examination of OSA when an MRI in the neck region is used with another indication other than OSA. Both OSA and velopharyngeal obstruction can be predicted using artificial intelligence. Compared to self-administered questionnaires, an essential advantage of our algorithm is that the location of obstruction can be identified with high precision and the examination is relatively fast compared to the home sleep test. Our results are especially crucial in cases where MRI was performed and was previously not diagnosed.

The present investigation had some limitations. First, the relatively low number of participants did not allow for the division of OSA into categories based on its severity. Moreover, the magnetic resonance examinations were performed on awake subjects and, therefore, did not present the situations during physiological sleep.

5. Conclusions

Based on the results of the present investigation, the MRI-based AT and anthropometric parameters were not significantly correlated with OSA severity; however, a significant correlation with BMI was detected. Parapharyngeal AT plays a significant role in the presence of velopharyngeal obstruction and OSA pathophysiology. However, it has a limited effect on oropharyngeal obstruction; moreover, it does not affect tongue-based obstruction. The BMI was defined as the most vital parameter of oropharyngeal and tongue-based obstruction; furthermore, its role in OSA pathophysiology is also significant. Neck circumference is essential to predict velopharyngeal obstruction and OSA, and abdominal circumference to predict oropharyngeal obstruction and OSA. In predicting OSA, age was determined as the most vital parameter, followed by the tongue fat midsagittal and neck circumference parameters, which had the same importance. In conclusion, using anthropometric and MRI AT parameters, by artificial intelligence, OSA and upper-airway obstruction can be predicted in 99% of cases, velopharyngeal obstruction in 80%, oropharyngeal obstruction in 77%, and tongue-based obstruction in 79%.

Author Contributions: Conceptualization, V.M., Z.L., D.L.T. and Á.D.T.; methodology, V.M., Z.L., D.L.T. and Á.D.T.; software, Z.L., D.L.T., Á.D.T. and Z.J.; validation, Z.L.; formal analysis: V.M.; investigation, V.M., L.K., D.L.T., Á.D.T., Z.J.; resources, L.T.; data curation, Z.L.; writing—original draft preparation, V.M.; writing—review and editing, V.M., Z.L., A.M., D.L.T., Á.D.T. and L.K.; visualization, V.M. and Z.J.; supervision, Z.L., A.M., D.L.T. and Á.D.T.; project administration, V.M. All authors have read and agreed to the published version of the manuscript.

Funding: This research received no external funding.

Institutional Review Board Statement: The study was conducted according to the guidelines of the Declaration of Helsinki and approved by the Hungarian Research Ethics Authority (National Institute of Pharmacy and Nutrition, approval reference number: 2788/2019).

Informed Consent Statement: Informed consent was obtained from all subjects involved in the study.

Data Availability Statement: Data supporting reported results can be provided upon reasonable request.

Acknowledgments: We would like to acknowledge the help of co-workers in organising (Fruzsina Németh, Emese Angyal, Helga Szabó, András Dienes) and additional medical staff of Semmelweis University in the study. We would like to thank the support of MTA (Hungarian Academy of Sciences) Bólyai scholarship program, ÚNKP-20-5 and ÚNKP-21-5 New National Excellence Program of the Ministry for Innovation and Technology, from the source of the National Research, Development and Innovation Fund (Á.D.T. and D.L.T.).

Conflicts of Interest: The authors declare no conflict of interest.

References

1. Lévy, P.; Kohler, M.; McNicholas, W.T.; Barbé, F.; McEvoy, R.D.; Somers, V.K.; Lavie, L.; Pépin, J.-L. Obstructive sleep apnoea syndrome. *Nat. Rev. Dis. Prim.* **2015**, *1*, 15015. [CrossRef]
2. Popkin, B.M.; Adair, L.S.; Ng, S.W. Global nutrition transition and the pandemic of obesity in developing countries. *Nutr. Rev.* **2012**, *70*, 3–21. [CrossRef] [PubMed]
3. World Health Organization (WHO). *Obesity and Overweight. Fact Sheet N 311*; WHO: Geneva, Switzerland, 2012; Available online: http://www.who.int/mediacentre/factsheets/fs311/en/ (accessed on 9 June 2021).
4. Prentice, A.; Jebb, S. Energy intake/physical activity interactions in the homeostasis of body weight regulation. *Nutr. Rev.* **2004**, *62*, S98–S104. [CrossRef] [PubMed]
5. Panuganti, K.K.; Nguyen, M.; Kshirsagar, R.K.; Doerr, C. *Obesity (Nursing)*; In StatPearls [Internet]; StatPearls Publishing: Tampa, FL, USA, 2021.
6. Dixon, A.E.; Peters, U. The effect of obesity on lung function. *Expert Rev. Respir. Med.* **2018**, *12*, 755–767. [CrossRef] [PubMed]
7. Antonaglia, C.; Passuti, G. Obstructive sleep apnea syndrome in non-obese patients. *Sleep Breath.* **2021**, *26*, 513–518. [CrossRef] [PubMed]
8. Wolfram-Gabel, R.; Kahn, J.; Bourjat, P. The parapharyngeal adipose corpus: Morphologic study. *Surg. Radiol. Anat.* **1997**, *19*, 249–255. [PubMed]
9. Santos, R.B.; Silva, W.A.; Parise, B.K.; Giatti, S.; Aielo, A.N.; Souza, S.P.; Santos, I.S.; Lotufo, P.A.; Bensenor, I.M.; Drager, L.F. Accuracy of global and/or regional anthropometric measurements of adiposity in screening sleep apnea: The ELSA-Brasil cohort. *Sleep Med.* **2019**, *63*, 115–121. [CrossRef]
10. Vana, K.D.; Silva, G.E.; Carreon, J.D.; Quan, S.F. Using anthropometric measures to screen for obstructive sleep apnea in the Sleep Heart Health Study cohort. *J. Clin. Sleep Med.* **2021**, *17*, 1635–1643. [CrossRef] [PubMed]
11. Polesel, D.N.; Nozoe, K.T.; Tufik, S.B.; Bezerra, A.G.; Fernandes, M.T.B.; Bittencourt, L.; Tufik, S.; Andersen, M.L.; Hachul, H. Gender differences in the application of anthropometric measures for evaluation of obstructive sleep apnea. *Sleep Sci.* **2019**, *12*, 2. [CrossRef]
12. Peppard, P.E.; Young, T.; Barnet, J.H.; Palta, M.; Hagen, E.W.; Hla, K.M. Increased prevalence of sleep-disordered breathing in adults. *Am. J. Epidemiol.* **2013**, *177*, 1006–1014. [CrossRef] [PubMed]
13. Mencar, C.; Gallo, C.; Mantero, M.; Tarsia, P.; Carpagnano, G.E.; Foschino Barbaro, M.P.; Lacedonia, D. Application of machine learning to predict obstructive sleep apnea syndrome severity. *Health Inform. J.* **2020**, *26*, 298–317. [CrossRef] [PubMed]
14. Bozkurt, S.; Bostanci, A.; Turhan, M. Can statistical machine learning algorithms help for classification of obstructive sleep apnea severity to optimal utilization of polysomno graphy resources? *Methods Inf. Med.* **2017**, *56*, 308–318. [PubMed]
15. Zhang, L.; Yan, Y.R.; Li, S.Q.; Li, H.P.; Lin, Y.N.; Li, N.; Sun, X.W.; Ding, Y.J.; Li, C.X.; Li, Q.Y. Moderate to severe OSA screening based on support vector machine of the Chinese population faciocervical measurements dataset: A cross-sectional study. *BMJ Open* **2021**, *11*, e048482. [CrossRef]
16. Zarei, A.; Asl, B.M. Automatic detection of obstructive sleep apnea using wavelet transform and entropy-based features from single-lead ECG signal. *IEEE J. Biomed. Health Inform.* **2018**, *23*, 1011–1021. [CrossRef]
17. Leelaarporn, P.; Wachiraphan, P.; Kaewlee, T.; Udsa, T.; Chaisaen, R.; Choksatchawathi, T.; Laosirirat, R.; Lakhan, P.; Natnithikarat, P.; Thanontip, K. Sensor-driven achieving of smart living: A review. *IEEE Sens. J.* **2021**, *21*, 10369–10391. [CrossRef]
18. Sun, L.M.; Chiu, H.-W.; Chuang, C.Y.; Liu, L. A prediction model based on an artificial intelligence system for moderate to severe obstructive sleep apnea. *Sleep Breath.* **2011**, *15*, 317–323. [CrossRef]
19. Tsuiki, S.; Nagaoka, T.; Fukuda, T.; Sakamoto, Y.; Almeida, F.R.; Nakayama, H.; Inoue, Y.; Enno, H. Machine learning for image-based detection of patients with obstructive sleep apnea: An exploratory study. *Sleep Breath.* **2021**, *25*, 2297–2305. [CrossRef]
20. Eastwood, P.; Gilani, S.Z.; McArdle, N.; Hillman, D.; Walsh, J.; Maddison, K.; Goonewardene, M.; Mian, A. Predicting sleep apnea from three-dimensional face photography. *J. Clin. Sleep Med.* **2020**, *16*, 493–502. [CrossRef]
21. Berry, R.B.; Budhiraja, R.; Gottlieb, D.J.; Gozal, D.; Iber, C.; Kapur, V.K.; Marcus, C.L.; Mehra, R.; Parthasarathy, S.; Quan, S.F. Rules for scoring respiratory events in sleep: Update of the 2007 AASM manual for the scoring of sleep and associated events: Deliberations of the sleep apnea definitions task force of the American Academy of Sleep Medicine. *J. Clin. Sleep Med.* **2012**, *8*, 597–619. [CrossRef]
22. Li, Y.; Lin, N.; Ye, J.; Chang, Q.; Han, D.; Sperry, A. Upper airway fat tissue distribution in subjects with obstructive sleep apnea and its effect on retropalatal mechanical loads. *Respir. Care* **2012**, *57*, 1098–1105. [CrossRef]
23. Shelton, K.E.; Woodson, H.; Gay, S.; Suratt, P.M. Pharyngeal fat in obstructive sleep apnea. *Am. Rev. Respir. Dis.* **1993**, *148*, 462–466. [CrossRef] [PubMed]
24. Turnbull, C.; Wang, S.; Manuel, A.; Keenan, B.; McIntyre, A.; Schwab, R.; Stradling, J. Relationships between MRI fat distributions and sleep apnea and obesity hypoventilation syndrome in very obese patients. *Sleep Breath.* **2018**, *22*, 673–681. [CrossRef] [PubMed]
25. Stanley, J.J. Drug-induced sleep endoscopy: Techniques, interpretation and implications. *Curr. Opin. Pulm. Med.* **2020**, *26*, 623–628. [CrossRef]

26. Midway, S.; Robertson, M.; Flinn, S.; Kaller, M. Comparing multiple comparisons: Practical guidance for choosing the best multiple comparisons test. *PeerJ* **2020**, *8*, e10387. [CrossRef] [PubMed]
27. Fernández-Delgado, M.; Cernadas, E.; Barro, S.; Amorim, D. Do we need hundreds of classifiers to solve real world classification problems? *J. Mach. Learn. Res.* **2014**, *15*, 3133–3181.
28. Boehmke, B.; Greenwell, B. *Hands-On Machine Learning with R*, 1st ed.; Chapman and Hall/CRC: New York, NY, USA, 2019; pp. 54–62.
29. Hastie, M.T. *Package 'mda'. CRAN R Proj*; Springer: New York, NY, USA, 2017.
30. Garson, G.D. A comparison of neural network and expert systems algorithms with common multivariate procedures for analysis of social science data. *Soc. Sci. Comp. Rev.* **1991**, *9*, 399–434. [CrossRef]
31. Milborrow, M.S. *Package 'Earth'. R Software Package*, 2019.
32. Shuster, J.J. *CRC Handbook of Sample Size Guidelines for Clinical Trials*; CRC Press: Boca Raton, FL, USA, 2019.
33. Sauerbrei, W.; Royston, P.; Schumacher, M.; Austin, P.C.; Tu, J.V. Bootstrap Methods for Developing Predictive Models. *Am. Stat.* **2004**, *58*, 131–137: Comment by Sauerbrei, Royston, and Schumacher and Reply. *Am. Stat.* **2005**, *59*, 116–118.
34. Chung, F.; Abdullah, H.R.; Liao, P. STOP-Bang questionnaire: A practical approach to screen for obstructive sleep apnea. *Chest* **2016**, *149*, 631–638. [CrossRef]
35. Chiu, H.-Y.; Chen, P.-Y.; Chuang, L.-P.; Chen, N.-H.; Tu, Y.-K.; Hsieh, Y.-J.; Wang, Y.-C.; Guilleminault, C. Diagnostic accuracy of the Berlin questionnaire, STOP-BANG, STOP, and Epworth sleepiness scale in detecting obstructive sleep apnea: A bivariate meta-analysis. *Sleep Med. Rev.* **2017**, *36*, 57–70. [CrossRef]
36. Rosen, I.M.; Kirsch, D.B.; Chervin, R.D.; Carden, K.A.; Ramar, K.; Aurora, R.N.; Kristo, D.A.; Malhotra, R.K.; Martin, J.L.; Olson, E.J. Clinical use of a home sleep apnea test: An American Academy of Sleep Medicine position statement. *J. Clin. Sleep Med.* **2017**, *13*, 1205–1207. [CrossRef]
37. Carlisle, T.; Carthy, E.R.; Glasser, M.; Drivas, P.; McMillan, A.; Cowie, M.R.; Simonds, A.K.; Morrell, M.J. Upper airway factors that protect against obstructive sleep apnoea in healthy older males. *Eur. Respir. J.* **2014**, *44*, 685–693. [CrossRef]
38. Chebbo, A.; Tfaili, A.; Ghamande, S. Anatomy and physiology of obstructive sleep apnea. *Sleep Med. Clin.* **2013**, *8*, 425–431. [CrossRef]
39. Worsnop, C.; Kay, A.; Kim, Y.; Trinder, J.; Pierce, R. Effect of age on sleep onset-related changes in respiratory pump and upper airway muscle function. *J. Appl. Physiol.* **2000**, *88*, 1831–1839. [CrossRef]
40. Malhotra, A.; Huang, Y.; Fogel, R.; Lazic, S.; Pillar, G.; Jakab, M.; Kikinis, R.; White, D.P. Aging influences on pharyngeal anatomy and physiology: The predisposition to pharyngeal collapse. *Am. J. Med.* **2006**, *119*, 72.e9–72.e14. [CrossRef] [PubMed]
41. Ahbab, S.; Ataoğlu, H.E.; Tuna, M.; Karasulu, L.; Çetin, F.; Temiz, L.Ü.; Yenigün, M. Neck circumference, metabolic syndrome and obstructive sleep apnea syndrome; evaluation of possible linkage. *Med. Sci. Monit.* **2013**, *19*, 111. [PubMed]
42. Kawaguchi, Y.; Fukumoto, S.; Inaba, M.; Koyama, H.; Shoji, T.; Shoji, S.; Nishizawa, Y. Different impacts of neck circumference and visceral obesity on the severity of obstructive sleep apnea syndrome. *Obesity* **2011**, *19*, 276–282. [CrossRef]
43. Pływaczewski, R.; Bieleń, P.; Bednarek, M.; Jonczak, L.; Górecka, D.; Śliwiński, P. Wpływ obwodu szyi i wskaźnika masy ciała na zaawansowanie obturacyjnego bezdechu sennego u mężczyzn. *Pneumonol. Alergol. Pol.* **2008**, *76*, 313–320.
44. Endeshaw, Y. Clinical characteristics of obstructive sleep apnea in community-dwelling older adults. *J. Am. Geriatr. Soc.* **2006**, *54*, 1740–1744. [CrossRef]
45. Kroll, C.; Mastroeni, S.S.; Czarnobay, S.A.; Ekwaru, J.P.; Veugelers, P.J.; Mastroeni, M.F. The accuracy of neck circumference for assessing overweight and obesity: A systematic review and meta-analysis. *Ann. Hum. Biol.* **2017**, *44*, 667–677. [CrossRef]
46. Ben-Noun, L.; Sohar, E.; Laor, A. Neck circumference as a simple screening measure for identifying overweight and obese patients. *Obes. Res.* **2001**, *9*, 470–477. [CrossRef]
47. Schwartz, A.R.; Patil, S.P.; Laffan, A.M.; Polotsky, V.; Schneider, H.; Smith, P.L. Obesity and obstructive sleep apnea: Pathogenic mechanisms and therapeutic approaches. *Proc. Am. Thorac. Soc.* **2008**, *5*, 185–192. [CrossRef] [PubMed]
48. Nashi, N.; Kang, S.; Barkdull, G.C.; Lucas, J.; Davidson, T.M. Lingual fat at autopsy. *Laryngoscope* **2007**, *117*, 1467–1473. [CrossRef] [PubMed]
49. Jugé, L.; Olsza, I.; Knapman, F.L.; Burke, P.G.; Brown, E.C.; Stumbles, E.; de Frescheville, A.F.B.; Gandevia, S.C.; Eckert, D.J.; Butler, J.E. Effect of upper airway fat on tongue dilation during inspiration in awake people with obstructive sleep apnea. *Sleep* **2021**, *44*, zsab192. [CrossRef] [PubMed]
50. Kim, A.M.; Keenan, B.T.; Jackson, N.; Chan, E.L.; Staley, B.; Poptani, H.; Torigian, D.A.; Pack, A.I.; Schwab, R.J. Tongue fat and its relationship to obstructive sleep apnea. *Sleep* **2014**, *37*, 1639–1648. [CrossRef]
51. Chen, H.-C.; Wang, C.-J.; Lo, Y.-L.; Hsu, H.-C.; Huang, C.-G.; Kuo, I.; Lu, Y.-A.; Hsin, L.-J.; Lin, W.-N.; Fang, T.-J. Parapharyngeal fat pad area at the subglosso-supraglottic level is associated with corresponding lateral wall collapse and apnea-hypopnea index in patients with obstructive sleep apnea: A pilot study. *Sci. Rep.* **2019**, *9*, 17722. [CrossRef]
52. Gao, F.; Li, Y.; Xu, W.; An, Y.; Wang, H.; Xian, J.; Han, D. Upper airway morphological changes in obstructive sleep apnoea: Effect of age on pharyngeal anatomy. *J. Laryngol. Otol.* **2020**, *134*, 354–361. [CrossRef]
53. Jang, M.-S.; Kim, H.Y.; Dhong, H.-J.; Chung, S.-K.; Hong, S.D.; Cho, H.-J.; Jung, T.-Y. Effect of parapharyngeal fat on dynamic obstruction of the upper airway in patients with obstructive sleep apnea. *Am. J. Respir. Crit. Care Med.* **2014**, *190*, 1318–1321. [CrossRef]
54. Pahkala, R.; Seppä, J.; Ikonen, A.; Smirnov, G.; Tuomilehto, H. The impact of pharyngeal fat tissue on the pathogenesis of obstructive sleep apnea. *Sleep Breath.* **2014**, *18*, 275–282. [CrossRef]

55. Kjaer, M. Role of extracellular matrix in adaptation of tendon and skeletal muscle to mechanical loading. *Physiol. Rev.* **2004**, *84*, 649–698. [CrossRef]
56. Shigeta, Y.; Ogawa, T.; Ando, E.; Clark, G.T.; Enciso, R. Influence of tongue/mandible volume ratio on oropharyngeal airway in Japanese male patients with obstructive sleep apnea. *Oral Surg. Oral Med. Oral Pathol. Oral Radiol. Endod.* **2011**, *111*, 239–243. [CrossRef]

life

Article

A Comparison of the Reliability of Five Sleep Questionnaires for the Detection of Obstructive Sleep Apnea

Šárka Solecka [1,2,*], Karel Matler [1], Tomáš Kostlivý [3], Vojtěch Kubec [3], Hana Tomášková [2] and Jaroslav Betka [4]

1 Department of Otorhinolaryngology, Hospital in Frýdek-Místek, 73801 Frýdek-Místek, Czech Republic
2 Department of Epidemiology and Public Health, Faculty of Medicine, University of Ostrava, 70103 Ostrava, Czech Republic
3 Department of Otorhinolaryngology, Faculty of Medicine in Pilsen, University Hospital in Pilsen, Charles University, 30000 Pilsen, Czech Republic
4 Department of Otorhinolaryngology and Head and Neck Surgery, First Faculty of Medicine, University Hospital Motol, Charles University, V Uvalu 84, 15006 Prague, Czech Republic
* Correspondence: sarkasolecka@gmail.com

Abstract: The aim of this study was to compare the reliability of five sleep questionnaires in detecting the occurrence of obstructive sleep apnea (OSA). The study was conducted on a group of 201 patients. The patients completed five sleep questionnaires: the Epworth Sleepiness Scale (ESS), the STOP-Bang questionnaire, the STOP questionnaire, the Berlin questionnaire (BQ) and the Pittsburgh Sleep Quality Index (PSQI). Subsequently, the patients were examined using limited polygraphy, and the sensitivity and specificity of the questionnaires were evaluated. The STOP-Bang, Berlin and STOP questionnaires had the highest sensitivity for OSA detection (81.6%, 78.7%, and 74.2%, respectively), while the sensitivities of PSQI and ESS were low (50.8% and 34.5%). The ESS, STOP-Bang, STOP and Berlin questionnaires had the highest specificity (82.6%, 75%, 61.9%, and 61.9%). In our sample, we found the STOP-Bang and Berlin questionnaires to be the most suitable for OSA screening with the highest sensitivities (81.6%, 78.7%) and satisfactory specificities (75%, 61.9%). The STOP questionnaire was also relatively reliable, especially given its time-saving nature; though short, it preserved satisfactory sensitivity (74.2%) and specificity (61.9%). The ESS and PSQI were unsuitable for OSA screening.

Keywords: obstructive sleep apnea; Berlin questionnaire; STOP-Bang questionnaire; STOP questionnaire; Epworth Sleepiness Scale; Pittsburgh Sleep Quality Index

Citation: Solecka, Š.; Matler, K.; Kostlivý, T.; Kubec, V.; Tomášková, H.; Betka, J. A Comparison of the Reliability of Five Sleep Questionnaires for the Detection of Obstructive Sleep Apnea. *Life* **2022**, *12*, 1416. https://doi.org/10.3390/life12091416

Academic Editor: Luca Steardo

Received: 18 August 2022
Accepted: 8 September 2022
Published: 10 September 2022

1. Introduction

Obstructive sleep apnea (OSA) is the single most important preventable medical cause of excessive daytime sleepiness and driving accidents. OSA may also adversely affect work performance through a decrease in productivity and an increase in the injury rate. The odds of having a work-related accident were found to be nearly double in workers with OSA in comparison to controls [1]. An appropriate screening questionnaire for OSA could help identify high-risk workers and reduce the risk of accidents at work through therapy.

The severity of OSA is determined by the apnea hypopnea index (AHI) value (number of apneas/hypopneas per hour) and is divided into three grades of severity. An AHI range of 5–14.9 (with the presence of subjective difficulties) is indicative of mild OSA in the adult population, while patients with an AHI of 15–29.9 are considered to have moderate OSA, and those with an AHI of 30 and above are considered to have severe OSA.

The prevalence of obstructive sleep apnea (OSA) is estimated at one billion people worldwide, including over 400 million who have moderate-to-severe symptoms [2]. A number of screening methods for OSA exist: questionnaires, clinical screening models, and blood biomarkers to help identify patients with OSA [3–10]; however, until now, the gold standard for the diagnosis of OSA remains overnight monitoring performed by limited polygraphy (PG) or polysomnography (PSG).

This study aimed at comparing five established sleep questionnaires regarding their predictive probabilities for OSA: the Epworth sleepiness scale (ESS), STOP-Bang questionnaire, STOP questionnaire, Berlin questionnaire (BQ) and Pittsburgh Sleep Quality Index (PSQI).

2. Materials and Methods

2.1. Materials

In a prospective study carried out between September 2018 and March 2020, we examined a cohort of 237 consecutive patients in an outpatient clinic for snoring and sleep-disordered breathing at the ENT department. Patients were most often referred by a general practitioner, cardiologist, or an ENT physician. Some of them requested an observation following their partner's complaints and/or their partners observing sleep apnea.

Thirty-six patients were excluded from the study: three patients due to the presence of central sleep apnea, 11 patients that did not undergo a limited polygraphy examination, and 22 patients that did not complete at least 3 of the 5 questionnaires. A total of 201 patients were included in the study. We present the inclusion/exclusion process in Figure 1.

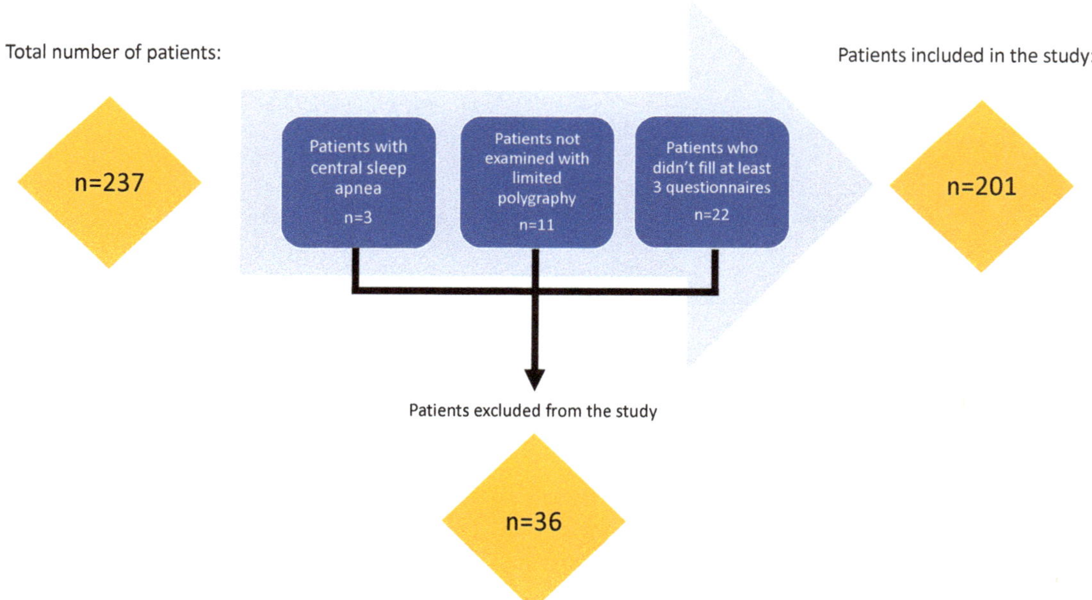

Figure 1. Flow chart—the inclusion/exclusion process.

2.2. *Descriptive Statistics and OSA of the Sample*

A total of 143 men and 58 women were enrolled in the study. The mean age in years was 51.56 and the median was 52. The mean age was higher for women: 55, in contrast to 50 for men. The youngest patient was 19 years old, and the oldest was 75 years old. The mean and median BMI of patients were 30.9 and 30.5 kg/m^2. The mean neck circumference in the patients was 41.8 cm, and the median 42 cm (for details, see Table 1).

Table 1. Basic indicators found in the whole group of patients (n = 201).

Indicators	Mean	Median	SD	Min	Max
Age (years)	51.6	52	12.32	19	75
BMI (kg/m^2)	30.9	30.5	5.53	17.6	53
Neck circumference (cm)	41.8	42	4.33	20	57

SD—standard deviation.

In our sample, OSA was not present in 11.9% of the patients (AHI below 5). We found mild OSA (AHI 5–14.9) in 13.9% of the patients, moderate OSA (AHI 15–29.9) in 32.3% and severe OSA (AHI 30 and over) in 41.8% (for details, see Table 2).

Table 2. OSA in the sample group examined by limited polygraphy.

AHI	Total		Men		Women	
	Number	%	Number	%	Number	%
Total	201	100	143	100	58	100
<5	24	11.9	10	7	14	24.1
>5	177	88.1	133	93.0	44	75.9
Mild OSA	28	13.9	19	13.3	9	15.5
Moderate OSA	65	32.3	48	33.6	17	29.3
Severe OSA	84	41.8	66	46.2	18	31.0

2.3. *Methods*

Patients completed five written sleep questionnaires individually and were subsequently examined by limited polygraphy at the Department of Neurology. Manual polygraphy validation was performed.

BMI (body mass index)—defined as body weight divided by the square of height.

AHI (apnea-hypopnea index)—defined as the total number of apnea and hypopnea episodes in the course of 1 h.

Mild OSA—defined as $5 \leq$ AHI < 14.9.

Moderate OSA—defined as $15 \leq$ AHI < 29.9.

Severe OSA—defined as AHI ≥ 30.

2.4. *Sleep Questionnaires Used in The Study*

2.4.1. The Epworth Sleepiness Scale (ESS)

The ESS was developed and validated by Johns [11] as a simple tool to assess excessive daytime sleepiness. The ESS consists of eight items that list various daily situations in which the patient evaluates the probability of falling asleep or napping using a scale of 0–3. The total score is the sum of the individual responses and is, therefore, in the range 0–24. Excessive daytime sleepiness and a greater likelihood of OSA are observed in patients with an ESS value > 10 [11,12]. In other studies, the sensitivity and specificity of the ESS vary, between 39–66% and 33–71%, respectively [13–16].

2.4.2. STOP-Bang Questionnaire

The STOP-Bang questionnaire was developed by Chung et al. as a screening questionnaire for OSA [17]. It contains eight questions related to snoring, fatigue during the day, sleep apnea, high blood pressure, BMI, age, neck circumference and gender. It is possible to receive 0–1 points for each question. The total score is the sum of the individual answers and ranges from 0 to 8.

A score of 0–2 points indicates a low risk of obstructive sleep apnea (OSA), whereas 3–4 points indicate a medium risk, and 5–8 points indicate a high risk. A high risk can alternatively be indicated by a score of 2 for the first four questions plus BMI >35 kg/m^2, or a score of 2 for the first four questions plus neck circumference (43 cm for men, 41 cm for women), or a score of 2 for the first four questions plus male gender.

Shrestha et al. found the sensitivity and specificity of the STOP-Bang questionnaire to be 92% and 33%, respectively. In a systematic review and meta-analysis by Bianca Pivetta et al., the sensitivity and specificity were found to be 91% and 28%, respectively. In the study by Costa et al., the sensitivity was lower, 68.4%, and the specificity was 85% [16,18,19].

2.4.3. STOP Questionnaire

The STOP questionnaire is a simpler version of the STOP-Bang questionnaire. It was developed in 2008 in an attempt to establish an easy-to-use questionnaire for OSA screening in surgical patients [17]. It contains four questions about snoring, fatigue during the day, sleep apnea and high blood pressure. It is possible to receive 0–1 points for each question. The total score is the sum of the individual answers and is, therefore, in the range of 0–4. A high risk of OSA is indicated by a score ≥2. In the studies of Chung et al. and Patel et al., the sensitivity of the STOP questionnaire varied from 66 to 89% [17,20].

2.4.4. Berlin Questionnaire (BQ)

The Berlin questionnaire was developed in 1996 at the Conference on Sleep in Primary Care in Berlin, Germany. It is a validated instrument that is used to identify individuals who are at risk for OSA in primary and some non-primary care settings. It contains 10 questions, which are divided into three categories. In the first category, there are five questions about snoring and breathing during sleep. In the second category, there are three questions about increased daily fatigue and drowsiness. In the last, third category, there are questions about hypertension and BMI. Each category is evaluated separately; the total score is calculated as the sum of points for each category and ranges from 0 to 3. A score of ≥2 indicates a risk for OSA [21,22]. Two previous studies found varying degrees of the sensitivity and specificity for the BQ: 73–83% and 22–44%, respectively [14,23].

2.4.5. Pittsburgh Sleep Quality Index (PSQI)

The PSQI was not originally designed to screen for OSA. Rather, it is focused on sleep quality (sleep latency, sleep duration, sleep efficiency, sleep interruptions, use of sleep-inducing drugs, and daily dysfunction related to poor sleep) [24]. It contains 10 questions, which are divided into seven categories. Each category is evaluated separately using 0 to 3 points, and the total score is calculated as the sum of points for each category and ranges from 0 to 21. Poor sleep quality, which is also expected in patients with OSA, is noted for scores >5. The sensitivity of PSQI was shown to be low in two different studies (38–51%), and the specificity was shown to be 67–76% [16,25].

Inclusion criteria: (1) age over 18 years, (2) OSA assessment (diagnosis, follow-up) using PG, (3) completed three or more sleep questionnaires.

Exclusion criteria: (1) diagnosed with central sleep apnea, (2) OSA assessment performed using methods other than PG, or incomplete data from PG, (3) completed less than 3 sleep questionnaires, or questionnaires that were not answered completely.

2.5. Statistical Methods

Descriptive statistics (numbers, arithmetic mean, median, standard deviation, min. and max. value) were used to describe the data. Correlations between the results were evaluated using Spearman's correlation coefficient. Furthermore, the sensitivity and specificity of individual screening questionnaires were evaluated. Statistical tests were evaluated at a significance level of 5%. The statistical program Stata version 13 was used for processing.

3. Results

For the Epworth Sleepiness Scale, 197 questionnaires were included and four excluded (for details, see Table 3). The sensitivity of ESS was 34.5%, and specificity 82.6%.

Table 3. Results of Epworth Sleepiness Scale.

Epworth Sleepiness Scale (n = 197)					
Score	0–10	11–12	13–15	16–24	
	133	25	23	16	
	Mean	Median	SD	Min.	Max.
	8.3	7	4.69	1	22
PG	0–4.9	5–14.9	15–29.9	30–	
	23	28	63	83	

In the case of the STOP-Bang questionnaire scale, 183 questionnaires were included and 18 not included, with the best sensitivity of 81.6% and specificity of 75% (for details, see Table 4).

Table 4. Results of STOP-Bang questionnaire.

STOP-Bang Questionnaire (n = 183)					
Score	Low Risk	Intermediate	High Risk		
	45	68	70		
	Mean	Median	SD	Min.	Max.
	4.3	4	1.71	1	8
PG	0–4.9	5–14.9	15–29.9	30–	
	20	25	60	78	

The STOP questionnaire scale had 184 included questionnaires and 17 not included questionnaires, with sensitivity of 74.2% and specificity of 61.9% (for details, see Table 5).

Table 5. Results of STOP questionnaire.

STOP Questionnaire (n = 184)					
Score	Low Risk	High Risk			
	55	129			
	Mean	Median	SD	Min.	Max.
	2.2	2	1.16	0	4
PG	0–4.9	5–14.9	15–29.9	30–	
	21	25	60	78	

For the Berlin Questionnaire Scale, there were 185 questionnaires included and 16 not included, with the second-highest sensitivity of 78.7% and specificity of 61.9% (for details see Table 6).

Table 6. Results of Berlin questionnaire.

Berlin Questionnaire (n = 185)					
Score	Low Risk	High Risk			
	48	137			
	Mean	Median	SD	Min.	Max.
	2.0	2	0.79	0	3
PG	0–4.9	5–14.9	15–29.9	30–	
	21	25	61	78	

The Pittsburgh Sleep Quality Index had 147 included and 54 not-included questionnaires, and had the worst results, sensitivity of 50.8%, and specificity of 47.4% (for details, see Table 7).

Table 7. Results of Pittsburgh Sleep Quality Index.

Pittsburgh Sleep Quality Index (*n* = 147)					
Score	0–5	6–21			
	72	75			
	Mean	Median	SD	Min.	Max.
	6.4	6	3.5	1	19
PG	0–4.9	5–14.9	15–29.9	30–	
	19	21	47	60	

The highest sensitivity was found in the STOP-Bang questionnaire, the Berlin questionnaire, and the STOP questionnaire (81.6%, 78.7%, and 74.2%, respectively). The ESS and the PSQI had the lowest sensitivity (34.5% and 50.8%, respectively).

The ESS had the highest specificity (82.6%), followed by the STOP-Bang, STOP and Berlin questionnaires (75%, 61.9%, and 61.9%, respectively). The PSQI has the lowest specificity (47.4%) (for details, see Table 8).

Table 8. Sensitivity and specificity of questionnaires.

Sensitivity and Specificity	**ESS**		**BQ**		**PSQI**		**STOP Bang**		**STOP**	
Test	Number	%	Number	%	Number	%	Number	%	Number	%
False neg.	114	65.5	35	21.3	63	49.2	30	18.4	42	25.8
True pos.	60	34.5	129	78.7	65	50.8	133	81.6	121	74.2
Total	174	100	164	100	128	100	163	100	163	100
Sensitivity	**34.5%**		**78.7%**		**50.8%**		**81.6%**		**74.2%**	
Test	Number	%	Number	%	Number	%	Number	%	Number	%
False neg.	19	82.6	13	61.9	9	47.4	15	75	13	61.9
True pos.	4	17.4	8	38.1	10	52.6	5	25	8	38.1
Total	23	100	21	100	19	100	20	100	21	100
Specificity	**82.6%**		**61.9%**		**47.4%**		**75%**		**61.9%**	

4. Discussion

The aim of this study was to compare the predictive capabilities of five established sleep questionnaires for OSA. The questionnaires tested in this study were the ESS, BQ, STOP and STOP-Bang, as well as the PSQI. All questionnaires were filled in by patients presenting sleep disorders. The scores were evaluated against limited polygraphy based on AHI.

One of the most commonly used questionnaires in sleep medicine, the Epworth Sleepiness Scale, deals with only one of the presumed risk factors for OSA: excessive daytime sleepiness [11,26]. The advantage of ESS is clarity; it is a simple evaluation method. According to Johns et al., ESS scores significantly distinguished patients with primary snoring from those with OSA, and ESS scores increased with the severity of OSA [27]. However, the association between AHI and ESS scores was not confirmed by Laub et al. According to Laub et al., ESS is not a good questionnaire for the evaluation of the presence or severity of obstructive sleep apnea [28]. Similarly, in a study by Mediano et al., excessive daytime sleepiness measured by ESS was not invariably present in patients with OSA. Patients with OSA and excessive daytime sleepiness were characterized by worse nocturnal oxygenation than those without excessive daytime sleepiness. Both groups exhibited a similar AHI [29].

In other studies the sensitivity and specificity of ESS varied between 39–66% and 33–71% [13–16,30]. The results of our study demonstrated that ESS had a lower sensitivity for OSA (34.5%) and higher specificity (82.6%) in comparison to the findings by other authors. The low sensitivity was not surprising given that the ESS is a standard questionnaire designed to measure subjective excessive daytime sleepiness, which can occur secondary to multiple causes other than OSA.

The STOP-Bang questionnaire is widely used worldwide. [30] It is quick and simple. According to a meta-analysis by Chiu et al. from 2017, it had a high sensitivity (88%), but the specificity was low (42%) [30]. In an earlier study, it was found that the STOP-Bang questionnaire had high sensitivity for detecting moderate and severe OSA (93% and 100%, respectively), but the specificity of the STOP-Bang questionnaire was still low: 47% and 37% for moderate and severe OSA, respectively, resulting in fairly high false-positive rates [17]. Silva et al. reported that the STOP-Bang questionnaire had the highest sensitivity for moderate-to-severe (87.0%) and severe (70.4%) OSA in comparison to the ESS and the STOP [13]. In other studies, the sensitivity and specificity of the STOP-Bang questionnaire varied between 91–92% and 28–33% [16,18]. In our study, the sensitivity of the STOP-Bang questionnaire for OSA was found to be 81.6%, and its specificity 75%, which was higher compared to the study by Kee et al. (60% and 69%, respectively) [31].

The STOP questionnaire contains the first four questions from the STOP-Bang questionnaire. According to a meta-analysis from 2016, it had a sensitivity of 87% and a specificity of 42% [30]. In other studies, the sensitivity of the STOP questionnaire varied between 66 and 89%. In a systematic review article, Abrishami et al. recommended the use of the STOP-Bang and STOP questionnaires for their high-quality methodology and accurate results, although the sensitivity and specificity were not significantly higher compared to other questionnaires [32]. In our sample, the sensitivity of the STOP questionnaire for OSA was found to be 74.2%, and the specificity 61.9%. According to the results of our study, the STOP-Bang and STOP questionnaires were relatively suitable screening tools in comparison with other questionnaires.

The Berlin questionnaire is more time-consuming compared to the ESS, STOP-Bang and STOP questionnaires. Ahmadi et al. [33] tested the BQ with patients in a sleep clinic, retrospectively. Out of the 130 individuals tested, only 26.2% had a respiratory disturbance index (RDI) >10, whereas the BQ identified 58.5% as being at high-risk of having sleep apnea, with a 62% sensitivity and 43% specificity. The discrepancy between these results and our study could be attributed to the use of RDI rather than AHI at a higher cut-off (i.e., >10). In other studies, the sensitivity and specificity of BQ varied between 73–83% and 22–59%, respectively [14,23,30,31]. In our study, the sensitivity of the BQ for OSA was found to be 78.7%, and its specificity was established as 61.9%. Due to its satisfactory sensitivity and specificity, the BQ appears to be a suitable tool for OSA screening.

The PSQI is one of the most frequently used sleep questionnaires worldwide. Completing and evaluating the questionnaire is complex and time-consuming. The PSQI addresses psychological symptoms and correlates OSA with the occurrence of depression, anxiety or stress [34,35]. The PSQI is unsuitable for OSA screening. According to a study by Scarlata et al., the sensitivity of the PSQI was only 37.8%, and its specificity 76.1% [25]. In a different study by Amado-Garzón, the sensitivity for OSA and central apnea was 80–85% [36]. Based on our results, the PSQI had lower sensitivity in comparison to the STOP-Bang, STOP and BQ (50.8%,). The specificity was the lowest among all our questionnaires (47.4%).

A certain limitation of the study can be its monocentricity and the fact that not all patients filled in all five questionnaires completely. Patients that completed less than (or did not completely answer) three sleep questionnaires were excluded (see exclusion criteria). Another limit of the study could be the missing gender differences evaluation for the relatively small number of respondents (143 men and 58 women).

5. Conclusions

The STOP-Bang and Berlin questionnaires, which had the highest sensitivity (81.6%, 78.7%) and satisfactory specificity (75%, 61.9%), were found to be the most suitable for OSA screening in our sample. The STOP questionnaire was also relatively reliable, especially given its time-saving nature, which did not impair its satisfactory sensitivity (74.2%) and specificity (61.9%). The Epworth Sleepiness Scale and the Pittsburgh Sleep Quality Index had the lowest sensitivity (34.5%, 50.8%) and are unsuitable for OSA screening.

Author Contributions: Conceptualization: Š.S., K.M. and T.K. Methodology: T.K., V.K. and H.T. Investigation: Š.S. Original Draft Preparation: Š.S. and H.T., Writing—Review and Editing: J.B. All authors have read and agreed to the published version of the manuscript.

Funding: Funding for the article was supported by the Ministry of Health, Czech Republic—Conceptual Development of Research Organization (Faculty Hospital in Pilsen-FNPl, 00669806).

Institutional Review Board Statement: The study was conducted in accordance with the Declaration of Helsinki, and approved by the Ethics Committee of the hospital in Frýdek-Místek (Ref. number 017/18, approval date: 21 June 2018).

Informed Consent Statement: Informed consent was obtained from all subjects involved in the study.

Data Availability Statement: Not applicable.

Acknowledgments: Supported by the Cooperatio Program, research area SURG.

Conflicts of Interest: The authors declare no conflict of interest. The funders had no role in the design of the study; in the collection, analyses, or interpretation of data; in the writing of the manuscript; or in the decision to publish the results.

References

1. Garbarino, S.; Guglielmi, O.; Sanna, A.; Mancardi, G.L.; Magnavita, N. Risk of Occupational Accidents in Workers with Obstructive Sleep Apnea: Systematic Review and Meta-analysis. *Sleep* **2016**, *39*, 1211–1218. [CrossRef] [PubMed]
2. Benjafield, A.V.; Ayas, N.T.; Eastwood, P.R.; Heinzer, R.; Ip, M.S.M.; Morrell, M.J.; Nunez, C.M.; Patel, S.R.; Penzel, T.; Pépin, J.-L.; et al. Estimation of the global prevalence and burden of obstructive sleep apnoea: A literature-based analysis. *Lancet Respir. Med.* **2019**, *7*, 687–698. [CrossRef]
3. Deegan, P.; McNicholas, W. Predictive value of clinical features for the obstructive sleep apnoea syndrome. *Eur. Respir. J.* **1996**, *9*, 117–124. [CrossRef] [PubMed]
4. Crocker, B.D.; Olson, L.G.; Saunders, N.A.; Hensley, M.J.; McKeon, J.L.; Allen, K.M.; Gyulay, S.G. Estimation of the Probability of Disturbed Breathing during Sleep before a Sleep Study. *Am. Rev. Respir. Dis.* **1990**, *142*, 14–18. [CrossRef] [PubMed]
5. Viner, S.; Szalai, J.P.; Hoffstein, V. Are History and Physical Examination a Good Screening Test for Sleep Apnea? *Ann. Intern. Med.* **1991**, *115*, 356. [CrossRef]
6. Bliwise, D.L.; Nekich, J.C.; Dement, W.C. Relative Validity of Self-Reported Snoring as a Symptom of Sleep Apnea in a Sleep Clinic Population. *Chest* **1991**, *99*, 600–608. [CrossRef]
7. Slouka, D.; Kucera, R.; Gal, B.; Betka, J.; Skalova, A.V. Biomarkers—A posibility for monitoring of obstructive sleep apnea syndrome. *Neuro. Endocrinol. Lett.* **2019**, *40*, 85–92.
8. Kump, K.; Whalen, C.; Tishler, P.V.; Browner, I.; Ferrette, V.; Strohl, K.P.; Rosenberg, C.; Redline, S. Assessment of the validity and utility of a sleep-symptom questionnaire. *Am. J. Respir. Crit. Care Med.* **1994**, *150*, 735–7411. [CrossRef]
9. Slouka, D.; Windrichova, J.; Rezackova, H.; Houfkova, K.; Kucera, R.; Cerna, V.; Kostlivy, T.; Topolcan, O.; Pesta, M. The potential of miR-499 plasmatic level as a biomarker of obstructive sleep apnea syndrome. *Biomarkers Med.* **2021**, *15*, 1011–1019. [CrossRef]
10. Lyons, M.M.; Kraemer, J.F.; Dhingra, R.; Keenan, B.T.; Wessel, N.; Glos, M.; Penzel, T.; Gurubhagavatula, I. Screening for Obstructive Sleep Apnea in Commercial Drivers Using EKG-Derived Respiratory Power Index. *J. Clin. Sleep Med.* **2019**, *15*, 23–32. [CrossRef]
11. Johns, M.W. A New Method for Measuring Daytime Sleepiness: The Epworth Sleepiness Scale. *Sleep* **1991**, *14*, 540–545. [CrossRef] [PubMed]
12. Johns, M.W. Sleepiness in Different Situations Measured by the Epworth Sleepiness Scale. *Sleep* **1994**, *17*, 703–710. [CrossRef] [PubMed]
13. Silva, G.E.; Vana, K.D.; Goodwin, J.L.; Sherrill, D.L.; Quan, S.F. Identification of Patients with Sleep Disordered Breathing: Comparing the Four-Variable Screening Tool, STOP, STOP-Bang, and Epworth Sleepiness Scales. *J. Clin. Sleep Med.* **2011**, *7*, 467–472. [CrossRef]
14. Ulasli, S.S.; Gunay, E.; Koyuncu, T.; Akar, O.; Halici, B.; Ulu, S.; Unlu, M. Predictive value of Berlin Questionnaire and Epworth Sleepiness Scale for obstructive sleep apnea in a sleep clinic population. *Clin. Respir. J.* **2014**, *8*, 292–296. [CrossRef] [PubMed]
15. Rosenthal, L.D.; Dolan, D.C. The Epworth Sleepiness Scale in the Identification of Obstructive Sleep Apnea. *J. Nerv. Ment. Dis.* **2008**, *196*, 429–431. [CrossRef]
16. Shrestha, S.K.; Shrestha, S.; Rajkarnikar, R.; Malla, J.; Rayamajhi, S.; Bhattarai, S.; Sharma, L.; Bhusal, Y. Comparing four standard Sleep Questionnaires to Polysomnography to predict possibility of Obstructive Sleep Apnea and its severity. *J. Adv. Intern. Med.* **2021**, *10*, 33–37. [CrossRef]
17. Chung, F.; Yegneswaran, B.; Liao, P.; Chung, S.A.; Vairavanathan, S.; Islam, S.; Khajehdehi, A.; Shapiro, C.M. STOP Questionnaire. *Anesthesiology* **2008**, *108*, 812–821. [CrossRef]
18. Pivetta, B.; Chen, L.; Nagappa, M.; Saripella, A.; Waseem, R.; Englesakis, M.; Chung, F. Use and Performance of the STOP-Bang Questionnaire for Obstructive Sleep Apnea Screening Across Geographic Regions. *JAMA Netw. Open* **2021**, *4*, e211009. [CrossRef]

19. Costa, J.C.; Rebelo-Marques, A.; Machado, J.P.N.; Valentim, B.M.F.; Ferreira, C.S.A.V.; Gonçalves, J.D.O.; Gama, J.M.D.R.; Teixeira, M.F.L.; Moita, J.J.M. STOP-Bang and NoSAS questionnaires as a screening tool for OSA: Which one is the best choice? *Rev. Assoc. Med. Bras.* **2020**, *66*, 1203–1209. [CrossRef]
20. Patel, M.D.; Tsang, B.J.; Saripella, M.A.; Nagappa, M.M.; Islam, M.S.; Englesakis, M.M.; Chung, M.F. Validation of the STOP questionnaire as a screening tool for OSA among different populations: A systematic review and meta-regression analysis. *J. Clin. Sleep Med.* **2022**, *18*, 1441–1453. [CrossRef]
21. Netzer, N.C.; Stoohs, R.A.; Netzer, C.M.; Clark, K.; Strohl, K.P. Using the Berlin Questionnaire to Identify Patients at Risk for the Sleep Apnea Syndrome. *Ann. Intern. Med.* **1999**, *131*, 485–491. [CrossRef] [PubMed]
22. Berlin questionnaire. *Sleep Breath* **2000**, *4*, 99. [CrossRef] [PubMed]
23. Karakoc, O.; Akcam, T.; Genc, H.; Yetkin, S.; Piskin, B.; Gerek, M. Use of the Berlin Questionnaire to screen at-risk patients for obstructive sleep apnea. *B-ENT* **2014**, *10*, 21–25. [PubMed]
24. Buysse, D.J.; Reynolds, C.F., III; Monk, T.H.; Berman, S.R.; Kupfer, D.J. The Pittsburgh sleep quality index: A new instrument for psychiatric practice and research. *Psychiatry Res.* **1989**, *28*, 193–213. [CrossRef]
25. Scarlata, S.; Pedone, C.; Curcio, G.; Cortese, L.; Chiurco, D.; Fontana, D.; Calabrese, M.; Fusiello, R.; Abbruzzese, G.; Santangelo, S.; et al. Pre-polysomnographic assessment using the Pittsburgh Sleep Quality Index questionnaire is not useful in identifying people at higher risk for obstructive sleep apnea. *J. Med Screen.* **2013**, *20*, 220–226. [CrossRef]
26. Johns, M.W. Reliability and Factor Analysis of the Epworth Sleepiness Scale. *Sleep* **1992**, *15*, 376–381. [CrossRef]
27. Johns, M.W. Daytime sleepiness, snoring, and obstructive sleep apnea. The Epworth Sleepiness Scale. *Chest* **1993**, *103*, 30–36. [CrossRef]
28. Laub, R.R.; Mikkelsen, K.L.; Tønnesen, P. Evaluation of the significance of Epworth sleepiness scale among 687 patients with suspected sleep apnea. *Eur. Respir. J. Sep.* **2015**, *46* (Suppl. S59). [CrossRef]
29. Mediano, O.; Barcelo, A.; de la Pena, M.; Gozal, D.; Agusti, A.; Barbe, F. Daytime sleepiness and polysomnographic variables in sleep apnoea patients. *Eur. Respir. J.* **2007**, *30*, 110–113. [CrossRef]
30. Chiu, H.-Y.; Chen, P.-Y.; Chuang, L.-P.; Chen, N.-H.; Tu, Y.-K.; Hsieh, Y.-J.; Wang, Y.-C.; Guilleminault, C. Diagnostic accuracy of the Berlin questionnaire, STOP-BANG, STOP, and Epworth sleepiness scale in detecting obstructive sleep apnea: A bivariate meta-analysis. *Sleep Med. Rev.* **2017**, *36*, 57–70. [CrossRef]
31. Kee, K.; Dixon, J.; Shaw, J.; Vulikh, E.; Schlaich, M.; Kaye, D.M.; Zimmet, P.; Naughton, M.T. Comparison of Commonly Used Questionnaires to Identify Obstructive Sleep Apnea in a High-Risk Population. *J. Clin. Sleep Med.* **2018**, *14*, 2057–2064. [CrossRef] [PubMed]
32. Abrishami, A.; Khajehdehi, A.; Chung, F. A systematic review of screening questionnaires for obstructive sleep apnea. *Can. J. Anaesth.* **2010**, *57*, 423–438. [CrossRef] [PubMed]
33. Ahmadi, N.; Chung, S.A.; Gibbs, A.; Shapiro, C.M. The Berlin questionnaire for sleep apnea in a sleep clinic population: Relationship to polysomnographic measurement of respiratory disturbance. *Sleep Breath.* **2007**, *12*, 39–45. [CrossRef] [PubMed]
34. Buysse, D.J.; Hall, M.L.; Strollo, P.J.; Kamarck, T.W.; Owens, J.; Lee, L. Relationships between the Pittsburgh Sleep Quality Index (PSQI), Epworth Sleepiness Scale (ESS), and clinical/polysomnographic measures in a community sample. *J. Clin. Sleep Med.* **2008**, *4*, 563–571, Erratum in *J. Clin. Sleep Med.* **2010**, *6*, table of contents. [CrossRef]
35. Carpenter, J.S.; Andrykowski, M.A. Psychometric evaluation of the pittsburgh sleep quality index. *J. Psychosom. Res.* **1998**, *45*, 5–13. [CrossRef]
36. Amado-Garzón, S.B.; Ruiz, A.J.; Rondón-Sepúlveda, M.A.; Hidalgo-Martínez, P. Sensitivity and specificity of four screening sleep-disordered breathing tests in patients with and without cardiovascular disease. *Sleep Sci.* **2021**, *14*, 311–318. [CrossRef]

life

MDPI

Article

Epiglottopexy Is a Treatment of Choice for Obstructive Sleep Apnea Caused by a Collapsing Epiglottis

Michaela Masárová [1,2], Martin Formánek [1,2,*], Ondřej Jor [3], Vilém Novák [4], Adéla Vrtková [5], Petr Matoušek [1,2], Pavel Komínek [1,2] and Karol Zeleník [1,2]

1 Department of Otorhinolaryngology and Head and Neck Surgery, University Hospital Ostrava, 708 52 Ostrava, Czech Republic
2 Faculty of Medicine, University of Ostrava, 708 52 Ostrava, Czech Republic
3 Department of Anesthesiology, Resuscitation and Intensive Medicine, University Hospital Ostrava, 708 52 Ostrava, Czech Republic
4 Department of Pediatric Neurology, University Hospital Ostrava, 708 52 Ostrava, Czech Republic
5 Department of Applied Mathematics, Faculty of Electrical Engineering and Computer Science, VSB-Technical University of Ostrava, 708 52 Ostrava, Czech Republic
* Correspondence: martin.formanek@fno.cz; Tel.: +42-0597375812

Abstract: Drug-induced sleep endoscopy (DISE) reveals epiglottic collapse to be a frequent cause of obstructive sleep apnea (OSA) and intolerance of positive airway pressure (PAP). These patients require different management. This prospective study aimed to compare transoral laser epiglottopexy outcomes in patients with OSA caused by epiglottic collapse with the patients' previous PAP outcomes. Fifteen consecutive adult patients with OSA and epiglottic collapse during DISE were included; ten were analyzed. Before inclusion, PAP was indicated and ineffective in six patients, one of whom underwent unsuccessful uvulopalatopharyngoplasty. PAP was performed during DISE in all patients before epiglottopexy and was uniformly ineffective. ENT control was performed at 1 week and 1 month, and control limited polygraphy to 6 months after surgery. The apnea–hypopnea index (AHI) and Epworth Sleepiness Scale (ESS) were significantly improved ($p < 0.001$ and $p = 0.003$, respectively) in all patients after epiglottopexy. Surgery was successful in 9/10 patients; the remaining patient had a significantly decreased AHI and could finally tolerate PAP. Transoral laser epiglottopexy is used to treat OSA in patients with epiglottic collapse. Unlike other methods, it significantly reduces both AHI and ESS and should be considered for these patients. An active search for OSA patients with epiglottic collapse is recommended to prevent treatment failure.

Keywords: obstructive sleep apnea; drug-induced sleep endoscopy; epiglottopexy; positive airway pressure

Citation: Masárová, M.; Formánek, M.; Jor, O.; Novák, V.; Vrtková, A.; Matoušek, P.; Komínek, P.; Zeleník, K. Epiglottopexy Is a Treatment of Choice for Obstructive Sleep Apnea Caused by a Collapsing Epiglottis. *Life* **2022**, *12*, 1378. https://doi.org/10.3390/life12091378

Academic Editors: David Slouka and Milan Štengl

Received: 2 August 2022
Accepted: 30 August 2022
Published: 5 September 2022

1. Introduction

The role of the epiglottis in the development of obstructive sleep apnea (OSA) in adults has been known for many years [1]. The most common type of epiglottic pathology is "closing door" epiglottis, when the epiglottis collapses on the posterior wall of the hypopharynx during inspiration [2].

In the past, the prevalence of epiglottic collapse evaluated by clinical examination was estimated to be 12% in OSA patients [3]. In recent years, however, drug-induced sleep endoscopy (DISE) has shown that the epiglottis is a much more common cause of OSA [4].

Currently, there is a lack of good understanding about the relationship between epiglottic collapse and its most effective treatment. The treatment of OSA with positive airway pressure (PAP) is considered the "gold standard" [5,6]. However, in cases of epiglottic collapse, several studies have suggested that PAP may rather aggravate airway obstruction by pushing the epiglottis further down into the laryngeal inlet (Figure 1) [1].

Thus, these patients could require different management and it seems logical for surgery to play a crucial role regardless of the severity of OSA.

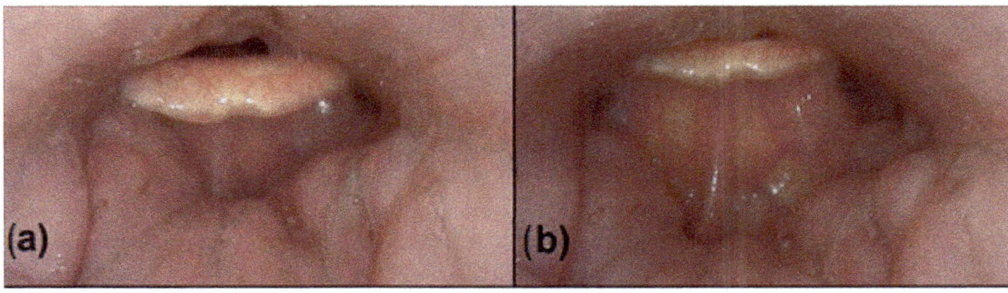

Figure 1. DISE with simultaneous PAP, endoscopic view. (**a**) Anterior–posterior obstruction of the epiglottis as a cause of upper-airway obstruction, without PAP. (**b**) PAP failure: the pressure is 18 hPa and aggravates the upper-airway obstruction by pushing the epiglottis further down into the laryngeal inlet.

Currently, there are insufficient data regarding the effect of epiglottis intervention [2,7,8].

Therefore, the aim of this prospective study was to evaluate the treatment outcomes of transoral laser epiglottopexy in patients with OSA at different severities caused by epiglottic collapse and to compare them with the outcomes of pre- and perioperative positive airway pressure treatments in these patients.

2. Materials and Methods

This prospective study was performed with the consent of the Ethics Committee of the University Hospital Ostrava and performed in accordance with the Declaration of Helsinki and the requirements of good clinical practice. Written informed consent was obtained from each patient before any procedure was initiated.

2.1. Study Design

This study was performed at the tertiary referral center.

2.2. Inclusion Criteria

Adult patients with OSA confirmed by limited polygraphy or polysomnography with an AHI $\geq$ 5 episodes/h and epiglottic collapse during DISE.

2.3. Exclusion Criteria

Major comorbidities representing excessive risk for general anesthesia, craniofacial malformations, neurological pathologies, patients who did not want surgery or did not agree to their inclusion in the study, and patients in need of multilevel surgery (epiglottopexy and soft palate or oropharyngeal surgery).

2.4. Clinical Evaluation

Patients were evaluated by obtaining their comprehensive history covering sleep habits and disturbances: shift work, sleep duration, sleep quality, lack of sleep, daily sleepiness, and snoring. Excessive daytime sleepiness was estimated by the ESS. Each patient's body mass index (BMI) was also recorded.

Clinical evaluation included a complete head and neck examination that evaluated the presence of soft palatal webbing, the presence of an elongated uvula, and the size of the tonsils according to Friedmann. The relation of the base of the tongue to the soft palate was evaluated using the Mallampati classification.

The upper airways and digestive tract were examined using a flexible videoendoscope with a diameter of 3.5 mm (Olympus, Tokyo, Japan) in each patient. Laterolateral narrowing of the oropharynx, narrowing of the base of the tongue, and the presence of pathology in the epiglottis were assessed according to Kezirian [9].

2.5. Drug-Induced Sleep Endoscopy

Sleep endoscopy was performed by an experienced otorhinolaryngologist and anesthesiologist in the operating room. After intramuscular administration of 5 mg of Dormicum (midazolam) and 0.5 mg of atropine 30 min before the examination, the patient in supine position without neck extension was induced to sleep with intravenous propofol (a 1 mg/kg bolus at the beginning, then 20–30 mg every 3–5 min). The depth of anesthesia was measured by the bispectral index throughout the examination and maintained in the range of 50–70. Vital signs were monitored and included blood pressure, heart rate, electrocardiogram, oxygen saturation, and respiratory rate. Sleep endoscopy was performed using a flexible videoendoscope with a diameter of 3.5 mm (Olympus, Tokyo, Japan), and the length of the examination was 15–20 min. The Kezirian VOTE classification was used to evaluate the results (Figure 2) [9].

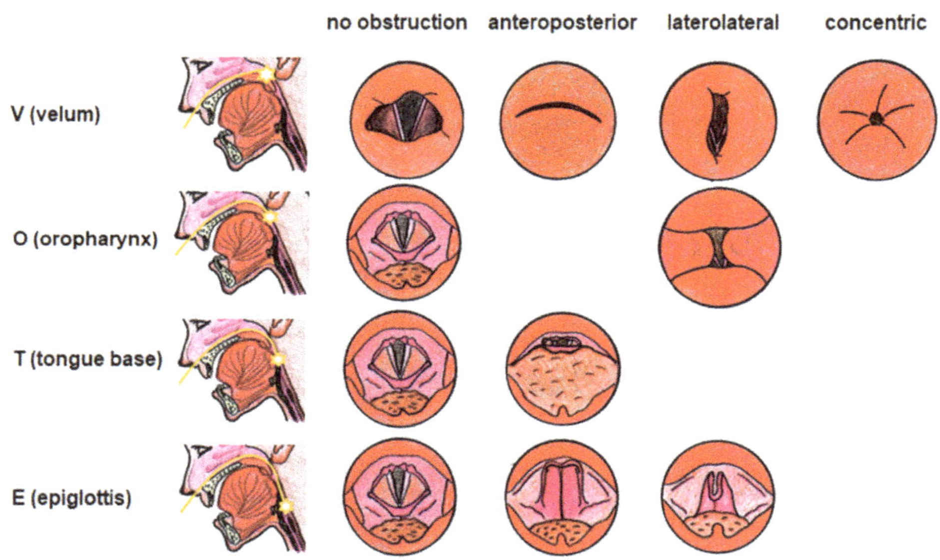

Figure 2. The VOTE classification according to Kezirian (2011). Obstruction in four upper-airway areas is evaluated: the area of the soft palate, the lateral pharynx walls and the tonsils, the tongue base, and the epiglottis. In each of these locations, the degree of obstruction (0, no obstruction; 1, partial obstruction; 2, complete obstruction) and obstruction configuration (anteroposterior, circular, laterolateral) are evaluated [7].

2.6. Positive Airway Pressure Titration during the Sleep Endoscopy

The BiPAP A40 (Philips Respironics, Murrysville, PA, USA) in PAP mode was used for titration. PAP titration was performed immediately after DISE with the patient in supine position. An overpressure ventilation mask (Respironics PerforMax Full-face mask, Philips Respironics, Murrysville, PA, USA) of appropriate size was applied to the patient's face. A special connecting valve (Philips Respironics, Murrysville, PA, USA) was inserted between the mask and the device hose, through which a flexible endoscope was inserted into the nose and nasopharynx. The mask was subsequently fixed with straps.

Sleep endoscopy was then performed under overpressure ventilation. The start of the examination was at a pressure of 6 hPa. Gradually, the PAP was elevated (always after at least 30 s) in the range of 6, 8, 10, 12, 14, and 18 hPa. At each pressure tested, an evaluation was performed by two physicians independently. The efficiency of overpressure ventilation was assessed visually, and at the same time the values of blood oxygen saturation measured on a finger of the upper limb were monitored.

The Kezirian VOTE classification was used for evaluation at each PAP tested to observe the effect of increasing pressure on the obstruction of the monitored localities.

2.7. Surgical Technique

In patients with an epiglottic collapse during sleep endoscopy, transoral laser epiglottopexy was subsequently performed under general anesthesia using an operating microscope (Zeiss, Oberkochen, Germany). The patient lay in supine position and their superior teeth were protected by a silicone protector. The base of the tongue and the entire epiglottic vallecula and epiglottis were exposed by a laryngoscope.

Removal of the mucosa at the base of the tongue, epiglottic vallecula, and the lingual surface of the epiglottis was performed using a Thulium laser (Revolix, Getz Healthcare, Singapore). The laser was set at 8–10 watts of delivered power. A rim of 5 mm of healthy mucosa was left intact along the entire lingual surface of the epiglottis. After de-epithelialization of the mucosa, the epiglottis was fixed to the base of the tongue with two absorbable stitches (Figure 3A).

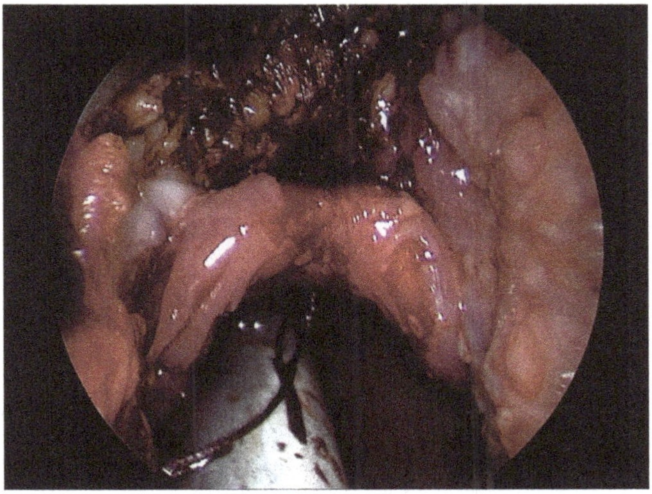

(A)

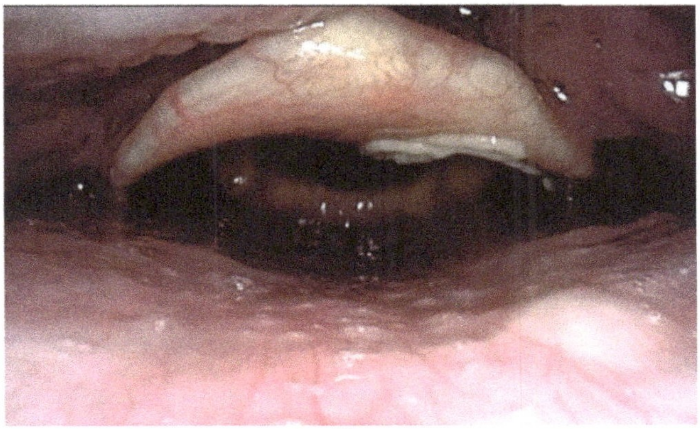

(B)

Figure 3. The principle of epiglottopexy. (**A**) Endoscopic peroperative view. (**B**) Follow-up videoendoscopic examination 1 month after surgery.

Moderate perioperative bleeding was treated with cauterization. Due to the laser's ability to coagulate vessels, it was usually not necessary to deal with major perioperative bleeding.

2.8. Postoperative Care

Antibiotics were administered intravenously for 24 h, and after that orally until the 7th postoperative day. Analgesics were also recommended for 1 week. If the patient had no problems with swallowing, they were allowed to resume oral intake including liquids from the first postoperative day.

2.9. Follow-Up

ENT control with a flexible videoendoscope (Olympus, Tokyo, Japan) was performed 1 week and 1 month after the surgery (Figure 3B). Control polysomnography or limited polygraphy was performed 6 months after the surgery. The presence of complications during follow-up was monitored.

2.10. Statistical Analysis

The primary analysis was performed using standard tools of exploratory data analysis. For the description of numerical variables, the mean or the median was used. Categorical variables were presented with absolute and/or relative frequencies. The success of surgery and healing were evaluated according to Sher's criteria [10,11]. Previous therapy was chosen as a control for each patient. Statistical analysis was performed using R software (version 3.6.0; R Core Team, R Foundation for Statistical Computing, Vienna, Austria). The one-tailed Wilcoxon signed-rank test was used with a 5% significance level.

3. Results

Altogether, 15 consecutive patients were included in this study. Four patients had to be excluded: three patients did not want surgery and were treated with a mandibular advancement with partial effect, and one patient did not come for surgery. Surgery was performed on 11 patients. One patient was lost to follow-up and had to be excluded from analysis.

A total of 10 patients (eight males and two females) were analyzed. The mean age of analyzed patients was 43.7 years. The patients had varying degrees of nonpositional OSA (mean AHI: 25.9; mean ESS: 15.6), and their mean BMI was 28.6. There were no cases of head and neck cancer or multiple system atrophy in these patients. Before their inclusion in the study, positive airway pressure treatment was primarily indicated as the first-line treatment in six patients and was not effective in any of them according to AHI and ESS. In one patient, uvulopalatopharyngoplasty was primarily performed in another hospital, also with no effect on OSA (Table 1).

According to the awake examination findings, including the use of a flexible endoscope, none of the patients had any epiglottic pathology. Positive airway pressure was not effective during the DISE in any patient, even at the strongest pressure tested, resulting in PAP treatment failure in all patients if it had been indicated, as was confirmed in those six patients who underwent preoperative positive airway pressure treatment. In all cases, positive airway pressure only aggravated the airway obstruction by pushing the epiglottis further down into the laryngeal inlet.

Postoperative AHI and ESS were significantly improved in all patients 6 months after epiglottopexy (Table 2; Figure 4). In total, 9/10 (90%) patients did not need further treatment after surgery, and quality of life was significantly improved. Therefore, we assume that the treatment was successful for them, which was also confirmed by Sher's criteria. There was an almost 40% decrease in AHI (from 46.5 to 28) in the one remaining patient with very severe OSA, who had not tolerated previous PAP treatment. Postoperatively, the patient could tolerate positive airway pressure treatment.

Table 1. Characteristics of the analyzed patients.

Patient	Age (Years)	Preoperative BMI	Preoperative AHI	Preoperative ESS	Preinclusion Failed Therapy	Awake Examination Findings (Obstruction)	DISE Findings (Obstruction)	Opening PAP Pressure (hPa)	Postoperative BMI	Postoperative AHI	Postoperative ESS
1	22	25.48	46.5	15	PAP	Tonsils	Epiglottis	>18	25.15	1.3	5
2	50	30.88	33	16	PAP UPPP	Tongue base	Epiglottis Tongue base	>18	31.25	5.3	4
3	44	29.32	8.3	11	-	Soft palate Tongue base	Epiglottis	>18	28.76	4.8	4
4	58	31.1	46.5	20	PAP	Soft palate Tongue base	Epiglottis, Tongue base	>18	31.15	28	12
5	49	26.59	18.3	18	-	Soft palate Tongue base	Epiglottis Tongue base	>18	25.48	6.1	6
6	48	27.65	28.83	19	PAP	Soft palate Tongue base	Epiglottis, Tongue base	>18	28.88	11.7	4
7	32	28.47	25.74	15	PAP	Soft palate Tongue base	Epiglottis Tongue base	>18	29.32	10.2	3
8	41	26.2	23.99	15	PAP	Soft palate	Epiglottis Tongue base	>18	29.7	10.8	6
9	38	32	11	13	-	Soft palate Tongue base	Epiglottis	>18	31.67	9	2
10	55	28	17.1	14	-	Soft palate Tongue base	Epiglottis Tongue base	>18	26.54	7.7	4

BMI, body mass index; AHI, apnea–hypopnea index; ESS, Epworth sleepiness scale; PAP, positive airway pressure.

Table 2. Analysis of AHI and ESS postoperative reduction.

	Median (Min; Max)			
	Preoperative	**Postoperative**	**Difference**	*p* *
AHI	24.9 (8.3; 46.5)	8.4 (1.3; 28.0)	14.4 (2.0; 45.2)	<0.001
ESS	15 (11; 20)	4 (2; 12)	11 (7; 15)	0.003

Difference = Preoperative–Postoperative; AHI, apnea–hypopnea index; ESS, Epworth sleepiness scale. * The *p*-value of the one-tailed Wilcoxon signed-rank test.

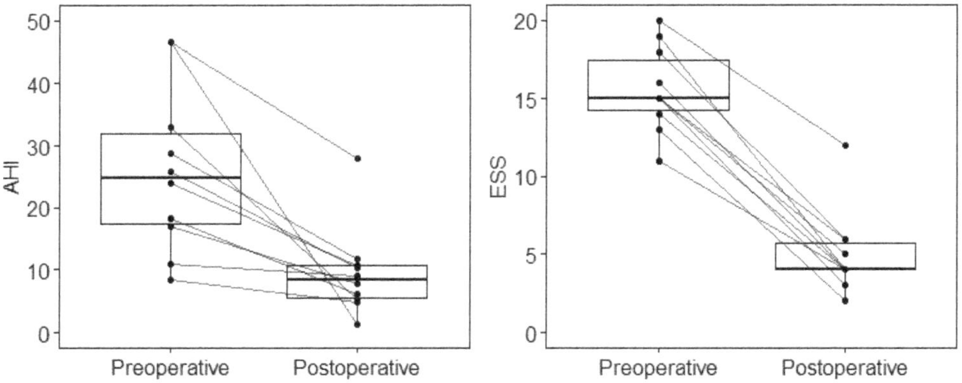

Figure 4. Visualization of postoperative AHI and ESS reduction (paired boxplots). AHI, apnea–hypopnea index; ESS, Epworth sleepiness scale.

Postoperative bleeding from the base of the tongue was noted in one patient 1 week after surgery, and a revision was performed under general anaesthesia. No other perioperative or postoperative complication was observed. No patient had postoperative swallowing problems. All patients started a liquid diet on the first postoperative day. A soft non-irritating diet was started one week after surgery. All patients were transitioned to a solid diet without any problems one month after surgery.

4. Discussion

The morphology of upper-airway structures plays a major role in the pathogenesis of OSA [7,12]. In recent years, DISE has gained popularity in identifying specific anatomical locations and patterns of obstruction. It allows examination of the dynamic status of upper airways during sleep and has become an important part of the armamentarium of surgeons to evaluate their OSA patients [12]. Specifically, it has a strength in identifying epiglottic collapse, which is not easily predictable without sedation [3,4,13].

In the past, the prevalence of epiglottic collapse was estimated to be around 12% in adult OSA patients. Later, and based on DISE, a significantly higher percentage of adult patients with OSA were found to exhibit epiglottic collapse [2,4]. Fernandez-Julian et al. compared surgical recommendations in 162 patients and found that the epiglottis was involved in the obstruction of 36.4% of patients according to DISE, but of only 24.1–28.4% according to awake examination [14]. Koutsourelakis et al. reported 49 OSA patients who were evaluated during DISE before upper-airway surgery and found 36 patients (73.5%) with some degree of epiglottic collapse [15]. Thus, epiglottic collapse seems to play an important role in airway obstruction in patients with OSA [14,15].

The treatment of OSA with positive airway pressure is currently considered as the "gold standard" for many patients [5,6,16]. Although PAP has high efficacy, the effectiveness of the therapy is limited, as demonstrated by 46–83% of patients being non-adherent to therapy (defined as >4 h of use per night) [16]. These patients very often lose interest in further solutions and hope for their healing or are usually offered another possible therapy that consists mainly of uvulopalatopharyngoplasty with or without surgery on the base of the tongue [9,17]. However, this type of surgery is very rarely successful in these patients. Therefore, these patients are forced to take the considerable risks associated with general anesthesia and uvulopalatopharyngoplasty and, ultimately, to live with their OSA. Velopharyngeal OSA surgery has an increased risk of postoperative complications, such as bleeding (sometimes life-threatening), foreign body sensation, dry throat, globus sensation, problems with phlegm, and velopharyngeal insufficiency [17]. Only single-site case series provide current estimates of the incidence of the perioperative complications of UPPP, with a pooled crude serious complication rate of 3.5% and a crude mortality rate of 0.4% [18]. Haavisto et al. reported that as late as 1 year after surgery, 57% of patients had some kind of problem in relation to the operation, the most common complaint being velopharyngeal insufficiency (24%) [19].

According to the literature, a collapsing epiglottis has been found in 15–31.4% of adult patients with OSA who did not tolerate positive airway pressure therapy or in whom positive airway pressure treatment was ineffective [1,3,4,6]. In the case of epiglottic collapse, several studies have shown that rising overpressure of positive airway pressure therapy only pushes the epiglottis onto the back wall of the pharynx more and thus makes the obstruction worse [2,11,20]. Torre et al. states that PAP does not resolve primary epiglottis collapse and too high pressures are required to open it. Furthermore, PAP has been found to be particularly ineffective in cases of latero-lateral epiglottis obstruction. In the case of the latero-lateral epiglottis obstruction the problem persists even at pressures higher than 15.0 hPa [2].

It can be stated that positive airway pressure therapy is not an ideal choice for patients with epiglottic collapse. Positional therapy or mandibular advancement seems to have slightly more promising results. It has been shown that both methods affect epiglottic collapse relatively positively, but only in patients with mild OSA. In patients with a higher AHI these methods often fail and additional treatment is required [21–23]. Kent et al. analyzed 35 consecutively screened adult patients with OSA with positive airway pressure therapy intolerance and incomplete response to oral appliance therapy and found 20% of them to have epiglottic collapse [21]. In addition, both treatment modalities (positional therapy and mandibular advancement) require high compliance and do not solve their OSA definitively [3,24]. Less compliant patients are thus at risk of not being properly treated and exposed to all OSA complications.

It is interesting that patients with epiglottis collapse seem to be thinner than standard patients with OSA and their AHI does not correlate with their BMI [25]. Likewise, epiglottis collapse is not usually related to craniofacial deformities, which can also predispose patients to OSA [17,25]. Patients with epiglottic collapse are often young and slim with a different degree of OSA, in our experience. Therefore, especially with these patients, it is very important to think about the possibility of a collapsing epiglottis [2,3].

Based on our results, transoral laser epiglottopexy seems to be an ideal therapeutic option for these patients. There are multiple choices for epiglottic surgery including epiglottidectomy (total, partial, and V-shaped) or transoral epiglottopexy [4,13,26,27]. However, transoral epiglottopexy (or glossoepiglottopexy) is the most gentle surgical technique in terms of impairing the fundamental functions of the epiglottis, complications, and postoperative morbidity [2,3,9,26–30].

Generally, many surgical approaches and techniques in this region are significantly invasive and associated with a high rate of complications, such as bleeding, edema, persistent dysphagia, dysgeusia, etc. However, new technologies (diathermy, CO_2 laser, thulium laser, and coblation) have allowed the introduction of innovative surgical techniques that are relatively safe [2,3,9].

The functions of the epiglottis must be taken into account. The epiglottis is involved in preventing food aspirations in two ways. It works as a mechanical closure of the laryngeal aditus during swallowing. Its sensitive receptors, distributed over its surface, participate in swallowing as well. In an effort to prevent unwanted complications such as aspirations, it is important from a technical point of view to leave a 3–4 mm rim of healthy mucosa along the entire profile of the epiglottis [3,4,27].

However, to date, no serious complications associated with epiglottopexy have been reported in the literature. In addition, it is a reversible method in case of unexpected complications [3,9]

The most risky procedure is an intervention at the base of the tongue (risking injury to the lingual artery with subsequent perioperative or postoperative bleeding). Therefore, it is very important not to reduce the base of the tongue or expose the mucosa too laterally, but to adhere strictly to the area around the midline [3,8,18,30].

Although our results are limited by the small number of patients, it can be stated that correctly applied epiglottopexy seems to be a very effective first-stage treatment for OSA caused by epiglottic collapse, with minimal risk of complications. What is more, epiglottopexy can significantly improve the tolerance of PAP treatment in some cases and prevent serious complications associated with other upper-airway surgeries (such as tonsillectomy, uvulopalatopharyngoplasty, etc.) that incorrectly diagnosed patients undergo unnecessarily [4,19,30]. A collapsing epiglottis is one of the very frequent reasons for PAP intolerance and cannot be properly evaluated during clinical examination with endoscopy. Thus, DISE should be considered in such cases.

5. Conclusions

A collapsing epiglottis is a relatively common cause of OSA and one of the reasons for positive airway pressure intolerance. Since it cannot be detected while awake, DISE should be recommended in these cases. Transoral laser epiglottopexy is an elegant technique that significantly ameliorates the severity of OSA, or in some cases even cures OSA in patients with epiglottic collapse. Unlike other methods, it significantly reduces both AHI and ESS and should be the method of choice for these patients. Despite the intervention in the larynx, it provides stable support to the epiglottis without affecting its function during swallowing. Transoral laser epiglottopexy is thus a safe surgical technique. It is essential to actively seek out patients with epiglottic collapse among other patients with OSA to prevent treatment failure and lessen the risk of potential complications.

Author Contributions: Conceptualization, K.Z., P.M. and P.K.; Data curation, M.M. and A.V.; Formal analysis, M.M. and A.V.; Funding acquisition, M.M., M.F., P.M. and P.K.; Investigation, M.M., O.J., V.N., K.Z. and P.M.; Methodology, M.F., K.Z. and P.M.; Supervision, M.F. and P.K.; Writing—original draft, M.M.; Writing—review & editing, M.F., K.Z., P.M. and P.K. All authors have read and agreed to the published version of the manuscript.

Funding: This work was supported by the project of University of Ostrava, SGS11/LF/2021 and by the Ministry of Health, Czech Republic, Conceptual Development of Research Organization, MH CZ–17/DRO-FNOs/2021.

Institutional Review Board Statement: The study was conducted in accordance with the Declaration of Helsinki, and approved by the Ethics Committee of University Hospital Ostrava (protocol code 259/2021, date of approval 25 March 2021).

Informed Consent Statement: Informed consent was obtained from the patients in the report.

Data Availability Statement: All available data presented.

Conflicts of Interest: The authors declare that they have no conflict of interest.

References

1. Dedhia, E.C.; Rosen, C.A.; Soose, R.J. What is the role of the larynx in adult obstructive sleep apnea? *Laryngoscope* **2014**, *124*, 1029–1034. [CrossRef] [PubMed]
2. Torre, C.; Camacho, M.; Liu, S.Y.; Huon, L.K.; Capasso, R. Epiglottis collapse in adult obstructive sleep apnea: A systematic review. *Laryngoscope* **2016**, *126*, 515–523. [CrossRef] [PubMed]

3. Roustan, V.; Barbieri, M.; Incandela, F.; Missale, F.; Camera, H.; Braido, F.; Mora, R.; Peretti, G. Transoral glossoepiglottopexy in the treatment of adult obstructive sleep apnoea: A surgical approach. *Acta Otorhinolaryngol. Ital.* **2018**, *38*, 38–44. [CrossRef] [PubMed]
4. Kwon, E.O.; Jung, S.Y.; Al-Dilaijan, K.; Min, J.Y.; Lee, K.H.; Kim, S.W. Is Epiglottis Surgery Necessary for Obstructive Sleep Apnea Patients with Epiglottis Obstruction? *Laryngoscope* **2019**, *129*, 2658–2662. [CrossRef]
5. Pavwoski, P.; Shelgikar, A.V. Treatment options for obstructive sleep apnea. *Neurol. Clin. Pract.* **2017**, *7*, 77–85. [CrossRef] [PubMed]
6. Waters, T. Alternative interventions for obstructive sleep apnea. *Clevel. Clin. J. Med.* **2019**, *86*, 34–41. [CrossRef] [PubMed]
7. Arora, A.; Chaidas, K.; Garas, G.; Amlani, A. Outcome of TORS to tongue base and epiglottis in patients with OSA intolerant of conventional treatment. *Sleep Breath.* **2016**, *20*, 739–747. [CrossRef] [PubMed]
8. Bourolias, C.; Hajiioannou, J.; Sobol, E.; Velegrakis, G.; Helidonis, E. Epiglottis reshaping using CO_2 laser: A minimally invasive technique and its potent applications. *Head Face Med.* **2008**, *4*, 15. [CrossRef] [PubMed]
9. Kezirian, E.J.; Hussey, H.M.; Brietzke, S.E.; Cohen, S.M.; Davis, G.E.; Shin, J.J.; Weinberger, D.G.; Cabana, M.D. Hypopharyngeal surgery in obstructive sleep apnea: Practice patterns, perceptions, and attitudes. *Otolaryngol. Head Neck Surg.* **2012**, *147*, 964–971. [CrossRef]
10. Yung-Chuan Liu, S.; Riley, R.W.; Yu, M.S. Surgical algorithm for obstructive sleep apnea: An update. *Clin. Exp. Otorhinolaryngol.* **2020**, *13*, 215–224. [CrossRef]
11. Hybášková, J.; Jor, O.; Novák, V.; Matoušek, P.; Komínek, P. Možné využití spánkové endoskopie pro zvýšení efektivity léčby (operační i neoperační) u pacientu s obstrukční spánkovou apnoí. *Česk Slov. Neurol. N* **2017**, *113*, 428–433. [CrossRef]
12. De Vito, A.; Cammaroto, G.; Chong, K.B.; Carrasco-Llatas, M.; Vicini, C. Drug-Induced Sleep Endoscopy: Clinical Application and Surgical Outcomes. *Healthcare* **2019**, *7*, 100. [CrossRef]
13. Mickelson, S.A.; Rosenthal, L. Midline glossectomy and epiglottidectomy for obstructive sleep apnea syndrome. *Laryngoscope* **1997**, *107*, 614–619. [CrossRef]
14. Fernández-Julián, E.; García-Pérez, M.A.; García-Callejo, J.; Ferrer, F.; Martí, F.; Marco, J. Surgical planning after sleep versus awake techniques in patients with obstructive sleep apnea. *Laryngoscope* **2014**, *124*, 1970–1974. [CrossRef]
15. Koutsourelakis, I.; Safiruddin, F.; Ravesloot, M.; Zakynthinos, S.; de Vries, N. Surgery for obstructive sleep apnea: Sleep endoscopy determinants of outcome. *Laryngoscope* **2012**, *122*, 2587–2591. [CrossRef]
16. Weaver, T.E.; Grunstein, R.R. Adherence to continuous positive airway pressure therapy: The challenge to effective treatment. *Proc. Am. Thorac. Soc.* **2008**, *5*, 173–178. [CrossRef]
17. Camacho, M.; Certal, V.; Capasso, R. Comprehensive review of surgeries for obstructive sleep apnea syndrome. *Braz. J. Otorhinolaryngol.* **2013**, *79*, 780–788. [CrossRef]
18. Kezirian, E.J.; Weaver, E.M.; Yueh, B.; Deyo, R.A.; Khuri, S.F.; Daley, J.; Henderson, W. Incidence of serious complications after uvulopalatopharyngoplasty. *Laryngoscope* **2004**, *114*, 450–453. [CrossRef] [PubMed]
19. Haavisto, L.; Suonpää, J. Complications of uvulopalatopharyngoplasty. *Clin. Otolaryngol. Allied Sci.* **1994**, *19*, 243–247. [CrossRef]
20. Verse, T.; Pirsig, W. Age-related changes in the epiglottis causing failure of nasal continuous positive airway pressure therapy. *J. Laryngol. Otol.* **1999**, *113*, 1022–1025. [CrossRef]
21. Kent, D.T.; Rogers, R.; Soose, R.J. Drug-induced sedation endoscopy in the evaluation of OSA patients with incomplete oral appliance therapy response. *Otolaryngol. Head. Neck Surg.* **2015**, *153*, 302–307. [CrossRef] [PubMed]
22. Kim, H.Y.; Sung, C.M.; Jang, H.B.; Kim, H.C.; Lim, S.C.; Yang, H.C. Patients with epiglottic collapse showed less severe obstructive sleep apnea and good response to treatment other than continuous positive airway pressure: A case-control study of 224 patients. *J. Clin. Sleep Med.* **2021**, *17*, 413–419. [CrossRef] [PubMed]
23. Azarbarzin, A.; Marques, M.; Sands, S.A.; de Beeck, S.O.; Genta, P.R.; Taranto-Montemurro, L.; de Melo, C.M.; Messineo, L.; Vanderveken, O.M.; White, D.P.; et al. Predicting epiglottic collapse in patients with obstructive sleep apnoea. *Eur. Respir. J.* **2017**, *50*, 1700345. [CrossRef]
24. De Almeida, F.R.; Lowe, A.A.; Tsuiki, S.; Otsuka, R.; Wong, M.; Fastlicht, S. Long-term compliance and side effects of oral appliances used for the treatment of snoring and obstructive sleep apnea syndrome. *J. Clin. Sleep Med.* **2005**, *1*, 143–152. [CrossRef]
25. Sung, C.M.; Kim, H.C.; Yang, H.C. The clinical characteristics of patients with an isolate epiglottic collapse. *Auris Nasus Larynx* **2020**, *47*, 450–457. [CrossRef]
26. Cassano, M. Endoscopic coblator-assisted epiglottoplasty in "obstructive sleep apnoea syndrome" patients. *Clin. Otolaryngol.* **2017**, *42*, 1112–1114. [CrossRef]
27. Oluwasanmi, A.F.; Mal, R.K. Diathermy epiglottectomy: Endoscopic technique. *J. Laryngol. Otol.* **2001**, *115*, 289–292. [CrossRef]
28. Lin, H.S.; Rowley, J.A.; Badr, M.S.; Folbe, A.J.; Yoo, G.H.; Victor, L.; Mathog, R.H.; Chen, W. Transoral robotic surgery for treatment of obstructive sleep apnea-hypopnea syndrome. *Laryngoscope* **2013**, *123*, 1811–1816. [CrossRef]
29. Catalfumo, F.J.; Golz, A.; Westerman, S.T.; Gilbert, L.M.; Joachims, H.Z.; Goldenberg, D. The epiglottis and obstructive sleep apnoea syndrome. *J. Laryngol. Otol.* **1998**, *112*, 940–943. [CrossRef]
30. Sorrenti, G.; Piccin, O.; Mondini, S.; Ceroni, A.R. One-phase management of severe obstructive sleep apnea: Tongue base reduction with hyoepiglottoplasty plus uvulopalatopharyngoplasty. *Otolaryngol. Head Neck Surg.* **2006**, *135*, 906–910. [CrossRef]

life MDPI

Communication

Cognition in Patients with Sleep-Disordered Breathing: Can Obstructive and Central Apneic Pauses Play a Different Role in Cognitive Impairment?

Patrik Karapin [1], Pavel Šiarnik [1,*], Bianka Suchá [1], Matúš Jurík [1], Miroslav Tedla [2,3], Michal Poddaný [4], Katarína Klobučníková [1], Stanislav Šutovský [1], Peter Turčáni [1] and Branislav Kollár [1]

1 1st Department of Neurology, Faculty of Medicine, Comenius University, 81499 Bratislava, Slovakia; patrikkarapin@gmail.com (P.K.); sucha50@uniba.sk (B.S.); jurik25@uniba.sk (M.J.); klobucnikova@gmail.com (K.K.); nilusuto@gmail.com (S.Š.); peter.turcani@sm.unb.sk (P.T.); cyklobrano@gmail.com (B.K.)
2 Department of ENT and HNS, Faculty of Medicine, University Hospital Bratislava, Comenius University, 81499 Bratislava, Slovakia; miro.tedla@gmail.com
3 Institute of Cancer and Genomic Sciences, University of Birmingham, Birmingham B15 2TT, UK
4 Department of Neurology, General Hospital, 03101 Liptovsky Mikulas, Slovakia; michal.poddany@gmail.com
* Correspondence: palo.siarnik@gmail.com; Tel.: +42-190-311-6499

Abstract: Background: There are increasing data linking sleep apnea with cognitive impairment. We aimed to clarify the relationship between sleep-disordered breathing (SDB) and cognition. Detailed attention was assigned to the potential role of central versus obstructive apneic pauses in cognitive impairment. Methods: Patients with suspected SDB were prospectively enrolled, and a complex sleep study was performed that included overnight polysomnography. A revised version of Addenbrooke's Cognitive Examination (ACE-R) was used to assess cognition, evaluating overall cognition and individual subdomains. Results: A total number of 101 participants were included in the study. In multivariate binary logistic regression analysis, obstructive apnea index ([OAI], 95% CI: 1.009–1.057, $p = 0.008$) was the only significant contributor to the model predicting attention deficit. The proportion of N1 stage of NREM sleep was the only significant contributor to the model predicting impaired verbal fluency (95% CI: 1.004–1.081, $p = 0.029$). No significant differences in sleep-related indices were observed in the remaining ACE-R subdomains. Conclusion: Except for verbal fluency and attention, we failed to find any significant association of sleep-related indices with the impairment in different cognitive subdomains. Our data suggest that impairment observed in verbal fluency is associated with a higher proportion of shallow NREM sleep, and attention deficit is associated with higher OAI. Obstructive respiratory episodes seem to play a more important role in cognitive impairment when compared to central ones.

Keywords: sleep-disordered breathing; cognition; cognitive deficit; polysomnography; central apneas; obstructive apneas

Citation: Karapin, P.; Šiarnik, P.; Suchá, B.; Jurík, M.; Tedla, M.; Poddaný, M.; Klobučníková, K.; Šutovský, S.; Turčáni, P.; Kollár, B. Cognition in Patients with Sleep-Disordered Breathing: Can Obstructive and Central Apneic Pauses Play a Different Role in Cognitive Impairment? *Life* **2022**, *12*, 1180. https://doi.org/10.3390/life12081180

Academic Editors: David Slouka and Milan Štengl

Received: 1 July 2022
Accepted: 1 August 2022
Published: 2 August 2022

1. Introduction

Sleep-disordered breathing (SDB) is a treatable disease that is frequent in the adult population [1]. The most common form of SDB is obstructive sleep apnea (OSA), which is characterized by recurrent upper airway obstruction, causing recurrent episodes of apnea or hypopnea during sleep, increased daytime sleepiness, sleep fragmentation, and intermittent hypoxia [2].

There is increased interest in investigating the impact of SDB on overall health and its association with numerous diseases [3–5]. Studies are also trying to elucidate the association between OSA and cognitive impairment (CI). Population studies evaluating the relationship between SDB and CI in healthy individuals have mixed results [6,7]. A recent meta-analysis [8] has shown that individuals with SDB have a 26% increased risk of cognitive deficit (risk ratio 1.26; 95% CI 1.05–1.50) when compared to the population without SDB. There are contradictory findings from studies that examined which cognitive domains are mostly impaired in patients with OSA [9]. Attention deficit and executive dysfunction are generally observed in these individuals. On the contrary, memory impairment is less prevalent in patients with OSA [10].

Furthermore, the role of SDB in the deterioration of overall cognition is not precisely recognized. Sleep fragmentation, sleep deprivation, intermittent nocturnal hypoxia, and disruption of sleep architecture are believed to be essential underlying mechanisms responsible for cognitive impairment [10]. Our study aimed to clarify the relationship between SDB and cognitive profile. Detailed attention was paid to the potential role of central versus obstructive apneic pauses in cognitive impairment, which has not been described in detail in previous works.

2. Materials and Methods

2.1. Participants

We consecutively enrolled patients who were suspected of suffering SDB and were hospitalized in the sleep laboratory of the 1st Department of Neurology, Comenius University, and University Hospital Bratislava (Old Town Hospital, Bratislava, Slovakia). From these patients, participants were selected for our study after meeting the inclusion criteria: (a) age between 18 and 80 years; (b) confirmed diagnosis of SDB by overnight polysomnography (defined as apnea/hypopnea index [AHI] $\geq$ 5). The exclusion criteria were: (a) presence of a cognitive deficit defined as Mini-Mental State Examination (MMSE) $\leq$ 24; (b) a known history of psychiatric illness, which may explain the presence of cognitive deficits (depression, psychotic disorders, attention deficit hyperactivity disorder, posttraumatic stress disorder); (c) use of drugs affecting sleep (hypnotics, sedatives, anxiolytics, steroids); (d) history of treatment for sleep disorders. The baseline evaluation of all patients included assessment of clinical and demographic characteristics including sex, age, neck circumference, and body mass index (BMI). Medical records of all patients were reviewed to search for medical conditions (arterial hypertension, diabetes mellitus, atrial fibrillation, stroke, dyslipidemia, and thyroid disease) that could contribute to CI.

All participants agreed to participate in the study. The agreement was confirmed by signing an informed consent prior to enrollment. The study was approved by the Ethics Committee of the Old Town Hospital, University Hospital Bratislava.

2.2. Sleep Study

Overnight polysomnography was performed on all participants (Alice 6, Philips-Respironics, Murrysville, PA, USA). The sleep specialist evaluated and scored polysomnography based on standardized criteria [11]. Observed parameters were apnea/hypopnea index (AHI), which was defined as the number of apneic or hypopneas pauses during an hour of sleep; other variables were arousal index ([AI], number of arousals per hour of sleep), desaturation index (number of all oxygen desaturations $\geq$3% per hour of sleep). We evaluated the obstructive apnea index (OAI), defined as the number of obstructive and mixed apneic pauses per hour of sleep, and the central apnea index (CAI), defined as the number of central apneic pauses per hour of sleep. We also assessed the hypopnea index, defined as the number of hypopneas per hour of sleep and the percentual duration of snoring during sleep. The hypnogram and the percentage of particular sleep stages (REM, NREM phases—N1, N2, and N3) were recorded. The diagnosis of SDB was defined as AHI $\geq$ 5. The Epworth Sleepiness Scale (ESS) questionnaire was used to assess daytime sleepiness [12], and the Pittsburgh Sleep Quality Index (PSQI) questionnaire was used to subjectively assess sleep quality [13].

2.3. Evaluation of Cognition

To assess the cognitive profile, we administered a psychometric examination of cognition in patients prior to polysomnography. An extended screening method was performed—a revised version of Addenbrooke's Cognitive Examination (ACE-R) [14]. We evaluated scores of overall cognition and particular cognitive subdomains: orientation/attention, memory, verbal fluency, language, and visuospatial functions. We classified patients according to the presence of an impairment in individual cognitive subdomains. Individual impairment in different subdomains was defined as a subtest score for orientation and attention $\leq$16; memory subtest score $\leq$17; verbal fluency subtest score $\leq$9; language subtest score $\leq$23; subtest score for visuospatial abilities $\leq$14 [14].

2.4. Statistical Analysis

Statistical analysis was performed using SPSS, version 21 (SPSS Inc, Chicago, IL, USA). Categorical variables were described as number and percentage (%), and continuous variables were described as mean $\pm$ standard deviation or median and IQR (interquartile range). To compare values of the continuous variables between two groups, we conducted Student's *t*-test if variables were normally distributed; otherwise, the Mann–Whitney test was conducted. The Chi-square test was used for categorical variables. *p*-value < 0.05 was considered statistically significant. In the binary logistic regression analysis, 95% confidence intervals were reported to declare the statistical significance and strength of association between deficit in particular cognitive subdomains (dependent variable) and the independent variables (demographic variables: age, gender, body mass index, neck circumference, presence of diseases: arterial hypertension, diabetes mellitus, atrial fibrillation, stroke, dyslipidemia, thyroid disease, and sleep parameters: AHI, AI, desaturation index, the proportion of particular sleep stages, OAI, CAI, hypopnea index, percentual duration of snoring, ESS, and PSQI) was assessed. Variables with a $p < 0.05$ in the bi-variable binary logistic regression analysis were considered for the multivariable analysis.

3. Results

3.1. Demographic and Cognitive Parameters

After meeting the inclusion criteria, 101 participants were included in the study. Baseline characteristics of the population are shown in Table 1. The decline in ACE-R subtest for orientation and attention was observed in six participants (5.9%). Memory deterioration was present in seven patients (6.9%). Impaired verbal fluency was confirmed in 11 participants (10.9%). Speech deficit was not observed in any individual in the study group. Deterioration of visuospatial functions was observed in 13 patients (12.9%). A CI found in at least one cognitive subdomain was present in 27 individuals (26.7%).

3.2. Comparison of Sleep Variables and Cognition

Our primary goal of the study was to compare sleep variables in groups divided according to the presence of CI in different cognitive subdomains. Statistically significant differences between observed sleep parameters were observed in individuals with attention deficit and impaired verbal fluency (detailed results are shown in Table 2). No significant differences in sleep parameters were found in CI subjects in other cognitive subdomains. In participants with attention deficit, we proved that they had statistically higher value of AHI ($p = 0.035$), AI ($p = 0.039$), and desaturation index ($p = 0.024$). In these patients, there were also statistically more frequent obstructive apneic pauses ($p = 0.025$), as well as central apneic pauses ($p = 0.036$). Furthermore, in participants with attention deficit, the ESS value was statistically higher ($p = 0.012$). In patients with impaired verbal fluency, a statistically higher percentage of the N1 NREM phase of sleep was observed ($p = 0.014$). In multivariate binary logistic regression analysis, we found that OAI (95% CI: 1.009–1.057, $p = 0.008$) was the only significant contributor to the model predicting attention deficit. The proportion of the N1 NREM phase was the only significant factor in the model that predicted impaired verbal fluency (95% CI: 1.004–1.081, $p = 0.029$).

Table 1. Baseline characteristics.

Demographic Variables		**Sleep Parameters**		**Cognitive Parameters**	
Number of participants (%)	101 (100%)	**AHI** (*n*/**hour**)	31.8; 36.7 (5.8–157.2)	**MMSE**	28; 2 (25–30)
Men's representation (%)	70 (69.3%)	**Arousal index** (*n*/**hour**)	18; 25.3 (1.4–89)	**ACE-R**	94; 7 (81–100)
Age (**years**)	55.08 ± 11.51	**Desaturation index** (*n*/**hour**)	29.3; 39.2 (1.2–149.6)	**Orientation-attention**	18; 1 (16–18)
BMI (**kg/m²**)	33.86 ± 6.67	**REM** (%)	8.5; 10.4 (0–30.4)	**Memory**	23; 3 (13–26)
Neck circumference (**cm**)	42.37 ± 4.15	**N1** (%)	34.1; 22.9 (1.9–74.6)	**Verbal fluency**	12; 3 (7–14)
Arterial hypertension (%)	69 (68.3%)	**N2** (%)	27.8; 18.7 (6.3–82.7)	**Language**	26; 0 (24–26)
Diabetes mellitus (%)	21 (20.8%)	**N3** (%)	23.5; 21.9 (0–58.1)	**Visuospatial functions**	16; 1 (12–16)
Atrial fibrillation (%)	9 (8.9%)	**OAI** (*n*/**hour**)	5.72; 18.24 (0–132.28)		
Stroke (%)	5 (5.0%)	**CAI** (*n*/**hour**)	5.44; 12.87 (0–92.75)		
Dyslipidemia (%)	31 (30.7%)	**Hypopnea index** (*n*/**hour**)	15.41; 18.17 (1.26–65.86)		
Thyroid disease (%)	14 (13.9%)	**Snoring** (%)	3.9; 10.65 (0–42.4)		
		ESS	7; 8 (0–24)		
		PSQI	5; 4 (0–17)		

BMI—body mass index; ESS—Epworth Sleepiness Scale; PSQI—Pittsburgh Sleep Quality Index; AHI—apnea/hypopnea index; REM—percentage of REM phase during the whole sleep; N1—percentage of N1 NREM phase during the whole sleep; N2—percentage of N2 NREM phase during the whole sleep; N3—percentage of N3 NREM phase during the whole sleep; OAI—obstructive apnea index; CAI—central apnea index; MMSE—Mini-Mental State Examination; ACE-R—Addenbrooke's Cognitive Examination-Revised; Snoring—percentual duration of snoring during sleep.

Table 2. Baseline characteristics in populations with and without a deficit in the particular cognitive subdomain.

Variables	Attention			Verbal Fluency		
	Without Deficit	With Deficit	*p*-Value	Without Deficit	With Deficit	*p*-Value
Number of participants (%)	95 (94.1%)	6 (5.9%)		90 (89.1%)	11 (10.9%)	
Men's representation (%)	65 (68.4%)	5 (83.3%)	0.442	65 (72.2%)	5 (45.5%)	0.069
Age (**years**)	55.41 ± 11.17	49.83 ± 16.38	0.252	54.82 ± 11.30	57.18 ± 13.51	0.524
BMI (**kg/m²**)	33.57 ± 6.59	38.58 ± 6.81	0.074	33.91 ± 6.59	33.48 ± 7.65	0.842
Neck circumference (**cm**)	42.21 ± 4.18	44.83 ± 2.99	0.134	42.32 ± 3.93	42.73 ± 5.93	0.762
AHI (*n*/**hour**)	31.8; 37.6 (6.7–153.2)	64.9; 67.9 (21.0–140.3)	0.035 *	34.9; 44.9 (6.7–153.2)	31.2; 16.3 (15.2–72.9)	0.819
Arousal index (*n*/**hod.**)	16.8; 24.7 (1.4–89)	38; 27.2 (16.8–71.8)	0.039 *	18.9; 26.4 (1.4–89)	16; 7.7 (10.8–38.9)	0.586
Desaturation index (*n*/**hour**)	27.4; 36.6 (1.2–149.6)	66.1; 60.4 (17.5–120.1)	0.024 *	31.3; 41.1 (1.2–149.6)	25.1; 13.9 (7.5–70.1)	0.42
REM (%)	8.4; 10.3 (0–25.7)	11.65; 15.4 (0–30.4)	0.551	8.55; 9.8 (0–30.4)	2.9; 11.9 (0–15.8)	0.138
N1 (%)	34; 23.1 (1.9–74.6)	37.35; 36.2 (8.2–66.4)	0.537	31.35; 20.7 (1.9–74.6)	45.4; 19 (23–72.3)	0.014 *
N2 (%)	28; 19.5 (6.3–82.7)	24.5; 11.1 (18.8–36.4)	0.413	28.15; 19.8 (6.3–82.7)	25.9; 8.1 (8.3–37.9)	0.206
N3 (%)	23.5; 23.2 (0–58.1)	22.9; 9.6 (14.8–36.2)	0.886	24; 23.5 (0–58.1)	21.8; 16.5 (11–39.2)	0.731
OAI (*n*/**hour**)	5.33; 13.76 (0–119.18)	31.6; 66.18 (0.77–132.28)	0.025 *	5.97; 19.22 (0–132.28)	4.66; 7.31 (0–20.45)	0.38
CAI (*n*/**hour**)	4.79; 12.12 (0–92.75)	23.25; 31.53 (4.26–37.57)	0.036 *	5.75; 13.54 (0–92.75)	3.35; 7.26 (0.72–37.57)	0.922
Hypopnea index (*n*/**hour**)	14.2; 40.6 (0–169.4)	19.45; 47.7 (1.2–133.5)	0.886	14.75; 39.25 (0–163.8)	16.3; 63.1 (0–169.4)	0.407
Snoring (%)	3.60; 10.70 (0–42.40)	4.65; 17.05 (0.3–34.9)	0.556	3.75; 10.58 (0–40.9)	4.0; 15.0 (0–42.4)	0.883
ESS	7; 8 (0–20)	13; 10 (7–24)	0.012 *	7; 8 (0–24)	6; 8 (1–16)	0.922
PSQI	5; 4 (0–14)	6; 9 (2–17)	0.319	5; 4 (0–17)	8; 5 (4–14)	0.079

* $p < 0.05$. BMI—body mass index; ESS—Epworth Sleepiness Scale; PSQI—Pittsburgh Sleep Quality Index; AHI—apnea/hypopnea index; REM—percentage of REM phase during the whole sleep; N1—percentage of N1 NREM phase during the whole sleep; N2—percentage of N2 NREM phase during the whole sleep; N3—percentage of N3 NREM phase during the whole sleep; OAI—obstructive apnea index; CAI—central apnea index; snoring—percentual duration of snoring during sleep.

4. Discussion

The results of our study support the findings of previous works, which emphasize the association between SDB and cognitive impairment. Our results suggest impaired sleep parameters in patients with sleep apnea who also have a CI present in attention and executive functions. Our findings are in agreement with a recent meta-analysis [8], which presents that individuals with OSA had significantly worse executive functions than the population without OSA. On the other hand, there was no association with impaired global cognition or memory decline.

The findings of our work support previous findings about the high prevalence of CI in individuals with SDB. We also found that alterations in cognitive functions were associated with the presence of sleep apnea, the severity of nocturnal desaturations, sleep fragmentation, and increased daytime sleepiness [10,15]. Our findings suggest that recurrent apneic pauses with subsequent intermittent brain hypoxia and sleep fragmentation could affect cognitive impairment. The consequences of these processes can cause structural and cerebrovascular damage to the brain, which could ultimately explain cognitive impairment in patients with SDB [16]. However, the design of our study does not allow us to elucidate the causal role of sleep apnea in the process of cognitive impairment.

Altered sleep architecture and sleep disruption could be important underlying mechanisms that link SDB with cognitive decline. Our results suggest that in individuals with SDB, alterations in cognition were associated with a higher proportion of shallow NREM sleep. It could indicate a deterioration in restorative sleep function and possible subsequent negative effects on cognitive function. This hypothesis is supported by the findings of previous studies, where the reduction in REM sleep and increased percentage of N1 NREM phase correlated with attention deficit and executive dysfunction [17,18]. In our study, the proportion of the N1 NREM phase was the only significant factor in the model predicting impaired verbal fluency.

A possible causal link could be increased amyloid production in disruption of NREM sleep. Sleep duration is disrupted and shortened during physiological aging and in the development of Alzheimer's disease (AD). Furthermore, a significant decrease in slow-wave sleep is observed, which correlates with the N3 stage of NREM sleep [19]. These changes are more prominent in patients with mild CI and individuals with AD [20,21]. A bidirectional relationship is suspected between the reduction of deep NREM sleep phases and pathomorphological changes in AD [22]. Animal studies support this hypothesis. Experimental increase in cortical amyloid β (Aβ) deposition causes NREM sleep disruption [23]. On the other hand, experimental NREM sleep reduction and prolonged wakefulness result in increased Aβ production [24]. In contrast, NREM sleep promotes the clearance of extracellular Aβ, which accumulates during wakefulness [25].

The strength of our work is the assessment of the impact of obstructive versus central apneic pauses on cognitive profile. Most studies deal with OSA and its possible impact on cognition. We are aware of only a few studies that took central sleep apnea into account. In a recent study, the authors assessed the prevalence of different types of SDB in elderly patients. In this study, the authors found that OSA was associated with lower general cognition, while CSA was only associated with executive dysfunction [26]. We found that individuals with attention deficit had statistically more frequent obstructive and central apneas. This finding could lead to the assumption that both obstructive and central apneas are linked to attention deficit. This hypothesis could be explained by the fact that obstructive and central apnea overlap in some clinical consequences [27]. On the other hand, after evaluating multivariate binary logistic regression analysis, only the OAI and not CAI was associated with attention deficit. This could support the superior importance of obstructive apneas in the pathogenesis of attention deficit. The exact underlying mechanisms need to be elucidated by future prospective studies. Increased cerebrovascular morbidity in OSA patients is well-known and could play a role. On the other hand, the causal role of CSA in the pathogenesis of the cerebrovascular disease remains controversial [28].

The conclusions of our study could be affected by several limitations. The main limitation of our work is a relatively small study population. We are aware that multiple factors could impact the results in performed cognitive tests, including premorbid intellect or achieved education that were not studied in detail in our study. Another factor that may have interfered with our results is the presence of undiagnosed depression, but we tried to influence this modifying factor by the exclusion of participants who had been diagnosed with depression or other significant psychiatric illnesses that could interfere with cognition.

5. Conclusions

Our study confirmed the high frequency of cognitive deficit in subjects with SDB. Except for verbal fluency and attention, we failed to find any significant association of sleep-related indices with the deficit in other cognition subdomains. Our data suggest that alteration of verbal fluency was associated with a higher proportion of N1 NREM sleep. Attention deficit was associated with higher OAI. Obstructive respiratory episodes seem to play a more important role in cognitive deficit when compared to central ones.

Author Contributions: Conceptualization, P.Š., B.K. and P.T.; methodology, K.K. and B.K.; investigation, P.K. and B.S.; resources, P.K., M.J. and M.P.; writing—original draft preparation, P.K. and P.Š.; writing—review and editing, P.T., M.T. and S.Š.; supervision, B.K.; project administration, P.K. and P.Š. All authors have read and agreed to the published version of the manuscript.

Funding: APVV-15-0253 grant supported this work.

Institutional Review Board Statement: This observational study was approved by the Ethics Committee of the Old Town Hospital, University Hospital Bratislava (Approval Code: 26/2021, Approval Date: 15 February 2021). The study was conducted according to the guidelines of the Declaration of Helsinki.

Informed Consent Statement: Informed consent was obtained from all subjects involved in the study.

Data Availability Statement: The data presented in this study are available on request from the corresponding author.

Conflicts of Interest: The authors declare no conflict of interest.

References

1. Young, T.; Peppard, P.E.; Gottlieb, D.J. Epidemiology of obstructive sleep apnea: A population health perspective. *Am. J. Respir. Crit. Care Med.* **2002**, *165*, 1217–1239. [CrossRef] [PubMed]
2. Gottlieb, D.J.; Punjabi, N.M. Diagnosis and management of obstructive sleep apnea: A review. *JAMA* **2020**, *323*, 1389–1400. [CrossRef] [PubMed]
3. Nieto, F.J.; Young, T.B.; Lind, B.K.; Shahar, E.; Samet, J.M.; Redline, S.; D'agostino, R.B.; Newman, A.B.; Lebowitz, M.D.; Pickering, T.G. Association of sleep-disordered breathing, sleep apnea, and hypertension in a large community-based study. Sleep Heart Health Study. *JAMA* **2000**, *283*, 1829–1836. [CrossRef]
4. Rajan, P.; Greenberg, H. Obstructive sleep apnea as a risk factor for type 2 diabetes mellitus. *Nat. Sci. Sleep* **2015**, *7*, 113–125. [PubMed]
5. McDermott, M.; Brown, D.L. Sleep apnea and stroke. *Curr. Opin. Neurol.* **2020**, *33*, 4–9. [CrossRef] [PubMed]
6. Haba-Rubio, J.; Marti-Soler, H.; Tobback, N.; Andries, D.; Marques-Vidal, P.; Waeber, G.; Vollenweider, P.; von Gunten, A.; Preisig, M.; Castelao, E.; et al. Sleep characteristics and cognitive impairment in the general population: The HypnoLaus study. *Neurology* **2017**, *88*, 463–469. [CrossRef] [PubMed]
7. Lutsey, P.L.; Bengtson, L.G.; Punjabi, N.M.; Shahar, E.; Mosley, T.H.; Gottesman, R.F.; Wruck, L.M.; MacLehose, R.F.; Alonso, A. Obstructive sleep apnea and 15-year cognitive decline: The atherosclerosis risk in communities (ARIC) study. *Sleep* **2016**, *39*, 309–316. [CrossRef] [PubMed]
8. Leng, Y.; McEvoy, C.T.; Allen, I.E.; Yaffe, K. Association of sleep-disordered breathing with cognitive function and risk of cognitive impairment: A systematic review and meta-analysis. *JAMA Neurol.* **2017**, *74*, 1237–1245. [CrossRef]
9. Devita, M.; Montemurro, S.; Ramponi, S.; Marvisi, M.; Villani, D.; Raimondi, M.C.; Rusconi, M.L.; Mondini, S. Obstructive sleep apnea and its controversial effects on cognition. *J. Clin. Exp. Neuropsychol.* **2017**, *39*, 659–669. [CrossRef]
10. Seda, G.; Han, T.S. Effect of obstructive sleep apnea on neurocognitive performance. *Sleep Med. Clin.* **2020**, *15*, 77–85. [CrossRef]
11. Iber, C.; Ancoli-Israel, S.; Chesson, A.; Quan, S.F. *The AASM Manual for the Scoring of Sleep and Associated Events: Rules. Terminology and Technical Specifications*; American Academy of Sleep Medicine: Westchester, IL, USA, 2007.
12. Johns, M.W. A new method for measuring daytime sleepiness: The Epworth sleepiness scale. *Sleep* **1991**, *14*, 540–545. [CrossRef] [PubMed]
13. Buysse, D.J.; Reynolds, C.F., 3rd; Monk, T.H.; Berman, S.R.; Kupfer, D.J. The Pittsburgh sleep quality index: A new instrument for psychiatric practice and research. *Psychiatry Res.* **1989**, *28*, 193–213. [CrossRef]
14. Mioshi, E.; Dawson, K.; Mitchell, J.; Arnold, R.; Hodges, J.R. The addenbrooke's cognitive examination revised (ACE-R): A brief cognitive test battery for dementia screening. *Int. J. Geriatr. Psychiatry* **2006**, *21*, 1078–1085. [CrossRef] [PubMed]
15. Bucks, R.S.; Olaithe, M.; Eastwood, P. Neurocognitive function in obstructive sleep apnoea: A meta-review. *Respirology* **2013**, *18*, 61–70. [CrossRef]

16. Yaffe, K.; Falvey, C.M.; Hoang, T. Connections between sleep and cognition in older adults. *Lancet Neurol.* **2014**, *13*, 1017–1028. [CrossRef]
17. Song, Y.; Blackwell, T.; Yaffe, K.; Ancoli-Israel, S.; Redline, S.; Stone, K.L. Relationships between sleep stages and changes in cognitive function in older men: The MrOS sleep study. *Sleep* **2015**, *38*, 411–421. [PubMed]
18. Parker, J.L.; Appleton, S.L.; Melaku, Y.A.; Stevens, D.; Wittert, G.A.; Martin, S.; Adams, R.J.; Vakulin, A. Sleep macroarchitecture but not obstructive sleep apnea is independently associated with cognitive function in only older men of a population-based cohort. *J. Sleep Res.* **2021**, *30*, e13370. [CrossRef] [PubMed]
19. Ohayon, M.M.; Carskadon, M.A.; Guilleminault, C.; Vitiello, M.V. Meta-analysis of quantitative sleep parameters from childhood to old age in healthy individuals: Developing normative sleep values across the human lifespan. *Sleep* **2004**, *27*, 1255–1273. [CrossRef]
20. Westerberg, C.E.; Mander, B.A.; Florczak, S.M.; Weintraub, S.; Mesulam, M.M.; Zee, P.C.; Paller, K.A. Concurrent impairments in sleep and memory in amnestic mild cognitive impairment. *J. Int. Neuropsychol. Soc.* **2012**, *18*, 490–500. [CrossRef] [PubMed]
21. Liguori, C.; Romigi, A.; Nuccetelli, M.; Zannino, S.; Sancesario, G.; Martorana, A.; Albanese, M.; Mercuri, N.B.; Izzi, F.; Bernardini, S.; et al. Orexinergic system dysregulation, sleep impairment, and cognitive decline in Alzheimer disease. *JAMA Neurol.* **2014**, *71*, 1498–1505. [CrossRef]
22. Mander, B.A.; Winer, J.R.; Jagust, W.J.; Walker, M.P. Sleep: A novel mechanistic pathway, biomarker, and treatment target in the pathology of Alzheimer's disease? *Trends Neurosci.* **2016**, *39*, 552–566. [CrossRef] [PubMed]
23. Roh, J.H.; Huang, Y.; Bero, A.W.; Kasten, T.; Stewart, F.R.; Bateman, R.J.; Holtzman, D.M. Disruption of the sleep-wake cycle and diurnal fluctuation of β-amyloid in mice with Alzheimer's disease pathology. *Sci. Transl. Med.* **2012**, *4*, 150ra122. [CrossRef] [PubMed]
24. Kang, J.E.; Lim, M.M.; Bateman, R.J.; Lee, J.J.; Smyth, L.P.; Cirrito, J.R.; Fujiki, N.; Nishino, S. Amyloid-beta dynamics are regulated by orexin and the sleep-wake cycle. *Science* **2009**, *326*, 1005–1007. [CrossRef] [PubMed]
25. Xie, L.; Kang, H.; Xu, Q.; Chen, M.J.; Liao, Y.; Thiyagarajan, M.; O'Donnell, J.; Christensen, D.J.; Nicholson, C.; Iliff, J.J.; et al. Sleep drives metabolite clearance from the adult brain. *Science* **2013**, *342*, 373–377. [CrossRef] [PubMed]
26. Attier-Zmudka, J.; Sérot, J.M.; Valluy, J.; Saffarini, M.; Douadi, Y.; Malinowski, K.P.; Balédent, O. Sleep apnea syndrome in an elderly population admitted to a geriatric unit: Prevalence and effect on cognitive function. *Front. Aging Neurosci.* **2019**, *11*, 361. [CrossRef]
27. Eckert, D.J.; Jordan, A.S.; Merchia, P.; Malhotra, A. Central sleep apnea: Pathophysiology and treatment. *Chest* **2007**, *131*, 595–607. [CrossRef]
28. Javaheri, S.; Barbe, F.; Campos-Rodriguez, F.; Dempsey, J.A.; Khayat, R.; Javaheri, S.; Malhotra, A.; Martinez-Garcia, M.A.; Mehra, R.; Pack, A.I.; et al. Sleep apnea: Types, mechanisms, and clinical cardiovascular consequences. *J. Am. Coll. Cardiol.* **2017**, *69*, 841–858. [CrossRef] [PubMed]

life

MDPI

Article

Risk of Cardiovascular Disease in Apnoeic Individuals: Role of Comorbid Insomnia Disorder

Matthieu Hein *, Benjamin Wacquier, Jean-Pol Lanquart and Gwenolé Loas

Erasme Hospital, Department of Psychiatry and Sleep Laboratory, Université Libre de Bruxelles, ULB, Route de Lennik, 808, Anderlecht, 1070 Brussels, Belgium; benjamin.wacquier@erasme.ulb.ac.be (B.W.); jplanquart@skynet.be (J.-P.L.); gwenole.loas@erasme.ulb.ac.be (G.L.)

* Correspondence: matthieu.hein@erasme.ulb.ac.be or secmed.psy@erasme.ulb.ac.be; Tel.: +32-2555-3741; Fax: +32-2555-6955

Abstract: Given the limited data available, the aim of this study was to examine the 10-year cardiovascular disease (CVD) risk associated with comorbid insomnia disorder and its specific subtypes in apnoeic individuals. Data from 1104 apnoeic individuals recruited from the database of the Erasme Hospital Sleep Laboratory were analysed. Only apnoeic individuals with a Framingham Risk Score ≥10% were included in the group at moderate-to-high 10-year CVD risk. Logistic regression analyses were conducted to examine the risk of 10-year CVD risk associated with comorbid insomnia disorder and its specific subtypes in apnoeic individuals. Moderate-to-high 10-year CVD risk was present in 59.6% of the apnoeic individuals in our sample. After adjustment for the main confounding factors, multivariate logistic regression analyses revealed that comorbid insomnia disorder and, more particularly, its subtype with short sleep duration were significantly associated with moderate-to-high 10-year CVD risk in apnoeic individuals. In this study, we demonstrate that comorbid insomnia disorder and, more specifically, its subtype with short sleep duration appear to have a negative cumulative effect on 10-year CVD risk in apnoeic individuals, which justifies more systematic research and adequate therapeutic management of this disorder to allow for better cardiovascular disease prevention in this particular subpopulation.

Keywords: cardiovascular risk; insomnia disorder; obstructive sleep apnoea syndrome; polysomnography

Citation: Hein, M.; Wacquier, B.; Lanquart, J.-P.; Loas, G. Risk of Cardiovascular Disease in Apnoeic Individuals: Role of Comorbid Insomnia Disorder. *Life* **2022**, *12*, 944. https://doi.org/10.3390/life12070944

Academic Editors: David Slouka and Milan Štengl

Received: 29 April 2022
Accepted: 20 June 2022
Published: 23 June 2022

Highlights

- Apnoeic patients are a subpopulation at high risk of CVD.
- Comorbid insomnia disorder is associated with higher CVD risk in apnoeic patients.
- This association seems to be mediated by short sleep duration in apnoeic patients.
- Appropriate management of this disorder is essential in apnoeic patients.

1. Introduction

In the literature, there are many arguments in support of the association between obstructive sleep apnoea syndrome (OSAS) and cardiovascular disease (CVD). Indeed, the prevalence of OSAS may reach 70.0% in individuals with CVD, and the incidence of CVD is high in apnoeic individuals [1,2]. In addition, OSAS is associated with a negative impact on cardiovascular prognosis, both in the general population and in individuals with CVD [3–5]. Pathophysiologically, this higher risk of CVD in apnoeic individuals seems to be mediated by some deleterious mechanisms induced by intermittent hypoxia related to obstructive events (hyperactivation of the sympathetic nervous system, alterations in endothelial function, activation of pro-inflammatory pathways, alterations in the renin–angiotensin system, and metabolic dysregulations) [6]. In addition, in apnoeic individuals, there seems to be a severity-dependent effect of OSAS on the occurrence of these deleterious mechanisms induced by intermittent hypoxia [7]. However, despite a potential beneficial effect of OSAS treatments on these pathophysiological mechanisms that negatively impact cardiovascular prognosis [8], OSAS treatments have been shown to have only a limited effect on reducing cardiovascular risk in apnoeic individuals [9,10]. Thus, it seems necessary to carry out additional investigations in order to identify the potential cofactors involved in this higher risk of CVD in apnoeic individuals.

In apnoeic individuals, insomnia disorder is a frequent comorbidity [11], since its prevalence is estimated at 38.0% in this particular subpopulation [12]. However, similar to OSAS [13], there seems to be a special relationship between insomnia disorder and CVD. Indeed, insomnia disorder is a frequent comorbidity in individuals with CVD, and the prevalence of CVD is not negligible in individuals with insomnia disorder [14,15]. In addition, insomnia disorder appears to promote higher cardiovascular morbidity and mortality both in the general population and in individuals with CVD [16]. Nevertheless, this negative impact of insomnia disorder on cardiovascular prognosis seems to be only associated with some specific subtypes of this disorder [17]. Indeed, some studies have shown that insomnia sufferers with short sleep duration presented a higher risk of CVD than those without short sleep duration [17]. However, despite this high prevalence of comorbid insomnia disorder in apnoeic individuals, and its potential involvement in the occurrence of CVD, the potential role played by this disorder and its specific subtypes in the 10-year CVD risk has been poorly studied in this particular subpopulation [18]. Indeed, most of the studies available in the literature have mainly investigated the impact of comorbid insomnia disorder on the occurrence of some conventional cardiovascular risk factors and some specific CVD in apnoeic individuals [19–21]. Thus, given the limited data available in the literature, it could be interesting to study the 10-year CVD risk associated with comorbid insomnia disorder and its specific subtypes in apnoeic individuals in order to better understand the poor cardiovascular prognosis of this particular subpopulation.

The aim of this study was, therefore, to empirically investigate the 10-year CVD risk associated with comorbid insomnia disorder and its specific subtypes in apnoeic individuals. Our hypothesis was that comorbid insomnia disorder and, more specifically, its subtype with short sleep duration are associated with higher 10-year CVD risk in apnoeic individuals. The objective of this approach was to provide healthcare professionals caring for apnoeic individuals with reliable data regarding the 10-year CVD risk associated with comorbid insomnia disorder and its specific subtypes in order to allow the development of more targeted cardiovascular prevention strategies in this particular subpopulation.

2. Materials and Methods

2.1. Population

A total of 1104 apnoeic individuals were recruited from the clinical database of the Erasme Hospital Sleep Laboratory, which contains demographic and polysomnographic data from individuals who performed a polysomnographic recording between 2002 and 2019. The inclusion and exclusion criteria applied for the recruitment of these 1104 apnoeic individuals are available in Table 1 [22]. In this study, we included only apnoeic individuals because our objective was to focus on this particular subpopulation, in which the occurrence of comorbid insomnia disorder may negatively impact the cardiovascular prognosis [18]. Finally, the detailed description of the outpatient recruitment procedure for the apnoeic patients included in this study is available in Supplementary Materials—Section S1 [23].

Table 1. Inclusion and exclusion criteria.

Inclusion Criteria	Exclusion Criteria
Age ≥ 40 years	Severe psychiatric disorder • Psychotic disorder • Bipolar disorder • Current or past substance abuse
OSAS according to the diagnostic criteria of the American Academy of Sleep Medicine [22]	Uncontrolled somatic disorder • Chronic hepatic disorder • Chronic pancreatic disorder • Chronic pulmonary disorder • Chronic cardiovascular disorder • Chronic renal disorder • Autoimmune disorder • Infectious or inflammatory disorder • Disorder altering the activity of the hypothalamic–pituitary–adrenal axis
Absence of previous CVD • Coronary heart disease • Cerebrovascular disease • Peripheral arterial disease • Heart failure	Sleep disorder • Central hypersomnia • Parasomnia • Predominantly central sleep apnoea syndrome • OSAS being treated before the sleep examination
	Lesions or malformations • Current or past cranial trauma • Current or past central nervous system lesions involving the respiratory centres • Craniofacial or thoracic cavity malformations Pregnancy

CVD = cardiovascular disease, OSAS = obstructive sleep apnoea syndrome.

2.2. *Method*

2.2.1. Medical and Psychiatric Assessment of Participants

During their admission to the Sleep Laboratory, all apnoeic individuals included in this study benefited from a medical interview and a somatic check-up (including blood test, electrocardiogram, daytime electroencephalogram, and urine analysis) in order to allow a systematic diagnosis of their potential somatic comorbidities. Following this comprehensive somatic assessment, a systematic diagnosis of conventional cardiovascular risk factors (type 2 diabetes (American Diabetes Association diagnostic criteria), hypertension (World Health Organization diagnostic criteria), dyslipidemia (International Diabetes Federation diagnostic criteria), and cardiovascular comorbidities) was performed for all apnoeic individuals included in this study (a detailed description of the diagnostic criteria used is available in Supplementary Materials—Section S2) [24–27].

Based on these different elements collected during this systematic somatic assessment, the Framingham Risk Score was used to estimate the 10-year risk of manifesting clinical CVD (cardiovascular mortality, coronary heart disease, cerebrovascular disease, peripheral arterial disease, or heart failure) in apnoeic individuals included in this study [28]. The prediction model of the Framingham Risk Score integrates age, sex, smoking status, systolic blood pressure, taking antihypertensive medication, total-cholesterol levels, HDL-cholesterol levels, and diabetes status [28]. A Framingham Risk Score <10% indicates a low 10-year CVD risk, whereas a Framingham Risk Score ≥10% indicates a moderate-to-high

10-year CVD risk [28]. Finally, the Framingham Risk Score is a cardiovascular risk score frequently used in the subpopulation of apnoeic individuals [18,29,30].

After this comprehensive somatic assessment, all apnoeic individuals included in this study benefited from a systematic psychiatric evaluation by a unit psychiatrist to diagnose their potential psychiatric comorbidities according to the diagnostic criteria of the DSM-IV-TR (before 2013) or DSM 5 (after 2013) [31,32].

Finally, all apnoeic individuals included in this study completed a series of self-questionnaires during their admission to the Sleep Laboratory to allow a first assessment of their subjective complaints of depression (Beck Depression Inventory (reduced to 13 items)), insomnia (Insomnia Severity Index), and daytime sleepiness (Epworth Sleepiness Scale) (a detailed description is available in Supplementary Materials—Section S3) [27].

2.2.2. Sleep Evaluation and Study

During their admission to the Sleep Laboratory, all apnoeic individuals included in this study benefited from a specific sleep interview by a unit psychiatrist to allow a systematic assessment of their sleep-related complaints, including sleeping habits, symptoms of insomnia disorders, symptoms of sleep-related breathing disorders, symptoms of central hypersomnia, symptoms of circadian rhythm sleep–wake disorders, symptoms of parasomnias, symptoms of restless leg syndrome (RLS), and abnormal nocturnal movements (such as periodic limb movements during sleep (PLMs)).

The participants benefited from a polysomnographic recording from which the data were collected for the analyses. These polysomnographic recordings performed in the Sleep Laboratory (accredited by the Belgian National Institute for Health and Disability Insurance for the diagnosis and treatment of OSAS) meet the recommendations of the American Academy of Sleep Medicine [33]. The detailed description of the stay conditions at the Sleep Laboratory and the applied polysomnography-montage are available in Supplementary Materials—Section S4 [23,27]. Finally, under the supervision of certified somnologists, these polysomnographic recordings were visually scored by specialised technicians according to the criteria of the American Academy of Sleep Medicine [34,35].

Through these different steps, all apnoeic individuals included in this study benefited from an assessment of their OSAS severity—mild (apnoea–hypopnoea index $\geq$5/hour and <15/hour), moderate (obstructive apnoea–hypopnoea index $\geq$15/hour and <30/hour), and severe (obstructive apnoea–hypopnoea index $\geq$30/hour), and a systematic diagnosis of their potential comorbid sleep disorder—moderate-to-severe PLMs (PLMs index was $\geq$15/hour), RLS (International Restless Legs Syndrome Study Group diagnostic criteria), insomnia disorder (American Academy of Sleep Medicine Work Group diagnostic criteria), and short sleep duration (<6 h) [36–40].

2.3. *Statistical Analyses*

Statistical analyses were performed using Stata 14. The normal distribution of the data was verified using histograms, boxplots, and quantile-quantile plots, and the equality of variances was checked using Levene's test.

We divided our sample of apnoeic individuals into a control group at low 10-year CVD risk and a patient group at moderate-to-high 10-year CVD risk. Only apnoeic individuals with a Framingham Risk Score $\geq$10% were included in the patient group at moderate-to-high 10-year CVD risk [28].

Given the asymmetric distribution of most continuous variables, non-parametric tests (Wilcoxon test) based on the medians (P25–P75) were used to demonstrate significant differences between the different groups of apnoeic individuals. Regarding categorical variables, percentages were used for descriptive analyses, and Chi^2 tests were used for comparative analyses.

Univariate logistic regression models were used to study the 10-year CVD risk associated with comorbid insomnia complaints (categorised: no, short sleep duration alone, comorbid insomnia disorder), comorbid insomnia subtypes (categorised: no, short sleep duration alone, comorbid insomnia disorder without short sleep duration, comorbid insomnia disorder with short sleep duration), and the potential confounding factors (detailed description available in Supplementary Materials—Section S5).

In the multivariate logistic regression models, the 10-year CVD risk associated with comorbid insomnia complaints and comorbid insomnia subtypes was only adjusted for significant confounding factors during the univariate analyses. The adequacy of these different models was verified by the Hosmer and Lemeshow test, whereas the specificity of the model was verified by the Link test.

The results were considered significant when the *p*-value was <0.05.

3. Results

3.1. Polysomnographic Data

Compared to those with low 10-year CVD risk, apnoeic individuals with moderate-to-high 10-year CVD risk showed:

- Reductions in sleep efficiency, sleep period time, total sleep time, % slow-wave sleep, and % REM sleep (Table 2).
- Increases in % stage 1, % wake after sleep onset, micro-arousal index, obstructive apnoea–hypopnoea index, oxygen desaturation index, total time under 90% of Sao2, and PLMs index (Table 2).

Table 2. Polysomnographic data (*n* = 1104).

	Whole Sample (*n* = 1104)	Subjects with Low 10-Year CVD Risk (*n* = 446)	Subjects with Moderate-to-High 10-Year CVD Risk (*n* = 658)	*p*-Value
Sleep latency (min)	25.0 (12.7–50.5)	23.2 (12.0–49.0)	26.6 (13.0–52.0)	0.154
Sleep efficiency (%)	77.0 (67.4–83.9)	79.4 (71.1–85.8)	75.0 (64.9–82.8)	<0.001
Sleep period time (min)	452.5 (414.4–485.3)	457.8 (422.0–492.0)	451.0 (407.0–482.0)	0.005
Total sleep time (min)	379.0 (332.8–422.3)	396.2 (348.7–430.7)	368.0 (321.0–413.3)	<0.001
% stage 1	9.0 (6.0–13.0)	8.0 (5.8–11.8)	9.3 (6.8–13.8)	<0.001
% stage 2	54.0 (47.1–60.0)	54.0 (48.0–59.6)	54.0 (46.0–60.0)	0.347
% slow-wave sleep	1.8 (0.0–7.0)	3.5 (0.4–9.0)	1.0 (0.0–5.3)	<0.001
% REM sleep	15.2 (11.0–19.7)	16.2 (12.4–20.3)	14.9 (10.0–19.0)	<0.001
REM latency (min)	83.5 (59.0–124.0)	85.0 (61.0–121.3)	81.5 (57.5–129.0)	0.631
% wake after sleep onset	14.6 (8.8–23.0)	12.9 (7.8–19.5)	15.4 (9.8–24.1)	<0.001
Number of awakenings	35 (24–52)	34 (23–49)	36 (25–53)	0.169
Micro-arousal index	17 (11–31)	15 (10–23)	19 (12–32)	<0.001
Apnoea–hypopnoea index	14 (8–30)	12 (7–22)	17 (9–35)	<0.001
Oxygen desaturation index	5 (2–14)	4 (1–8)	6 (2–18)	<0.001
Total time under 90% of SaO2 (min)	14.6 (1.7–69.0)	6.0 (0.5–32.5)	21.5 (3.5–96.0)	<0.001
PLMs index	2 (0–11)	1 (0–8)	3 (0–14)	0.001
	Median (P25–P75)	**Median (P25–P75)**	**Median (P25–P75)**	**Wilcoxon Test**

CVD = cardiovascular disease, OSAS = obstructive sleep apnoea syndrome, PLMs = periodic limb movements during sleep, REM = rapid eye movement.

There were no significant differences between the two groups for sleep latency, % stage 2, REM latency, or number of awakenings (Table 2).

3.2. Demographic Data

Moderate-to-high 10-year CVD risk was present in 59.6% (*n* = 658) of apnoeic individuals from our sample (Table 3). Male sex, body mass index ≥25 and <30 kg/m^2, body mass index ≥30 kg/m^2, age ≥54 years, smoking, alcohol consumption, moderate-to-severe OSAS, RLS alone or combined with PLMs, short sleep duration alone, comorbid insomnia disorder, type 2 diabetes, untreated hypertension, controlled hypertension, uncontrolled hypertension, dyslipidaemia without statin therapy, dyslipidaemia with statin therapy, cardiovascular comorbidities, aspirin therapy, and CRP levels ≥1 mg/L were more frequent in apnoeic individuals with moderate-to-high 10-year CVD risk than in those with low 10-year CVD risk (Table 3). In addition, apnoeic individuals with moderate-to-high 10-year

CVD risk had higher body mass index, age, and CRP levels than those with low 10-year CVD risk (Table 3). There were no significant differences between the two groups for snoring, excessive daytime sleepiness, depression status, Epworth Sleepiness Scale scores, Beck Depression Inventory scores, and Insomnia Severity Index scores (Table 3). Finally, in apnoeic individuals, comorbid insomnia disorder was very frequent, since its prevalence was 40.0% in this particular subpopulation (Table 3).

Table 3. Sample description (*n* = 1104).

Variables	Categories	%	Subjects with Low 10-Year CVD Risk	Subjects with Moderate-to-High 10-Year CVD Risk	*p*-ValueChi^2
Gender	Female (*n* = 263)	23.8%	65.0%	35.0%	<0.001
	Male (*n* = 841)	76.2%	32.7%	67.3%	
BMI (kg/m^2)	<25 (*n* = 194)	17.6%	55.7%	44.3%	<0.001
	≥25 and <30 (*n* = 432)	39.1%	39.8%	60.2%	
	≥30 (*n* = 478)	43.3%	34.7%	65.3%	
Age (years)	<54 (*n* = 535)	48.5%	57.4%	42.6%	<0.001
	≥54 (*n* = 569)	51.5%	24.4%	75.6%	
Smoking	No (*n* = 900)	81.5%	45.8%	54.2%	<0.001
	Yes (*n* = 204)	18.5%	16.7%	83.3%	
Alcohol	No (*n* = 725)	65.7%	42.6%	57.4%	0.037
	Yes (*n* = 379)	34.3%	36.2%	63.8%	
Snoring	Non (*n* = 173)	15.7%	38.7%	61.3%	0.626
	Yes (*n* = 931)	84.3%	40.7%	59.3%	
OSAS severity	Mild (*n* = 559)	50.6%	49.4%	50.6%	<0.001
	Moderate (*n* = 269)	24.4%	35.7%	64.3%	
	Severe (*n* = 276)	25.0%	26.8%	73.2%	
Sleep movement disorders	No (*n* = 861)	78.0%	42.6%	57.4%	0.008
	Moderate-to-severe PLMs alone (*n* = 70)	6.3%	38.6%	61.4%	
	RLS alone or combined with PLMs (*n* = 173)	15.7%	30.1%	69.9%	
Comorbid insomnia complaints	No (*n* = 441)	39.9%	46.0%	54.0%	0.002
	Short sleep duration alone (*n* = 221)	20.1%	32.1%	67.9%	
	Comorbid insomnia disorder (*n* = 442)	40.0%	38.9%	61.1%	
Excessive daytime sleepiness	No (*n* = 649)	58.8%	38.4%	61.6%	0.100
	Yes (*n* = 455)	41.2%	43.3%	56.7%	
Type 2 diabetes	No (*n* = 890)	80.6%	48.1%	51.9%	<0.001
	Yes (*n* = 214)	19.4%	8.4%	91.6%	
Hypertension	No (*n* = 496)	44.9%	60.1%	39.9%	<0.001
	Untreated (*n* = 169)	15.3%	29.6%	70.4%	
	Controlled (*n* = 301)	27.3%	28.6%	71.4%	
	Uncontrolled (*n* = 138)	12.5%	8.7%	91.3%	
Dyslipidaemia	No (*n* = 460)	41.7%	56.1%	43.9%	<0.001
	Without statin therapy (*n* = 366)	33.1%	32.2%	67.8%	
	With statin therapy (*n* = 278)	25.2%	25.2%	74.8%	
Cardiovascular comorbidities	No (*n* = 945)	85.6%	42.3%	57.7%	0.001
	Yes (*n* = 159)	14.4%	28.9%	71.1%	
Aspirin therapy	No (*n* = 913)	82.7%	45.1%	54.9%	<0.001
	Yes (*n* = 191)	17.3%	17.8%	82.2%	
CRP (mg/L)	<1 (*n* = 294)	26.6%	49.3%	50.7%	<0.001
	≥1 (*n* = 810)	73.4%	37.2%	62.8%	
Depression	No (*n* = 649)	58.8%	39.8%	60.2%	0.349
	Remitted (*n* = 208)	18.8%	44.7%	55.3%	
	Current (*n* = 247)	22.4%	38.5%	61.5%	

Table 3. *Cont.*

Variables	Categories	%	Subjects with Low 10-Year CVD Risk	Subjects with Moderate-to-High 10-Year CVD Risk	*p*-ValueChi^2
10-year CVD Risk	Low (*n* = 446) Moderate-to-high (*n* = 658)	40.4% 59.6%			
	Median (P25–P75)				**Wilcoxon Test**
BMI (kg/m^2)	29.0 (26.1–32.9)		28.1 (25.1–31.7)	29.6 (26.8–33.6)	<0.001
Age (years)	54 (48–61)		49 (45–55)	57 (52–63)	<0.001
ESS	9 (6–13)		9 (6–13)	9 (6–12)	0.397
BDI	3 (1–7)		3 (1–7)	3 (2–7)	0.415
ISI	13 (8–17)		13 (8–17)	13 (9–17)	0.846
CRP (mg/L)	1.7 (1.0–3.6)		2.2 (1.0–6.9)	2.9 (1.2–7.7)	0.030
Framingham Risk Score (%)	11.9 (7.2–19.7)		6.3 (4.6–8.2)	17.9 (13.0–25.2)	<0.001

CVD = cardiovascular disease, OSAS = obstructive sleep apnoea syndrome, BMI = body mass index, CRP = C-reactive protein, PLMs = periodic limb movements during sleep, RLS = restless legs syndrome, ESS = Epworth Sleepiness Scale, BDI = Beck Depression Inventory, ISI = Insomnia Severity Index.

3.3. Univariate Analyses for 10-Year CVD Risk Associated with Comorbid Insomnia Complaints and Potential Confounding Factors in Apnoeic Individuals

Male sex, overweight, obesity, older age, smoking, alcohol consumption, moderate-to-severe OSAS, RLS alone or combined with PLMs, short sleep duration alone, comorbid insomnia disorder, type 2 diabetes, untreated hypertension, controlled hypertension, uncontrolled hypertension, dyslipidaemia without statin therapy, dyslipidaemia with statin therapy, cardiovascular comorbidities, aspirin therapy, and CRP levels ≥1 mg/L were significantly associated with moderate-to-high 10-year CVD risk in apnoeic individuals (Table 4).

Table 4. Univariate analyses for 10-year CVD risk associated with comorbid insomnia complaints and potential confounding factors in apnoeic individuals (*n* = 1104).

Variables	OR (CI 95%)	*p*-Value
Gender		<0.001
Female	1	
Male	3.83 (2.86 to 5.12)	
BMI (kg/m^2)		
<25	1	<0.001
≥25 and <30	1.90 (1.35 to 2.67)	
≥30	2.36 (1.68 to 3.32)	
Age (years)		
<54	1	<0.001
≥54	4.16 (3.22 to 5.38)	
Smoking		
No	1	<0.001
Yes	4.22 (2.86 to 6.24)	
Alcohol		
No	1	0.038
Yes	1.31 (1.02 to 1.70)	
Snoring		
No	1	0.626
Yes	0.92 (0.66 to 1.28)	

Table 4. *Cont.*

Variables	OR (CI 95%)	*p*-Value
OSAS severity		<0.001
Mild	1	
Moderate	1.76 (1.30 to 2.37)	
Severe	2.66 (1.95 to 3.64)	
Sleep movement disorders		0.009
No	1	
Moderate-to-severe PLMs	1.18 (0.72 to 1.95)	
RLS alone or combined with PLMs	1.73 (1.22 to 2.46)	
Comorbid insomnia complaints		0.002
No	1	
Short sleep duration alone	1.80 (1.28 to 2.53)	
Comorbid insomnia disorder	1.34 (1.02 to 1.75)	
Excessive daytime sleepiness		0.101
No	1	
Yes	0.82 (0.64 to 1.04)	
Type 2 diabetes		<0.001
No	1	
Yes	10.08 (6.12 to 16.64)	
Hypertension		<0.001
No	1	
Untreated	3.58 (2.46 to 5.22)	
Controlled	3.76 (2.77 to 5.12)	
Uncontrolled	15.80 (8.51 to 29.34)	
Dyslipidaemia		<0.001
No	1	
Without statin therapy	2.68 (2.02 to 3.57)	
With statin therapy	3.80 (2.74 to 5.27)	
Cardiovascular comorbidities		0.002
No	1	
Yes	1.80 (1.25 to 2.60)	
Aspirin therapy		<0.001
No	1	
Yes	3.80 (2.56 to 5.63)	
CRP (mg/L)		<0.001
<1	1	
≥1	1.65 (1.26 to 2.15)	
Depression		0.35
Non	1	
Remitted	0.82 (0.60 to 1.12)	
Current	1.06 (0.78 to 1.43)	

OSAS = obstructive sleep apnoea syndrome, BMI = body mass index, CRP = C-reactive protein, PLMs = periodic limb movements during sleep, RLS = restless legs syndrome.

3.4. *Multivariate Analyses for 10-Year CVD Risk Associated with Comorbid Insomnia Complaints in Apnoeic Individuals*

After adjustment for the main confounding factors associated with cardiovascular risk highlighted during the univariate analyses, multivariate logistic regression analyses revealed that unlike short sleep duration alone, only comorbid insomnia disorder was significantly associated with moderate-to-high 10-year CVD risk in apnoeic individuals (Table 5).

Table 5. Multivariate analyses for 10-year CVD risk associated with comorbid insomnia complaints in apnoeic individuals (n = 1104).

Variables	OR Adjusted (CI 95%)	p-Value
Comorbid insomnia complaints		0.037
No	1	
Short sleep duration alone	0.99 (0.60 to 1.65)	
Comorbid insomnia disorder	1.64 (1.09 to 2.45)	

Model adjusted for gender, BMI, age, smoking, alcohol, OSAS severity, sleep movement disorders, type 2 diabetes, hypertension, dyslipidaemia, cardiovascular comorbidities, aspirin therapy, and CRP levels.

3.5. Additional Univariate and Multivariate Analyses for 10-Year CVD Risk Associated with Comorbid Insomnia Subtypes in Apnoeic Individuals

Unlike comorbid insomnia disorder without short sleep duration, only short sleep duration alone and comorbid insomnia disorder with short sleep duration were more frequent in apnoeic individuals with moderate-to-high 10-year CVD risk than in those with low 10-year CVD risk (Table 6). In addition, during univariate logistic regression analyses, unlike comorbid insomnia disorder without short sleep duration, only short sleep duration alone and comorbid insomnia disorder with short sleep duration were significantly associated with moderate-to-high 10-year CVD risk in apnoeic individuals (Table 6). Finally, after adjustment for the main confounding factors associated with cardiovascular risk highlighted during the univariate analyses, multivariate logistic regression analyses revealed that unlike short sleep duration alone and comorbid insomnia disorder without short sleep duration, only comorbid insomnia disorder with short sleep duration was significantly associated with moderate-to-high 10-year CVD risk in apnoeic individuals (Table 6).

Table 6. Additional univariate and multivariate analyses for 10-year CVD risk associated with comorbid insomnia subtypes in apnoeic individuals (n = 1104).

Variables	%	Subjects with Low 10-Year CVD Risk	Subjects with Moderate-to-High 10-Year CVD Risk	Model 1 OR Unadjusted (CI 95%)	p-Value	Model 2 OR Adjusted (CI 95%)	p-Value
Comorbid insomnia subtypes					<0.001		0.018
No	39.9% (n = 441)	46.0%	54.0%	1		1	
Short sleep duration alone	20.1% (n = 221)	32.1%	67.9%	1.80 (1.28 to 2.53)		0.99 (0.60 to 1.65)	
Without short sleep duration	22.4% (n = 247)	44.9%	55.1%	1.05 (0.76 to 1.43)		1.26 (0.78 to 2.04)	
With short sleep duration	17.6% (n = 195)	31.3%	68.7%	1.87 (1.31 to 2.67)		2.22 (1.33 to 3.72)	

Model 1 = model unadjusted. Model 2 = model adjusted for gender, BMI, age, smoking, alcohol, OSAS severity, sleep movement disorders, type 2 diabetes, hypertension, dyslipidaemia, cardiovascular comorbidities, aspirin therapy, and CRP levels. CVD = cardiovascular disease.

4. Discussion

In this study, we demonstrated that 59.6% of apnoeic individuals had a moderate-to-high 10-year CVD risk, which is significantly higher than in the general population [41]. However, the rate of apnoeic individuals with moderate-to-high 10-year CVD risk highlighted in our study seems to be higher than that of the study by Li et al. (2020) [42]. Indeed, in this previous study, only 34.0% of apnoeic individuals had a moderate-to-high 10-year CVD risk [42]. However, compared to our study, the apnoeic individuals recruited in the study by Li et al. (2020) had a better demographic (younger age and lower body mass index) and cardiometabolic profile (lower prevalence of type 2 diabetes, hypertension, and dyslipidaemia) [42], which may have led to an underestimation of the 10-year CVD risk in their study given the major role played by these demographic and cardiometabolic factors in the development of CVD [28]. On the other hand, the rate of apnoeic individuals with moderate-to-high 10-year CVD risk demonstrated in our study seems to be smaller than that of the study by Luyster et al. (2014) (66.4%) [18]. However, unlike our study, where OSAS was diagnosed during polysomnographic recordings, the use of the Multivariable Apnoea Prediction Questionnaire to identify individuals at high risk of OSAS could explain the overestimation of the 10-year CVD risk in their study given that the algorithm of this screening tool seems to favour the recruitment of individuals with higher cardiovascular risk (higher body mass index, older age, and more severe cardiometabolic comorbidities) in the groups at high risk of OSAS [43,44]. Finally, the rate of apnoeic individuals with moderate-to-high 10-year CVD risk from our study seems to be consistent with that of the studies by Matthews et al. (2011) and Cao et al. (2022) (57.1%), which recruited apnoeic individuals with demographic and cardiometabolic features more similar to those of our sample of apnoeic individuals [29,30]. Thus, regardless of some methodological differences with other studies available in the literature, we have confirmed that apnoeic individuals are a subpopulation at high risk of CVD, which justifies a better identification of the cardiovascular risk factors specific to this particular subpopulation.

Similar to the data available in the literature [12], we confirmed that comorbid insomnia disorder is common in apnoeic individuals. Indeed, in our study, 40% of apnoeic individuals had comorbid insomnia disorder, which highlights the importance of the co-occurrence of insomnia disorder and OSAS. In addition, we have shown that comorbid insomnia disorder and, more particularly, its subtype with short sleep duration were significantly associated with moderate-to-high 10-year CVD risk in apnoeic individuals. Pathophysiologically, several elements could help to better understand this frequent occurrence of comorbid insomnia disorder and its potential involvement in the 10-year CVD risk in apnoeic individuals. First, repeated nocturnal awakenings related to OSAS may induce the development of psychophysiological conditioning processes promoting dysfunctional sleep behaviours [45]. However, since dysfunctional sleep behaviours are one of the main pathophysiological mechanisms involved in the acute onset and maintenance of insomnia disorder [46], the development of these dysfunctional sleep behaviours related to OSAS could explain the frequent co-occurrence of insomnia disorder in our sample of apnoeic individuals. Secondly, in the literature, there are arguments in support of a potential synergistic effect of the co-occurrence between insomnia disorder and OSAS on some pathophysiological mechanisms (deregulation of the hypothalamic–pituitary–adrenal axis, hyperactivation of the sympathetic nervous system, and activation of pro-inflammatory mechanisms) [47,48]. However, since these pathophysiological mechanisms play a central role in the development of CVD in both apnoeic and insomniac individuals [6,49], the potential negative cumulative effect on the cardiovascular outcome of this pathophysiological synergy between insomnia disorder and OSAS could explain the higher 10-year CVD risk associated with comorbid insomnia disorder highlighted in our sample of apnoeic individuals. Third, in our study, we found that this higher 10-year CVD risk associated with comorbid insomnia disorder in apnoeic individuals appears to be mediated by sleep duration. Indeed, unlike comorbid insomnia disorder without short sleep duration, only comorbid insomnia disorder with short sleep duration was significantly associated with moderate-to-high 10-year CVD risk in apnoeic individuals. However, in insomnia sufferers with short sleep duration, the pathophysiological mechanisms favouring the development of CVD are more marked than in insomnia sufferers without short sleep duration [50], which could potentially explain this mediating effect of sleep duration on the 10-year CVD risk associated with comorbid insomnia disorder demonstrated in our sample of apnoeic individuals. Thus, based on these different elements, it seems essential to systematically screen and adequately treat comorbid insomnia disorder and, more particularly, its subtype with short sleep duration in apnoeic individuals in order to allow better cardiovascular prevention in this particular subpopulation.

The demonstration of this higher 10-year CVD risk associated with comorbid insomnia disorder and, more particularly, its subtype with short sleep duration in apnoeic individuals could allow a better understanding of the limited effect of OSAS treatments on reducing cardiovascular risk in this particular subpopulation [51]. Indeed, in apnoeic individuals, the absence of appropriate management of comorbid insomnia disorder could induce the persistence of pathophysiological mechanisms favouring the emergence of CVD [16], both by the direct negative effect of insomnia disorder on cardiovascular outcome and by the indirect negative effect of insomnia disorder on compliance with OSAS treatments [52,53]. However, although the implementation of an adequate combined treatment of comorbid insomnia disorder could open new perspectives to allow a better cardiovascular outcome in apnoeic individuals [54], it seems essential to take into account the specific features of this particular subpopulation for the choice of this combined treatment in order to avoid the establishment of treatments with a negative impact for the management of OSAS [55]. Indeed, since most pharmacological treatments for comorbid insomnia disorder may have a deleterious effect on respiratory parameters in apnoeic individuals, cognitive–behavioural therapy for insomnia combined with optimal treatment of OSAS (lifestyle changes plus continuous positive airway pressure therapy/mandibular advancement devices/surgery) seems to be the best therapeutic option for apnoeic individuals with comorbid insomnia disorder [56,57]. Finally, alongside this combined treatment of comorbid insomnia disorder in apnoeic individuals, it is essential to establish adequate therapeutic strategies for conventional cardiovascular risk factors in order to allow integrated cardiovascular management in this particular subpopulation [58].

Limitations

The results obtained in our study come from retrospective data that, even if they have been encoded in a systematic manner, cannot be verified directly with the subject in most cases, which means that our results need to be replicated in prospective studies. Furthermore, we only focused on OSAS, which means that our results cannot be generalised to other types of sleep-related breathing disorders (such as central sleep apnoea, sleep-related hypoventilation, or sleep-related hypoxemia disorder). In addition, although the Framingham Risk Score is a cardiovascular risk score frequently used in apnoeic individuals, it only allows an indirect measurement of the 10-year CVD risk, which may potentially limit the interpretation of our results. Finally, our database only contains apnoeic individuals who had agreed to undergo a Sleep Laboratory evaluation, which may also limit the generalisability of our results.

5. Conclusions

In this study, we confirmed that insomnia disorder was a frequent comorbidity in apnoeic individuals. Indeed, the prevalence of comorbid insomnia disorder was 40.0% in our sample of apnoeic individuals. In addition, we demonstrated a moderate-to-high 10-year CVD risk in 59.6% of apnoeic individuals from our sample, which confirms that apnoeic individuals are a subpopulation at high risk of CVD. Finally, we highlighted that comorbid insomnia disorder and, more specifically, its subtype with short sleep duration appear to have a negative cumulative effect on 10-year CVD risk in apnoeic individuals, which justifies more systematic research and adequate therapeutic management of this disorder in order to allow for better cardiovascular prevention in this particular subpopulation.

Supplementary Materials: The following supporting information can be downloaded at: https://www.mdpi.com/article/10.3390/life12070944/s1, Section S1 Detailed description of the outpatient recruitment procedure for the apnoeic patients included in this study; Section S2: Detailed description of the diagnostic criteria used for the conventional cardiovascular risk factors; Section S3: Detailed description of self-questionnaires used; Section S4: Description of the stay conditions at the Sleep Laboratory and description of the applied polysomnography-montage; Section S5: Description of the confounding factors included in the univariate analyses. References [24–26,59–68] are cited in the Supplementary Materials.

Author Contributions: M.H.: Conceptualization, Methodology, Formal Analysis, Investigation, Data Curation, Writing—Original Draft Preparation. B.W.: Methodology, Software, Data Curation, Writing—Original Draft Preparation. J.-P.L.: Methodology, Software, Data Curation, Writing—Original Draft Preparation. G.L.: Writing—Original Draft Preparation, Supervision. All authors have read and agreed to the published version of the manuscript.

Funding: This research received no external funding.

Institutional Review Board Statement: This research protocol was approved by the Hospital and Medical School Ethics Committee of the Erasme Hospital (Brussels University Clinics) (Erasme Reference: P2022/093). At Erasme Hospital, all patients are informed that their data could be used retrospectively for scientific research. If patients do not wish for their data to be used, they must inform the hospital, at which time, this directive is indicated in their medical records, and any future use of their data is prohibited.

Informed Consent Statement: Not applicable.

Data Availability Statement: The datasets used and/or analysed during the current study are available from the corresponding author on reasonable request.

Acknowledgments: I thank Christelle Bouchart for their support, as well as the Sleep Laboratory team from the Erasme Hospital for technical support.

Conflicts of Interest: The authors have no conflict of interest with the work carried out in this study.

Abbreviations

CRP: C-Reactive Protein; CVD: Cardiovascular Disease; DSM: Diagnostic and Statistical Manual of Mental Disorders; OSAS: Obstructive Sleep Apnoea Syndrome; PLMs: Periodic Limb Movements during Sleep; REM: Rapid Eye Movement; RLS: Restless Legs Syndrome.

References

1. Le Grande, M.R.; Beauchamp, A.; Driscoll, A.; Jackson, A.C. Prevalence of obstructive sleep apnoea in acute coronary syndrome patients: Systematic review and meta-analysis. *BMC Cardiovasc. Disord.* **2020**, *20*, 147. [CrossRef] [PubMed]
2. Tang, M.; Wang, Y.; Wang, M.; Tong, R.; Shi, T. Risk for Cardiovascular Disease and One-Year Mortality in Patients with Chronic Obstructive Pulmonary Disease and Obstructive Sleep Apnea Syndrome Overlap Syndrome. *Front. Pharmacol.* **2021**, *12*, 767982. [CrossRef] [PubMed]
3. Xie, C.; Zhu, R.; Tian, Y.; Wang, K. Association of obstructive sleep apnoea with the risk of vascular outcomes and all-cause mortality: A meta-analysis. *BMJ Open* **2017**, *7*, e013983. [CrossRef] [PubMed]
4. Qu, H.; Guo, M.; Zhang, Y.; Shi, D.Z. Obstructive sleep apnea increases the risk of cardiac events after percutaneous coronary intervention: A meta-analysis of prospective cohort studies. *Sleep Breath* **2018**, *22*, 33–40. [CrossRef] [PubMed]
5. Lavie, P.; Lavie, L. Cardiovascular morbidity and mortality in obstructive sleep apnea. *Curr. Pharm. Des.* **2008**, *14*, 3466–3473. [CrossRef] [PubMed]
6. Salman, L.A.; Shulman, R.; Cohen, J.B. Obstructive Sleep Apnea, Hypertension, and Cardiovascular Risk: Epidemiology, Pathophysiology, and Management. *Curr. Cardiol. Rep.* **2020**, *22*, 6. [CrossRef]
7. Lavie, L.; Lavie, P. Molecular mechanisms of cardiovascular disease in OSAHS: The oxidative stress link. *Eur. Respir. J.* **2009**, *33*, 1467–1484. [CrossRef]
8. Ning, Y.; Zhang, T.S.; Wen, W.W.; Li, K.; Yang, Y.X.; Qin, Y.W.; Zhang, H.N.; Du, Y.H.; Li, L.Y.; Yang, S.; et al. Effects of continuous positive airway pressure on cardiovascular biomarkers in patients with obstructive sleep apnea: A meta-analysis of randomized controlled trials. *Sleep Breath* **2019**, *23*, 77–86. [CrossRef]
9. Labarca, G.; Dreyse, J.; Drake, L.; Jorquera, J.; Barbe, F. Efficacy of continuous positive airway pressure (CPAP) in the prevention of cardiovascular events in patients with obstructive sleep apnea: Systematic review and meta-analysis. *Sleep Med. Rev.* **2020**, *52*, 101312. [CrossRef]
10. Khan, S.U.; Duran, C.A.; Rahman, H.; Lekkala, M.; Saleem, M.A.; Kaluski, E. A meta-analysis of continuous positive airway pressure therapy in prevention of cardiovascular events in patients with obstructive sleep apnoea. *Eur. Heart J.* **2018**, *39*, 2291–2297. [CrossRef]
11. Valipour, A.; Lothaller, H.; Rauscher, H.; Zwick, H.; Burghuber, O.C.; Lavie, P. Gender-related differences in symptoms of patients with suspected breathing disorders in sleep: A clinical population study using the sleep disorders questionnaire. *Sleep* **2007**, *30*, 312–319. [CrossRef]

12. Zhang, Y.; Ren, R.; Lei, F.; Zhou, J.; Zhang, J.; Wing, Y.K.; Sanford, L.D.; Tang, X. Worldwide and regional prevalence rates of co-occurrence of insomnia and insomnia symptoms with obstructive sleep apnea: A systematic review and meta-analysis. *Sleep Med. Rev.* **2019**, *45*, 1–17. [CrossRef]
13. Silvani, A. Sleep disorders, nocturnal blood pressure, and cardiovascular risk: A translational perspective. *Auton. Neurosci.* **2019**, *218*, 31–42. [CrossRef]
14. Frøjd, L.A.; Munkhaugen, J.; Moum, T.; Sverre, E.; Nordhus, I.H.; Papageorgiou, C.; Dammen, T. Insomnia in patients with coronary heart disease: Prevalence and correlates. *J. Clin. Sleep Med.* **2021**, *17*, 931–938. [CrossRef]
15. Benbir, G.; Demir, A.U.; Aksu, M.; Ardic, S.; Firat, H.; Itil, O.; Ozgen, F.; Yılmaz, H.; Karadeniz, D. Prevalence of insomnia and its clinical correlates in a general population in Turkey. *Psychiatry Clin. Neurosci.* **2015**, *69*, 543–552. [CrossRef]
16. Javaheri, S.; Redline, S. Insomnia and Risk of Cardiovascular Disease. *Chest* **2017**, *152*, 435–444. [CrossRef]
17. Bertisch, S.M.; Pollock, B.D.; Mittleman, M.A.; Buysse, D.J.; Bazzano, L.A.; Gottlieb, D.J.; Redline, S. Insomnia with objective short sleep duration and risk of incident cardiovascular disease and all-cause mortality: Sleep Heart Health Study. *Sleep* **2018**, *41*, zsy047. [CrossRef]
18. Luyster, F.S.; Kip, K.E.; Buysse, D.J.; Aiyer, A.N.; Reis, S.E.; Strollo, P.J., Jr. Traditional and nontraditional cardiovascular risk factors in comorbid insomnia and sleep apnea. *Sleep* **2014**, *37*, 593–600. [CrossRef]
19. Vozoris, N.T. Sleep apnea-plus: Prevalence, risk factors, and association with cardiovascular diseases using United States population-level data. *Sleep Med.* **2012**, *13*, 637–644. [CrossRef]
20. Gupta, M.A.; Knapp, K. Cardiovascular and psychiatric morbidity in obstructive sleep apnea (OSA) with insomnia (sleep apnea plus) versus obstructive sleep apnea without insomnia: A case-control study from a Nationally Representative US sample. *PLoS ONE* **2014**, *9*, e90021. [CrossRef]
21. Rissling, M.B.; Gray, K.E.; Ulmer, C.S.; Martin, J.L.; Zaslavsky, O.; Gray, S.L.; Hale, L.; Zeitzer, J.M.; Naughton, M.; Woods, N.F.; et al. Sleep Disturbance, Diabetes, and Cardiovascular Disease in Postmenopausal Veteran Women. *Gerontologist* **2016**, *56* (Suppl. 1), S54–S66. [CrossRef]
22. American Academy of Sleep Medicine. *International Classification of Sleep Disorders*, 3rd ed.; American Academy of Sleep Medicine: Darien, IL, USA, 2014.
23. Al Faker, M.; Wacquier, B.; Willame, H.; Point, C.; Dosogne, M.; Loas, G.; Hein, M. The association between type 2 diabetes and major depression in apnoeic individuals. *Sleep Biol. Rhythms* **2022**, *20*, 219–228. [CrossRef]
24. American Diabetes Association. Diagnosis and classification of diabetes mellitus. *Diabetes Care* **2014**, *37* (Suppl. 1), S81–S90. [CrossRef]
25. Mills, K.T.; Bundy, J.D.; Kelly, T.N.; Reed, J.E.; Kearney, P.M.; Reynolds, K.; Chen, J.; He, J. Global disparities of hypertension prevalence and control: A systematic analysis of population-based studies from 90 countries. *Circulation* **2016**, *134*, 441–450. [CrossRef]
26. Alberti, K.G.; Eckel, R.H.; Grundy, S.M.; Zimmet, P.Z.; Cleeman, J.I.; Donato, K.A.; Fruchart, J.C.; James, W.P.; Loria, C.M.; Smith, S.C., Jr.; et al. Harmonizing the metabolic syndrome: A joint interim statement of the International Diabetes Federation Task Force on Epidemiology and Prevention; National Heart, Lung, and Blood Institute; American Heart Association; World Heart Federation; International Atherosclerosis Society; and International Association for the Study of Obesity. *Circulation* **2009**, *120*, 1640–1645.
27. Hein, M.; Lanquart, J.P.; Mungo, A.; Loas, G. Cardiovascular risk associated with co-morbid insomnia and sleep apnoea (COMISA) in type 2 diabetics. *Sleep Sci.* **2022**, *15*, 184–194. [CrossRef]
28. D'Agostino, R.B., Sr.; Vasan, R.S.; Pencina, M.J.; Wolf, P.A.; Cobain, M.; Massaro, J.M.; Kannel, W.B. General cardiovascular risk profile for use in primary care: The Framingham Heart Study. *Circulation* **2008**, *117*, 743–753. [CrossRef] [PubMed]
29. Matthews, K.A.; Strollo, P.J., Jr.; Hall, M.; Mezick, E.J.; Kamarck, T.W.; Owens, J.F.; Buysse, D.J.; Reis, S.E. Associations of Framingham risk score profile and coronary artery calcification with sleep characteristics in middle-aged men and women: Pittsburgh SleepSCORE study. *Sleep* **2011**, *34*, 711–716. [PubMed]
30. Cao, W.; Luo, J.; Huang, R.; Xiao, Y. The Association between Sleep Breathing Impairment Index and Cardiovascular Risk in Male Patients with Obstructive Sleep Apnea. *Nat. Sci. Sleep* **2022**, *14*, 53–60. [CrossRef] [PubMed]
31. American Psychiatric Association. *Diagnostic and Statistical Manual of Mental Disorders*, 4th ed.; Text Rev; American Psychiatric Publishing: Arlington, VA, USA, 2000.
32. American Psychiatric Association. *Diagnostic and Statistical Manual of Mental Disorders*, 5th ed.; American Psychiatric Association: Washington, DC, USA, 2013.
33. Kushida, C.A.; Littner, M.R.; Morgenthaler, T.; Alessi, C.A.; Bailey, D.; Coleman, J., Jr.; Friedman, L.; Hirshkowitz, M.; Kapen, S.; Kramer, M.; et al. Practice Parameters for the Indications for Polysomnography and Related Procedures: An Update for 2005. *Sleep* **2005**, *28*, 499–521. [CrossRef]
34. Iber, C.; Ancoli-Israel, S.; Chesson, A.; Quan, S.F. For the American Academy of Sleep Medicine. In *The AASM Manual for the Scoring of Sleep and Associated Events: Rules, Terminology and Technical Specifications*, 1st ed.; American Academy of Sleep Medicine: Westchester, IL, USA, 2007.

35. Berry, R.B.; Budhiraja, R.; Gottlieb, D.J.; Gozal, D.; Iber, C.; Kapur, V.K.; Marcus, C.L.; Mehra, R.; Parthasarathy, S.; Quan, S.F.; et al. Rules for scoring respiratory events in sleep: Update of the 2007 AASM Manual for the Scoring of Sleep and Associated Events. Deliberations of the Sleep Apnea Definitions Task Force of the American Academy of Sleep Medicine. *J. Clin. Sleep Med.* **2012**, *8*, 597–619. [CrossRef]
36. Fleetham, J.; Ayas, N.; Bradley, D.; Ferguson, K.; Fitzpatrick, M.; George, C.; Hanly, P.; Hill, F.; Kimoff, J.; Kryger, M.; et al. Canadian Thoracic Society guidelines: Diagnosis and treatment of sleep disordered breathing in adults. *Can. Respir. J.* **2006**, *13*, 387–392. [CrossRef]
37. Haba-Rubio, J.; Marti-Soler, H.; Tobback, N.; Andries, D.; Marques-Vidal, P.; Vollenweider, P.; Preisig, M.; Heinzer, R. Clinical significance of periodic limb movements during sleep: The HypnoLaus study. *Sleep Med.* **2018**, *41*, 45–50. [CrossRef]
38. Allen, R.P.; Picchietti, D.L.; Garcia-Borreguero, D.; Ondo, W.G.; Walters, A.S.; Winkelman, J.W.; Zucconi, M.; Ferri, R.; Trenkwalder, C.; Lee, H.B.; et al. Restless legs syndrome/Willis-Ekbom disease diagnostic criteria: Updated International Restless Legs Syndrome Study Group (IRLSSG) consensus criteria–history, rationale, description, and significance. *Sleep Med.* **2014**, *15*, 860–873. [CrossRef]
39. Edinger, J.D.; Bonnet, M.H.; Bootzin, R.R.; Doghramji, K.; Dorsey, C.M.; Espie, C.A.; Jamieson, A.O.; McCall, W.V.; Morin, C.M.; Stepanski, E.J.; et al. Derivation of research diagnostic criteria for insomnia: Report of an American Academy of Sleep Medicine Work Group. *Sleep* **2004**, *27*, 1567–1596. [CrossRef]
40. Hein, M.; Lanquart, J.P.; Loas, G.; Hubain, P.; Linkowski, P. Insomnia with short sleep duration as risk factor for type 2 diabetes: A systematic review of the literature. *Rev. Med. Brux.* **2020**, *41*, 98–104. [CrossRef]
41. Cintra, F.; Bittencourt, L.R.; Santos-Silva, R.; Andersen, M.; de Paola, A.; Poyares, D.; Tufik, S. The association between the Framingham risk score and sleep: A São Paulo epidemiological sleep study. *Sleep Med.* **2012**, *13*, 577–582. [CrossRef]
42. Li, X.; Wang, F.; Xu, H.; Qian, Y.; Zou, J.; Yang, M.; Zhu, H.; Yi, H.; Guan, J.; Yin, S. Interrelationships among common predictors of cardiovascular diseases in patients of OSA: A large-scale observational study. *Nutr. Metab. Cardiovasc. Dis.* **2020**, *30*, 23–32. [CrossRef]
43. Vasheghani-Farahani, A.; Kazemnejad, F.; Sadeghniiat-Haghighi, K.; Saadat, S.; Tavakolipoor, P.; Yazdani, T.; Alidoosti, M.; Ghasem-Amooeian, V.; Ashraf, H. Obstructive sleep apnea and severity of coronary artery disease. *Caspian J. Intern. Med.* **2018**, *9*, 276–282.
44. Kariuki, J.K.; Yang, K.; Scott, P.W.; Chasens, E.R.; Godzik, C.; Luyster, F.S.; Imes, C.C. Obstructive Sleep Apnea Risk Is Associated with Severity of Metabolic Syndrome: A Secondary Analysis of the 2015–2018 National Health and Nutrition Examination Survey. *J. Cardiovasc. Nurs.* **2021**. [CrossRef]
45. Sweetman, A.; Lack, L.; Bastien, C. Co-Morbid Insomnia and Sleep Apnea (COMISA): Prevalence, Consequences, Methodological Considerations, and Recent Randomized Controlled Trials. *Brain Sci.* **2019**, *9*, 371. [CrossRef]
46. Hein, M.; Hubain, P.; Linkowski, P.; Loas, G. Support for insomnia: Recommendations for practice in general medicine. *Rev. Med. Brux.* **2016**, *37*, 235–241.
47. Benetó, A.; Gomez-Siurana, E.; Rubio-Sanchez, P. Comorbidity between sleep apnea and insomnia. *Sleep Med. Rev.* **2009**, *13*, 287–293. [CrossRef]
48. Luyster, F.S.; Buysse, D.J.; Strollo, P.J., Jr. Comorbid insomnia and obstructive sleep apnea: Challenges for clinical practice and research. *J. Clin. Sleep Med.* **2010**, *6*, 196–204. [CrossRef]
49. Khan, M.S.; Aouad, R. The Effects of Insomnia and Sleep Loss on Cardiovascular Disease. *Sleep Med. Clin.* **2017**, *12*, 167–177. [CrossRef]
50. Vgontzas, A.N.; Fernandez-Mendoza, J.; Liao, D.; Bixler, E.O. Insomnia with objective short sleep duration: The most biologically severe phenotype of the disorder. *Sleep Med. Rev.* **2013**, *17*, 241–254. [CrossRef]
51. Yu, J.; Zhou, Z.; McEvoy, R.D.; Anderson, C.S.; Rodgers, A.; Perkovic, V.; Neal, B. Association of Positive Airway Pressure with Cardiovascular Events and Death in Adults With Sleep Apnea: A Systematic Review and Meta-analysis. *JAMA* **2017**, *318*, 156–166. [CrossRef]
52. Sofi, F.; Cesari, F.; Casini, A.; Macchi, C.; Abbate, R.; Gensini, G.F. Insomnia and risk of cardiovascular disease: A meta-analysis. *Eur. J. Prev. Cardiol.* **2014**, *21*, 57–64. [CrossRef]
53. Drakou, T.; Steiropoulos, P.; Saroglou, M.; Georgopoulou, A.; Kazis, D.; Papagiannopoulos, S.; Porpodis, K.; Tryfon, S. The presence of insomnia and depression contributes to the acceptance of an initial treatment trial of continuous positive airway pressure therapy in patients with obstructive sleep apnea. *Sleep Breath* **2021**, *25*, 1803–1812. [CrossRef]
54. Grandner, M.A.; Alfonso-Miller, P.; Fernandez-Mendoza, J.; Shetty, S.; Shenoy, S.; Combs, D. Sleep: Important considerations for the prevention of cardiovascular disease. *Curr. Opin. Cardiol.* **2016**, *31*, 551–565. [CrossRef]
55. Meira, E.; Cruz, M.; Kryger, M.H.; Morin, C.M.; Palombini, L.; Salles, C.; Gozal, D. Comorbid Insomnia and Sleep Apnea: Mechanisms and implications of an underrecognized and misinterpreted sleep disorder. *Sleep Med.* **2021**, *84*, 283–288. [CrossRef] [PubMed]
56. Ragnoli, B.; Pochetti, P.; Raie, A.; Malerba, M. Comorbid Insomnia and Obstructive Sleep Apnea (COMISA): Current Concepts of Patient Management. *Int. J. Environ. Res. Public Health* **2021**, *18*, 9248. [CrossRef] [PubMed]
57. Lavie, P. Insomnia and sleep-disordered breathing. *Sleep Med.* **2007**, *8* (Suppl. 4), S21–S25. [CrossRef]
58. Bonsignore, M.R.; Baiamonte, P.; Mazzuca, E.; Castrogiovanni, A.; Marrone, O. Obstructive sleep apnea and comorbidities: A dangerous liaison. *Multidiscip. Respir. Med.* **2019**, *14*, 8. [CrossRef]

59. Beck, A.T.; Steer, R.A.; Ball, R.; Ranieri, W. Comparison of Beck Depression Inventories -IA and -II in psychiatric outpatients. *J. Pers Assess.* **1996**, *67*, 588–597. [CrossRef]
60. Johns, M.W. A new method for measuring daytime sleepiness: The Epworth sleepiness scale. *Sleep* **1991**, *14*, 540–545. [CrossRef]
61. Morin, C.M. *Insomnia: Psychological Assessment and Management*; Guilford Press: New York, NY, USA, 1993.
62. Loke, Y.K.; Brown, J.W.; Kwok, C.S.; Niruban, A.; Myint, P.K. Association of obstructive sleep apnea with risk of serious cardiovascular events: A systematic review and meta-analysis. *Circ. Cardiovasc. Qual. Outcomes* **2012**, *5*, 720–728. [CrossRef]
63. Kendzerska, T.; Gershon, A.S.; Hawker, G.; Leung, R.S.; Tomlinson, G. Obstructive sleep apnea and risk of cardiovascular events and all-cause mortality: A decade-long historical cohort study. *PLoS Med.* **2014**, *11*, e1001599. [CrossRef]
64. Kendzerska, T.; Mollayeva, T.; Gershon, A.S.; Leung, R.S.; Hawker, G.; Tomlinson, G. Untreated obstructive sleep apnea and the risk for serious long-term adverse outcomes: A systematic review. *Sleep Med Rev.* **2014**, *18*, 49–59. [CrossRef]
65. Bouloukaki, I.; Fanaridis, M.; Stathakis, G.; Ermidou, C.; Kallergis, E.; Moniaki, V.; Mauroudi, E.; Schiza, S.E. Characteristics of Patients with Obstructive Sleep Apnea at High Risk for Cardiovascular Disease. *Medicina* **2021**, *57*, 1265. [CrossRef]
66. Ayas, N.T.; Hirsch Allen, A.J.; Fox, N.; Peres, B.; Mehrtash, M.; Humphries, K.H.; Jen, R.; Taylor, C.M.; van Eeden, S.F. C-Reactive Protein Levels and the Risk of Incident Cardiovascular and Cerebrovascular Events in Patients with Obstructive Sleep Apnea. *Lung* **2019**, *197*, 459–464. [CrossRef]
67. Niiranen, T.J.; Kronholm, E.; Rissanen, H.; Partinen, M.; Jula, A.M. Self-reported obstructive sleep apnea, simple snoring, and various markers of sleep-disordered breathing as predictors of cardiovascular risk. *Sleep Breath.* **2016**, *20*, 589–596. [CrossRef]
68. Gottlieb, D.J.; Somers, V.K.; Punjabi, N.M.; Winkelman, J.W. Restless legs syndrome and cardiovascular disease: A research roadmap. *Sleep Med.* **2017**, *31*, 10–17. [CrossRef]

life

MDPI

Article

Pharyngeal Airspace Alterations after Using the Mandibular Advancement Device in the Treatment of Obstructive Sleep Apnea Syndrome

Pedro Dias Ferraz [1,*], Inês Francisco [2,3], Maria Inês Borges [1], Adriana Guimarães [2], Fátima Carvalho [1], Francisco Caramelo [3,4,5,6], José Pedro Figueiredo [1,6] and Francisco Vale [2,3,*]

1. Coimbra Hospital and University Centre (CHUC), 3000-075 Coimbra, Portugal; 11789@chuc.min-saude.pt (M.I.B.); mfatimacarvalho@live.com.pt (F.C.); jpfigueiredo@fmed.uc.pt (J.P.F.)
2. Faculty of Medicine, Institute of Orthodontics, University of Coimbra, 3000-075 Coimbra, Portugal; ifrancisco@fmed.uc.pt (I.F.); uc45131@uc.pt (A.G.)
3. Area of Environment Genetics and Oncobiology (CIMAGO), Faculty of Medicine, Coimbra Institute for Clinical and Biomedical Research (iCBR), University of Coimbra, 3000-075 Coimbra, Portugal; fcaramelo@fmed.uc.pt
4. Laboratory of Biostatistics and Medical Informatics (LBIM), Faculty of Medicine, University of Coimbra, 3004-531 Coimbra, Portugal
5. Centre for Innovative Biomedicine and Biotechnology (CIBB), University of Coimbra, 3000-075 Coimbra, Portugal
6. Clinical Academic Center of Coimbra (CACC), 3030-370 Coimbra, Portugal

* Correspondence: 11585@chuc.min-saude.pt (P.D.F.); fvale@fmed.uc.pt (F.V.)

Citation: Ferraz, P.D.; Francisco, I.; Borges, M.I.; Guimarães, A.; Carvalho, F.; Caramelo, F.; Figueiredo, J.P.; Vale, F. Pharyngeal Airspace Alterations after Using the Mandibular Advancement Device in the Treatment of Obstructive Sleep Apnea Syndrome. *Life* **2022**, *12*, 835. https://doi.org/10.3390/life12060835

Academic Editors: David Slouka and Milan Štengl

Received: 3 May 2022
Accepted: 1 June 2022
Published: 2 June 2022

Publisher's Note: MDPI stays neutral with regard to jurisdictional claims in published maps and institutional affiliations.

Abstract: Background: Mandibular Advancement Devices (MADs), inserted in non-surgical treatments for obstructive sleep apnea and hypopnea syndrome (OSAHS), are used intra-orally during the sleep period, with the aim of promoting mandibular protrusion. The aim of the study is to analyze the changes in the upper airway after the use of an MAD in the treatment of OSAHS. Methods: 60 patients diagnosed with OSAHS, as established by the Sleep Medicine Service, underwent treatment with the Silensor SL device at the Stomatology Service of the University Hospital Center of Coimbra, from January 2018 to January 2019. All patients completed two polysomnographies and two lateral teleradiographies: one before starting treatment (T0) and one after 1 year of treatment (T1). In the lateral teleradiography performed after one year of treatment, the patient had the MAD placed intra-orally. The linear measurements of the airspace proposed by the Arnett/Gunson FAB Surgery cephalometric analysis were measured at four craniometric points: A, MCI, B, Pog. Results: The results demonstrate an anteroposterior airway enlargement in two of the four points studied with the MAD placed intra-orally (B and Pog point). The greatest average increase is observed at point Pog (3 mm), followed by B (1 mm), and finally, point A (0.6 mm). Conclusions: This study proved that there is an improvement in anteroposterior measurements at various points in the upper airways after treatment with MAD.

Keywords: sleep apnea; obstructive; sleep apnea syndromes; occlusal splints; mandibular advancement devices; sleep disorders; intrinsic; cephalometry

1. Introduction

Obstructive sleep apnea and hypopnea syndrome (OSAHS) is a chronic and progressive respiratory disorder characterized by recurrent, total or partial collapse of the upper airways (UA) during sleep. This condition causes cessation of breathing for 10 or more seconds, with changes in the normal sleep pattern and changes in normal pulmonary ventilation, resulting in a deficit of oxygenation [1].

The severity of OSAHS is associated with the number of documented apnea and hypopnea events per hour of sleep (Apnea–Hypopnea Index (AHI)), and, therefore, it can be classified as mild ($5 \leq$ AHI <15), moderate ($15 \leq$ AHI < 30) or severe (AHI ≥ 30) [2].

Given the multifactorial characteristics of OSAHS, several possible treatments for its correction are described. Mandibular Advancement Devices (MADs), inserted in non-surgical treatments for OSAHS, are used intra-orally during the sleep period, with the aim of promoting mandibular protrusion. They are seen as a conservative and non-invasive treatment, with high tolerance and adherence to treatment [3–5].

It has been described that the use of MADs reduces the collapse of the UA during sleep [6,7]. It is thought that the advancement of maxillary bones leads to an increase in the caliber of the oropharynx and laryngopharynx and provides tension to the muscles involved in this anatomical region. These allow a reduction in UA collapse during deep sleep stages. However, despite more and more studies, the mechanisms that lead to an improvement in MAD are not fully known [7].

OSAHS is usually diagnosed based on physical examination and polysomnographic study (PSG). PSG is currently considered the gold standard method for diagnosing OSAHS and consists of recording chest and abdominal movements during sleep, combined with various parameters, such as AHI, during each hour of sleep. PSG findings are considered normal as long as the AHI is less than one event per hour, with an apnea episode duration of less than 5 s, oxyhemoglobin saturation greater than 90% and less than 10% of the carbon dioxide value at the end of expiration [5].

Profile teleradiography is a standardized and widely available radiographic technique, commonly used in orthodontics to study craniofacial structures. The increase in the dimensions of the UA seems to have an important effect on the success of MAD and, as such, the comparative analysis of craniofacial structures is one of the possible ways to assess the effect of this therapy.

There have been several studies that evaluate the use of cephalometry, which can predict the success of the treatment of OSAHS with an MAD [8,9].

The aim of the study is to evaluate the changes observed in the UA during treatment with the Silensor SL device in previously diagnosed OSAHS.

2. Materials and Methods

This study was conducted according to the 1964 Helsinki declaration and its later amendments or comparable ethical standards as well as approved by the Ethics Committee of the Faculty of Medicine at the University of Coimbra (CE-145/2020 on 25 November 2020). All patients gave their written informed consent prior to the start of the study.

First, 60 patients, 38 male and 22 female, with a diagnosis of OSAHS, established by the Sleep Medicine Service, underwent treatment with the Silensor SL device at the Stomatology Service at the Coimbra Hospital and University Centre, from January 2018 to January 2019. The mean age was 52.13 (from 21 to 75 years) and mean body mass index (BMI) was 27.4 kg/m^2 (from 18.4 to 35.3). Further, 26.6% of patients suffered from obesity (BMI > 30 kg/m^2). All patients had at least eight teeth in each dental arch, and all presented tolerance to the MAD throughout the entire treatment. Patients with advanced periodontal disease, temporomandibular joint pathology and head and neck malformations syndrome were excluded.

The Silensor SL device consists of two acrylic plates that cover the upper and lower arch, respectively. These two splints are joined by a fixed connector, in the upper part at the level of the canine and in the lower part at the level of the lower first and second molars. This connector orientation allows us to define the mandibular protrusion movement. This device has six connectors of different lengths with the aim of inducing mandibular advancements of 65% to 75% in the maximum protrusion of each patient. This device also allows limited laterality and mouth opening movements [10] (Figure 1).

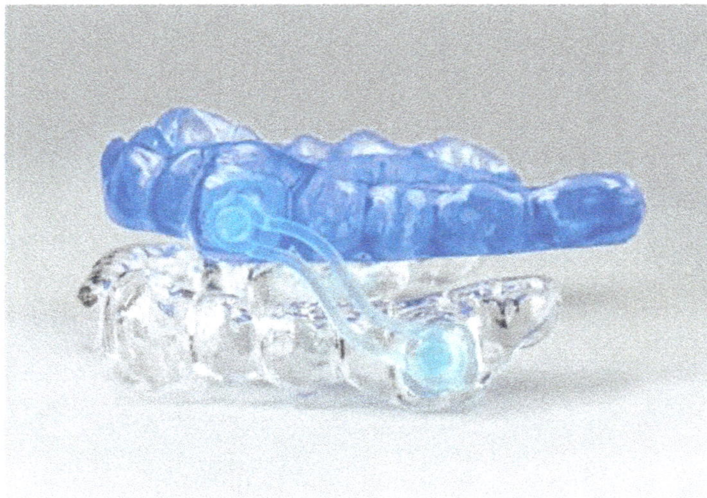

Figure 1. Silensor SL device.

Patients were classified according to the severity of their disease: mild (AHI of 5–15/h), moderate (AHI of 15–30/h) and severe (AHI > 30/h) [11].

In all patients, two lateral cephalograms were taken, one before starting treatment (T0) and the other after 1 year of treatment (T1). In the final teleradiography (T1) the patients had the MAD placed intra-orally (Figure 2). Moreover, during the evaluation of the results, all patients underwent cardiorespiratory study polysomnography (AASM level III).

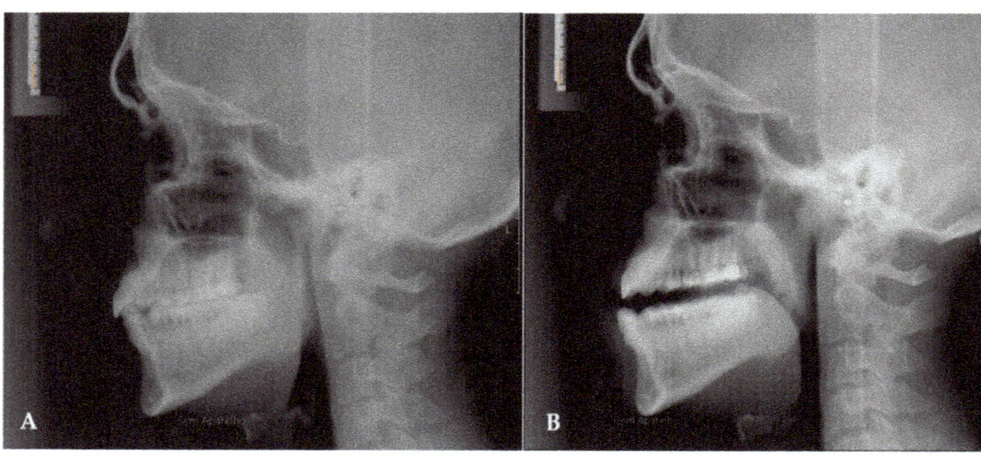

Figure 2. Teleradiography without the device in place (**A**) and teleradiography one year after treatment with the device in place (**B**). The orange line measures 10 mm.

To carry out this study, cephalometric analysis was performed using the Dolphin Image software, version 11.9 (DolphinImage & Management Solutions®®, Chatsworth, CA, USA), using the Arnett/Gunson FAB Surgery method. To transfer the acquired images to the virtual environment, the spatial orientation was previously obtained according to the Frankfurt Horizontal Plan (FHP).

The actual movement presented by each patient after the placement of the MAD was quantified at the Pogonion point (Pog), after the superimposition of the pre- and post-treatment radiographs of each patient. The superimposition method adopted was based on "The structural method" developed by Arne Björk [10,12].

The linear measurements of the airspace proposed by the Arnett/Gunson FAB Surgery cephalometric analysis were measured at four craniometric points: A, MCI, B, Pog (Figure 3):

- SPAS at point A (SPAS at A): A line is drawn perpendicular to the true vertical line that passes through point A and extends posteriorly, intersecting the anterior (A/G anterior SPAS at A) and posterior (A/G posterior SPAS at A) limits of the superior posterior airway space. The dimension of the UA is given by the distance between these two points.
- SPAS at point MCI (SPAS at MCI): A line is drawn perpendicular to the true vertical line that passes through point MCI (point located on the incisal edge of the maxillary central incisor) and extends posteriorly, intersecting the anterior (A/G anterior SPAS at MCI) and posterior (A/G posterior SPAS at MCI) limits of the superior posterior airway (Figure 3—Points 8 and 9).
- SPAS at point B (SPAS at B): A line is drawn perpendicular to the true vertical line that passes through point B and extends posteriorly, intersecting the anterior (A/G anterior SPAS at B) and posterior (A/G posterior SPAS at B) limits of the superior posterior airway space.
- SPAS at point Pog (SPAS at Pog): A line is drawn perpendicular to the true vertical line that passes through point Pog at the anterior (A/G anterior SPAS at Pog) and posterior (A/G posterior SPAS at Pog) wall of the superior posterior airway.

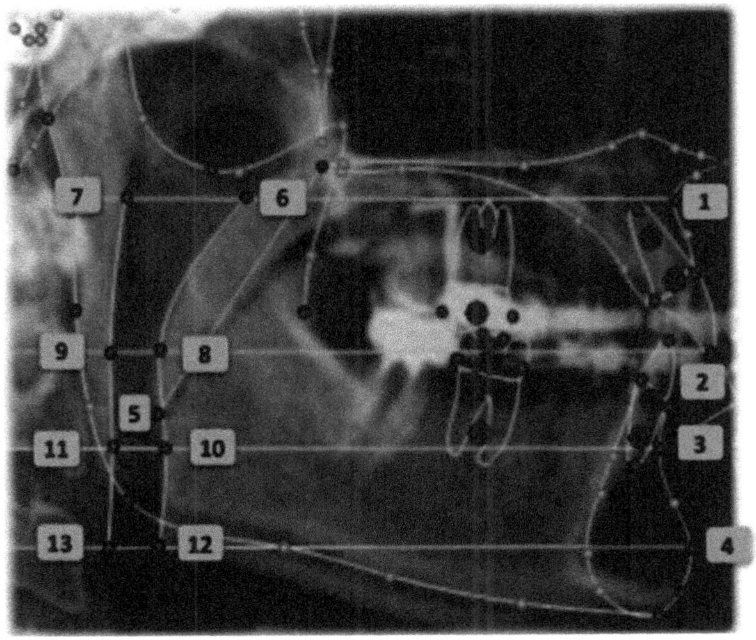

Figure 3. Cephalometric points: (1) A-point, (2) MCI-point, (3) B-point, (4) Pog-point, (5) A/G tip of soft Palate, (6) A/G anterior SPAS at A, (7) A/G posterior SPAS at A, (8) A/G anterior SPAS at MCI, (9) A/G posterior SPAS at MCI, (10) A/G anterior SPAS at B, (11) A/G posterior SPAS at B, (12) A/G anterior SPAS at Pog, (13) A/G posterior SPAS at Pog.

A power analysis was performed for the variables studied using the G*Power 3.1.9.7 software and the bilateral Student's t-test. The observed potency was 99.9% for Pog point, 74.5% for B point, 99.9% for SNB, and 99.9% for ANB. The variables under study were described as mean, standard deviation, maximum and minimum. All statistical tests were performed after evaluating the normality of distribution of quantitative variables with the Shapiro–Wilk test. To assess the difference between the changes observed in the AHI index and oxygen saturation after MAD treatment the Wilcoxon test was performed. BMI difference after MAD treatment was analyzed using the Student's t-test. Spearman correlation was performed in order to verify whether the variation in BMI could explain the variation observed in the AHI. Statistical analysis was performed on the IBM®® SPSS®® v26 platform, adopting a significance level of 0.05.

3. Results

The sample selected consisted of 60 individuals (38 males and 22 females) with an average of 52.1 (SD = 9.92). The distribution of patients according to AHI classification before and after MAD treatment is presented in Table 1. All patients showed an improvement in the AHI index. Patients with severe AHI in T0 achieved a moderate AHI after MAD treatment. Only one patient with a moderate AHI index maintained the severity, however, with improvements in the AHI index (21 to 17). Table 2 presents the results of the BMI, AHI and oxygen saturation changes.

Table 1. Distribution of patients according to AHI classification and sex.

	AHI	Total	Female	Male
T0	Mild	39	12	27
	Moderate	19	10	9
	Severe	2	0	2
T1	Mild	57	21	36
	Moderate	3	1	2
	Severe	0	0	0

Table 2. Descriptive statistics for the measured variables.

	Variables	Mean	SD	Minimum	Maximum
T0	BMI	27.40	3.32	18.30	35.30
	AHI	14.10	6.01	5.20	34.00
	SaO_2 (%)	94.30	1.78	84.90	98.00
T1	BMI	27.50	3.43	18.40	39.10
	AHI	6.50	5.46	0.00	27.90
	SaO_2 (%)	94.40	2.20	86.00	98.00

Regarding the AHI index, an improvement was found with statistically significant differences after MAD treatment ($p < 0.001$). Comparisons between TO and T1 showed no significant differences for SaO_2 and BMI variables after MAD treatment ($p = 0.754$ and $p = 0.550$, respectively).

There was no statistically significant correlation (r = 0.000; $p = 0.998$) between the difference in BMI (TO and T1) and the difference in AHI (TO and T1), which shows that the two variables are independent of each other (Figure 4).

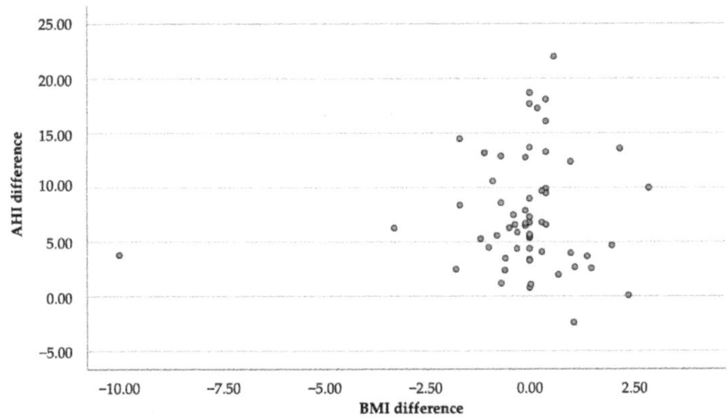

Figure 4. Correlation between AHI and BMI.

Table 3 shows the results of changes in the posterior UA: an anteroposterior increase was observed with the MAD placed intra-orally. The greatest average increase was observed at point Pog, followed by B, and finally point A.

Table 3. Cephalometric measurements.

Variable	T0	T1	Difference	p §
A	14.67 ± 4.16 (7.40/23.00)	15.27 ± 4.39 (7.20/27.20)	0.59 ± 3.48 (−9.90/10.60)	0.238
MCI	12.40 ± 11.77 (4.00/88.00)	12.06 ± 4.03 (4.40/20.90)	−0.34 ± 12.06 (−80.30/9.60)	0.844
B	10.89 ± 3.51 (4.20/23.00)	11.88 ± 3.66 (5.50/21.90)	0.99 ± 2.60 (−3.70/8.00)	0.010
Pog	12.92 ± 4.67 (5.60/24.00)	15.72 ± 4.64 (4.90/25.90)	2.80 ± 4.24 (−7.40/14.90)	<0.001
SNA	82.48 ± 5.31 (72.50/92.30)	82.48 ± 5.37 (71.50/93.40)	0.00 ± 1.35 (−4.40/2.70)	0.992
SNB	79.64 ± 4.70 (68.20/90.30)	80.77 ± 4.91 (71.50/94.60)	1.13 ± 1.85 (−5.20/5.10)	<0.001
ANB	2.77 ± 2.49 (−2.30/8.80)	1.69 ± 2.90 (−4.70/7.00)	−1.07 ± 1.58 (−4.80/2.00)	<0.001

§ Student's *t*-test.

Regarding point A, there was a slight advance from an average value of 14.67 mm to 15.27 mm. In B, the anteroposterior advancement was 1 mm from 10.89 mm to 11.88 mm. The biggest difference was found at the Pog point level (practically 3 mm from 12.92 mm to 15.72 mm). At the MCI point, there were no significant changes (Table 3).

Through the graphs presented (Figures 5–8), it is possible to interpret the individual variation in each patient. Dispersion diagrams of the measurements before and after use of the device are presented. In each of the diagrams, the line y = x is shown so that one can compare visually if the values increase (the points are above the line) or decrease (the points are below the line).

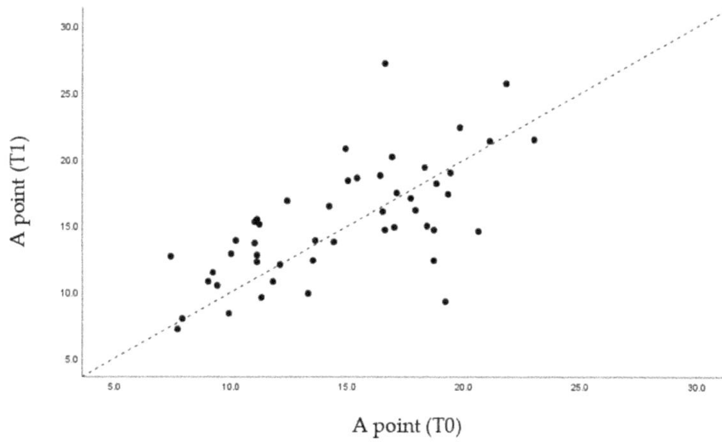

Figure 5. Dispersion of measurements before and after using the MAD at point A.

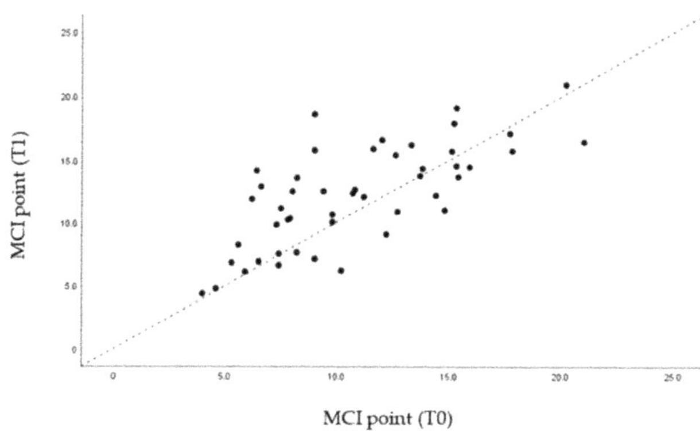

Figure 6. Dispersion of measurements before and after using the MAD at point MCI.

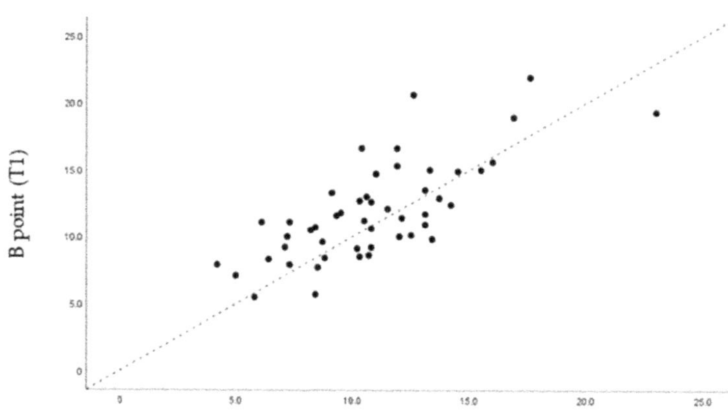

Figure 7. Dispersion of measurements before and after using the MAD at point B.

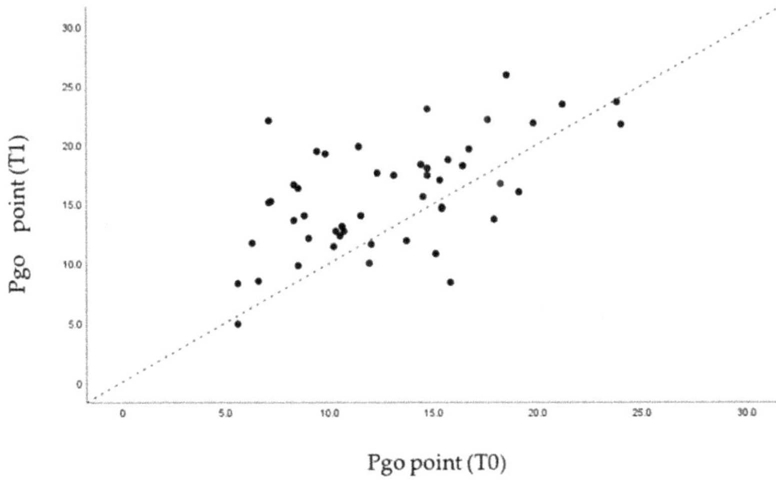

Figure 8. Dispersion of measurements before and after using the MAD at point Pog.

Regarding the skeletal reference points, the SNA values remained unchanged due to the fact that there was no movement of the jaw in the MAD mechanism (Figure 9). Due to the mandibular advance caused by the MAD, as expected, the SNB values rose from an average of 79.64 mm to 80.77 mm (Figure 10). Likewise, the value of the angle formed by ANB decreased slightly from 2.77° to 1.69°, also due to the mandibular advance caused by MAD (Table 3).

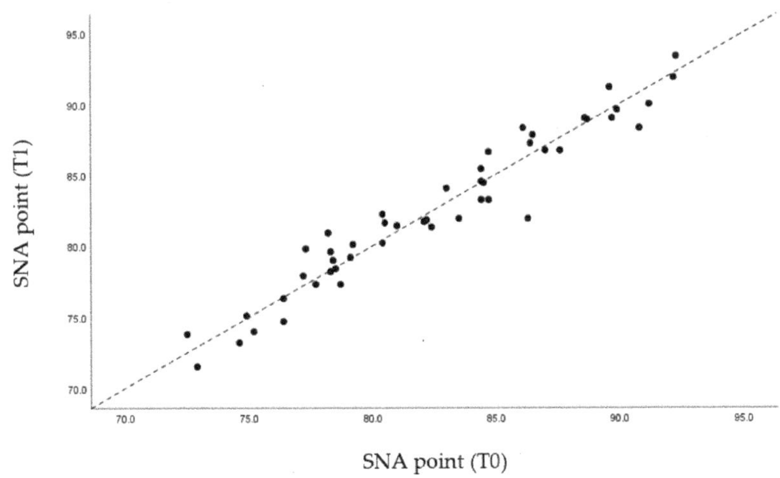

Figure 9. Dispersion of measurements before and after using the MAD with SNA angle.

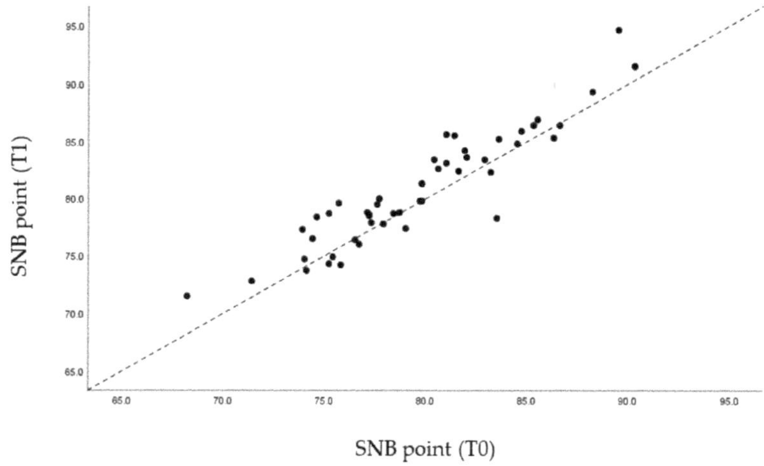

Figure 10. Dispersion of measurements before and after using the MAD with SNB angle.

In relation to the skeletal analysis, it was found that, according to the SNA angle, the values were very close to the original values. The fact that MAD causes the advance of the mandible results in point A remaining in the same position before and after treatment (Figure 5).

Regarding the SNB angle, the results shown in Figure 10 are predictable, due to the mandibular advancement caused by the MAD, which provoke the advancement of point B in space.

As for the ANB angle, the vast majority of the population studied presents an angle variation in negative values also due to mandibular advancement.

4. Discussion

MAD allow the mandible to be repositioned and stay stable with a minimal mouth opening, so that it keeps the tongue against the floor of the oral cavity, contributing to an increase in the caliber of the UA [13,14]. The sagittal pharyngeal cross-sectional area increased after the application of MAD. The results confirm the mechanical effectiveness of the Silensor SL appliance.

Through statistical analysis, we verified that in all the graphs presented (Figures 5–8), most of the points representing each patient studied are above the x = y line, which generally confirms the increase in the sagittal dimensions of the airway.

Most of the published studies emphasize that the main mechanism of action of the MAD consists of the mechanical advancement of the mandible and the increase in the anteroposterior dimensions of the oropharynx, thus, avoiding the collapse of the UA during sleep [6,15]. Other studies report that mandibular advancement improves the caliber of the UA, and that it occurs predominantly both due to the increased volume of the velopharynx and due to the increase in its lateral dimensions [16].

A possible mechanism leading to the widening of the airways in the palatal plane is related to the tension transmitted along the palatoglossus muscles to the soft palate. As the soft palate advances, tension is transmitted along the palatopharyngeal muscle to the posterior pharyngeal wall. This tension on the posterior pharyngeal wall increases the lateral volume of part of the oropharynx [17].

Other studies demonstrate that, with the placement of the MAD, the total area of the sagittal cross-section of the tongue significantly increased, which means that the shape and posture of the tongue change after insertion of MAD [16]. The sagittal sectional area of the tongue is enlarged and positioned inferiorly in the supine position. The explanation found is related to gravitational attraction and the mechanical effectiveness of the device [13].

This study presents an evaluation of the effectiveness of MADs through a widely used method for the study of UA. On the other hand, it is a valuable sample compared to existing studies. The constant and rigorous method was used when performing a profile teleradiography before and after a year of treatment [14]. Finally, the results of this study should make clinicians aware of the need to examine the oral cavity since it can be increasingly involved in the pathophysiology of OSAHS, so they can provide a valuable contribution to the screening of this pathology.

Nonetheless, this study also included some difficulties and limitations. It is not always easy to identify all the cephalometric points, due to distortions in the radiographic image or to the overlapping of structures, as this is a region with several anatomical structures involved, and, in fact, more than one image exam is needed to assess the dimensions of the airway. Although cephalometry can be considered an important tool for performing comparative skeletal measurements, it may have its known limitations. The lateral teleradiographies represent a bi-dimensional image of a three-dimensional structure, which makes its accuracy controversial. The alternative to the method used for cephalometric assessment would be cone beam computed tomography. However, the potential benefits of diagnosis and treatment planning do not outweigh the potential risks of an increase in radiation dose [10,18]. Moreover, further studies should be carried out with larger samples and follow-up.

5. Conclusions

Obstructive sleep apnea and hypopnea syndrome is considered a public health problem due to its high prevalence, being responsible for several short- and long-term comorbidities. This study proved that there is improvement in anteroposterior measurements, with statistically significant values, at various points in the UA, namely in the treatment of mild to moderate OSAHS.

Author Contributions: Conceptualization, F.V. and J.P.F.; methodology, P.D.F. and A.G.; software, A.G. and I.F.; validation, P.D.F. and M.I.B.; formal analysis, F.C. (Francisco Caramelo); investigation, F.C. (Fátima Carvalho), P.D.F. and A.G.; resources, F.C. (Francisco Caramelo); data curation, F.C. (Fátima Carvalho); writing—original draft preparation, P.D.F., M.I.B. and F.C. (Fátima Carvalho); writing—review and editing, F.V., I.F. and J.P.F.; visualization, I.F.; supervision, F.V. and J.P.F. All authors have read and agreed to the published version of the manuscript.

Funding: This research received no external funding.

Institutional Review Board Statement: The study was conducted according to the guidelines of the Declaration of Helsinki and approved by the Ethics Committee of Faculty of Medicine at the University of Coimbra.

Informed Consent Statement: Informed consent was obtained from all subjects involved in the study.

Data Availability Statement: The data presented in this study are available on request from the corresponding author.

Conflicts of Interest: The authors declare no conflict of interest.

References

1. Ferraz, P.; Borges, M.; Silva, M.; Carvalho, F.; Moita, J.; Figueiredo, J. Efficacy of the mandibular advancement device on obstructive sleep apnea syndrome—A retrospective study. *Rev. Port. Estomatol. Med. Dent. Cir. Maxilofac.* **2020**, *61*, 3–7.
2. Carberry, J.C.; Amatoury, J.; Eckert, D.J. Personalized Management Approach for OSA. *Chest* **2018**, *153*, 744–755. [CrossRef] [PubMed]
3. Cayla, G.V.D.; Collet, J.M.; Attali, V.; Kerbrat, J.B.; Benslama, L.; Goudot, P. Long-term effectiveness and side effects of mandibular advancement devices on dental and skeletal parameters. *J. Stomatol. Oral Maxillofac. Surg.* **2019**, *120*, 7–10. [CrossRef] [PubMed]
4. Vanderveken, O.M.; Devolder, A.; Marklund, M.; Boudewyns, A.N.; Braem, M.J.; Okkerse, W.; Verbraecken, J.A.; Franklin, K.A.; de Backer, W.A.; van de Heyning, P.H. Comparison of a Custom-made and a Thermoplastic Oral Appliance for the Treatment of Mild Sleep Apnea. *Am. J. Respir. Crit. Care Med.* **2008**, *178*, 197–202. [CrossRef] [PubMed]

5. Basyuni, S.; Barabas, M.; Quinnell, T. An update on mandibular advancement devices for the treatment of obstructive sleep apnoea hypopnoea syndrome. *J. Thorac. Dis.* **2018**, *10*, S48–S56. [CrossRef] [PubMed]
6. Shen, H.L.; Wen, Y.W.; Chen, M.H.; Liao, Y.F. Craniofacial morphologic predictors of oral appliance outcomes in patients with obstructive sleep apnea. *J. Am. Dent. Assoc.* **2012**, *143*, 1209–1217. [CrossRef] [PubMed]
7. Mehta, A.; Qian, J.; Petocz, P.; Ali Darendeliler, M.; Cistulli, P.A. A randomized, controlled study of a mandibular advancement splint for obstructive sleep apnea. *Am. J. Respir. Crit. Care Med.* **2001**, *163*, 1457–1461. [CrossRef] [PubMed]
8. Hoekema, A.; Doff, M.H.J.; De Bont, L.G.M.; Hoeven, J.H.V.D.; Wijkstra, P.J.; Pasma, H.R.; Stegenga, B. Predictors of obstructive sleep apnea-hypopnea treatment outcome. *J. Dent. Res.* **2007**, *86*, 1181–1186. [CrossRef] [PubMed]
9. Duran-Cantolla, J.; Crovetto-Martinez, R.; Alkhraisat, M.; Crovetto, M.; Municio, A.; Kutz, R.; Aizpuru, F.; Miranda, E.; Anitua, E. Efficacy of mandibular advancement device in the treatment of obstructive sleep apnea syndrome: A randomized controlled crossover clinical trial. *Med. Oral Patol. Oral Cirugía Bucal* **2015**, *20*, e605–e615. [CrossRef] [PubMed]
10. Vale, F.; Rodrigues, M.L.; Francisco, I.; Roseiro, A.; Santos, I.; Caramelo, F.; Rodrigues, M.J. Short-term pharyngeal airway space changes after mandibular advancement surgery in Class II patients—A two-dimensional retrospective study. *Orthod. Craniofac. Res.* **2019**, *22*, 81–86. [CrossRef] [PubMed]
11. Hudgel, D.W. Sleep Apnea Severity Classification Revisited. *Sleep* **2016**, *39*, 1165–1166. [CrossRef] [PubMed]
12. Björk, A. *Guide to Superimposition of Profile Radiographs by "The Structural Method"*; Allen Press: Lawrence, KS, USA, 2010; pp. 154–179.
13. Flores-Orozco, E.I.; Tiznado-Orozco, G.E.; Díaz-Peña, R.; Orozco EI, F.; Galletti, C.; Gazia, F.; Galletti, F. Effect of a Mandibular Advancement Device on the Upper Airway in a Patient with Obstructive Sleep Apnea. *J. Craniofac. Surg.* **2020**, *31*, e32–e35. [CrossRef] [PubMed]
14. Ingman, T.; Nieminen, T.; Hurmerinta, K. Cephalometric comparison of pharyngeal changes in subjects with upper airway resistance syndrome or obstructive sleep apnoea in upright and supine positions. *Eur. J. Orthod.* **2004**, *26*, 321–326. [CrossRef] [PubMed]
15. Skinner, M.A.; Robertson, C.J.; Kingshott, R.N.; Jones, D.R.; Taylor, D.R. The efficacy of a mandibular advancement splint in relation to cephalometric variables. *Sleep Breath.* **2002**, *6*, 115–124. [CrossRef] [PubMed]
16. Chan, A.S.L.; Sutherland, K.; Schwab, R.J.; Zeng, B.; Petocz, P.; Lee, R.W.W.; Darendeliler, M.A.; Cistulli, P.A. The effect of mandibular advancement on upper airway structure in obstructive sleep apnoea. *Thorax* **2010**, *65*, 726–732. [CrossRef] [PubMed]
17. Schendel, S.A.; Broujerdi, J.A.; Jacobson, R.L. Three-dimensional upper-airway changes with maxillomandibular advancement for obstructive sleep apnea treatment. *Am. J. Orthod. Dentofac. Orthop.* **2014**, *146*, 385–393. [CrossRef] [PubMed]
18. Haskell, J.A.; Mccrillis, J.; Haskell, B.S.; Scheetz, J.P.; Scarfe, W.C.; Farman, A.G. Effects of Mandibular Advancement Device (MAD) on Airway Dimensions Assessed with Cone-Beam Computed Tomography. *YSODO* **2009**, *15*, 132–158. [CrossRef]

life

Review

Diagnostic and Therapeutic Approach to Children and Adolescents with Obstructive Sleep Apnea Syndrome (OSA): Recommendations in Emilia-Romagna Region, Italy

Susanna Esposito [1,*], Giampiero Ricci [2], Riccardo Gobbi [3], Claudio Vicini [3], Fabio Caramelli [4], Silvia Pizzi [5], Agatina Fadda [6], Salvatore Ferro [6] and Giuseppe Plazzi [7,8]

1 Pediatric Clinic, Department of Medicine and Surgery, University Hospital, University of Parma, 43126 Parma, Italy
2 Pediatric Unit, IRCCS Azienda Ospedaliero-Universitaria di Bologna, 40138 Bologna, Italy; giampiero.ricci@unibo.it
3 Head-Neck and Oral Surgery Unit, Department of Head-Neck Surgery, Otolaryngology, Morgagni Piertoni Hospital, 47121 Forlì, Italy; riccardo.gobbi@auslromagna.it (R.G.); claudio@claudiovicini.com (C.V.)
4 Pediatric Intensive Care Unit, IRCCS Azienda Ospedaliero-Universitaria di Bologna, 40138 Bologna, Italy; fabio.caramelli@aosp.bo.it
5 Dentistry Unit, Department of Medicine and Surgery, Centro Universitario di Odontoiatria, University of Parma, 43126 Parma, Italy; s.pizzi@unipr.it
6 Direzione Generale Cura della Persona, Salute e Welfare, Emilia-Romagna Region, 40128 Bologna, Italy; agatina.fadda@regione.emilia-romagna.it (A.F.); salvatore.ferro@regione.emilia-romagna.it (S.F.)
7 IRCCS Istituto delle Scienze Neurologiche di Bologna, 40139 Bologna, Italy; giuseppe.plazzi@unibo.it
8 Department of Biomedical, Metabolic and Neural Sciences, University of Modena and Reggio Emilia, 41125 Modena, Italy
* Correspondence: susannamariaroberta.esposito@unipr.it; Tel.: +39-0521-903524

Citation: Esposito, S.; Ricci, G.; Gobbi, R.; Vicini, C.; Caramelli, F.; Pizzi, S.; Fadda, A.; Ferro, S.; Plazzi, G. Diagnostic and Therapeutic Approach to Children and Adolescents with Obstructive Sleep Apnea Syndrome (OSA): Recommendations in Emilia-Romagna Region, Italy. *Life* **2022**, *12*, 739. https://doi.org/10.3390/life12050739

Academic Editors: David Slouka and Milan Štengl

Received: 18 March 2022
Accepted: 13 May 2022
Published: 16 May 2022

Publisher's Note: MDPI stays neutral with regard to jurisdictional claims in published maps and institutional affiliations.

Abstract: Obstructive sleep apnoea syndrome (OSA) in paediatrics is a rather frequent pathology caused by pathophysiological alterations leading to partial and prolonged obstruction (hypoventilation) and/or intermittent partial (hypopnoea) or complete (apnoea) obstruction of the upper airways. Paediatric OSA is characterised by daytime and night-time symptoms. Unfortunately, there are few data on shared diagnostic-therapeutic pathways that address OSA with a multidisciplinary approach in paediatric age. This document summarizes recommendations from the Emilia-Romagna Region, Italy, developed in order to provide the most appropriate tools for a multidisciplinary approach in the diagnosis, treatment and care of paediatric patients with OSA. The multidisciplinary group of experts distinguished two different 'step' pathways, depending on the age group considered (i.e., under or over two years). In most cases, these pathways can be carried out by the primary care paediatrician, who represents the first filter for approaching the problem. For this reason, it is essential that the primary care paediatrician receives adequate training on how to formulate the diagnostic suspicion of OSA and on what criteria to use to select patients to be sent to the hospital centre. The relationship between the paediatrician of the patient and her/his parents must see a synergy of behaviour between the various players in order to avoid uncertainty about the diagnostic and therapeutic decisions as well as the follow-up phase. The definition and evaluation of the organizational process and outcome indicators of the developed flow-chart, and the impact of its implementation will remain fundamental.

Keywords: breathing disorder; hypoventilation; obstructive sleep apnea syndrome; OSA; paediatrics

1. Introduction

Obstructive sleep apnoea syndrome (OSA) in paediatrics is a rather frequent pathology caused by pathophysiological alterations leading to partial and prolonged obstruction (hypoventilation) and/or intermittent partial (hypopnoea) or complete (apnoea) obstruction of the upper airways [1–3]. This condition causes an absence of airflow despite continuous respiratory effort and is usually associated with reduced peripheral oxygen saturation and/or hypercapnia. Children with OSA have a respiratory effort that tends to overcome resistance in the upper airway, which is normally a trigger for producing arousals. Since obstructive events are mainly typical of REM sleep, which is characterised by muscular hypotonia, children with OSA usually have more arousals than those without OSA [1–3].

Paediatric OSA is characterised by daytime and night-time symptoms [4,5]. Diurnal symptoms include shortness of breath, irritability, nasal voice, chronic rhinitis, morning headache, poor school concentration, growth retardation and, more rarely, arterial hypertension and cardiac changes. During the night, snoring, pauses in breathing during sleep, shortness of breath, changes in hearth rate, changes in skin colour, choking sensation, fear or nocturnal agitation, abnormal sleeping positions, paradoxical movements of the chest and abdomen, intense sweating, insomnia, nocturnal enuresis, sleepwalking and/or bruxism are present. If not treated properly, OSA can lead to complications, sometimes serious, even in childhood, mainly due to intermittent hypoxemia [4–6]. These can cause a chronic inflammatory state leading to increased production of free radicals and other reactive oxygen species, which are responsible for oxidative stress. In addition to this, the fragmentation of the normal sleep pattern to which the paediatric patient with OSA is exposed deserves special consideration. These consequences have a major impact on cognitive development, especially if OSA is present from the earliest years of life. Several studies have, in fact, shown that untreated OSA patients more often present cognitive or neuropsychological function deficits (i.e., general intelligence, verbal intelligence, executive functions, learning, memory, visuospatial skills, language, mathematical skills, abstract and analytical thinking) [7]. Behavioural alterations, particularly hyperactivity, in younger children, and emotional lability, anxiety and depression are other frequent morbidity factors in paediatric patients with sleep disorders. However, although it is still controversial, some authors demonstrated that early treatment of paediatric OSA can improve the patient's cognitive ability and school and social performance [7]. Unfortunately, there are few data on shared diagnostic-therapeutic pathways that address OSA with a multidisciplinary approach in paediatric age.

Due to the lack of standardized multidisciplinary protocols on OSAS in paediatric age, this document summarizes recommendations from Emilia-Romagna Region, Italy, developed in order to provide the most appropriate tools for a multidisciplinary approach in the diagnosis, treatment and care of paediatric patients with OSA.

2. Epidemiology

OSA can occur throughout childhood, with a peak incidence between the second and sixth years of age [2,3]. Estimates of its prevalence in children are quite variable due to the different inclusion criteria adopted and the different pulse oximetry and polysomnographic parameters used for diagnosis. Currently, the prevalence of OSA in children is thought to be between 2% and 5.7% [2,3]. Obesity, male sex, the degree of severity of OSA and persistent adenotonsillar hypertrophy with mandibular hypoplasia have been shown to be risk factors for the persistence of the disease [2,3].

3. Aetiopathogenesis

By far the most frequent cause of OSA in paediatric age, particularly over two years of age, is adenotonsillar hypertrophy [1–3]. Other less frequent aetiological factors are excess weight and craniofacial dysmorphisms [8]. On the other hand, children aged <23 months are more prone to obstruction due to anatomical factors (i.e., shape of the face, small airways, forced nasal breathing), pulmonary mechanics (i.e., low residual functional capacity), immature and variable ventilatory control, prevalence of the REM phase of sleep (which can exacerbate obstruction due to reduced muscle tone) and easy collapsibility of the upper airways. In this age group, the most common causes of OSA [9] are:

- Craniofacial anomalies (particularly mandibular hypoplasia, as in the Pierre Robin sequence);
- Genetic syndromes (i.e., achondroplasia, Down's syndrome, Prader-Willi syndrome);
- Nasal obstruction (i.e., respiratory infections, coana atresia);
- Laryngeal obstruction (i.e., laryngeal malformations, laryngomalacia, vocal cord paralysis);
- Neurological and neuromuscular diseases (i.e., cerebral palsy, mitochondrial diseases, spinal muscular atrophy);
- Gastroesophageal reflux;
- Adenotonsillar hypertrophy after the age of six months.

Comorbidities include pulmonary hypertension, growth retardation, prematurity, behavioural problems, and feeding difficulties.

4. Diagnosis

4.1. Anamnestic Data

The suspicion of OSA must be raised by means of anamnestic and clinical criteria already during routine visits by the primary care paediatrician. In more details, the anamnestic information that the primary care paediatrician should note is:

- Presence of habitual snoring (three or more nights a week);
- Presence of breathing difficulties during sleep (apnoeas, noisy breathing or gasping);
- Enuresis;
- Preferred position during sleep (sitting or with hyperextended neck); cyanosis;
- Headache on awakening; daytime sleepiness;
- Cognitive and behavioural deficits (Table 1);
- In the first two years of life, poor growth or feeding difficulties; family history of adenotonsillectomy.

Table 1. Cognitive and behavioural deficits associated with sleep disorders in paediatric patients with OSA.

Cognitive Deficits	Behavioral Deficits
Intellectual deficits, verbalization disorder, poor vocabulary, behavioural anormalities	Hyperactivity/attention deficit hyperactivity disorder
Learning deficits	Somatisation
Poor school performance	Aggression and social problems
Poor impulse control	Excessive daytime sleepiness
Attention deficits	Anxiety

From Marcus CL et al. (2012), modified [7].

The use of clinical and anamnestic questionnaires (Pediatric Sleep Questionaire [10] or Teenager STOP-BANG [11], available as Supplementary Materials S1 and S2) is useful for a more accurate picture in support of the clinic. In children less than two years of age, suspicion of OSA may be advanced by the presence of a previous episode of acute life threatening event (ALTE)/brief resolved unexplained event (BRUE).

4.2. Clinical Picture

During the physical examination, particular attention should be paid to the possible presence of:

- Weight loss or gain, obesity;
- Growth deficit;
- Tonsillar hypertrophy;
- Adenoid facies;
- Micro/retrognancy;
- Ogival palate;
- Dental malocclusions;

- Craniofacial malformations (i.e., Pierre Robin sequence, craniostenosis, Apert syndrome, Crouzon syndrome, Treacher Collins syndrome, cleft lip and palate);
- Genetic and metabolic diseases (i.e., down syndrome, Prader-Willi syndrome, mucopolysaccharidosis);
- Infantile cerebral palsy;
- Neuromuscular disease;
- Arterial hypertension;
- Laryngomalacia with inspiratory stridor (in younger children).

Paediatric OSA recognises three phenotypes [8] based on objective examination:

- Classic phenotype, characterised by adenotonsillar hypertrophy with or without dental and skeletal malocclusion;
- Adult phenotype, characterised by obesity associated or not with aspects of the classical phenotype;
- Congenital phenotype, characterised by craniofacial anomalies associated with genetic syndromes (i.e., Pierre Robin sequence, Crouzon syndrome, Apert syndrome, Down syndrome).

4.3. Instrumental Examinations

After an accurate clinical and anamnestic assessment, possibly associated with an otorhinolaryngological examination with rhinofibrolaryngoscopy (under sedation in spontaneous breathing in the case of assessment of the lower airways), the diagnosis of OSA must necessarily be confirmed by means of instrumental examinations [12].

The instrumental methods for the diagnosis of OSA in paediatric age [12] are:

1. Night-time pulse oximetry with memory;
2. Night-time cardiorespiratory monitoring (polygraph);
3. Nocturnal polysomnography (PSG), which is the gold standard.

Abbreviated polysomnography (nappolygraphy) during afternoon sleep tends, due to its brevity, is associated with the risk to underestimate the prevalence and severity of OSA and in the case of negativity does not allow the exclusion of OSA [12]. Therefore, it is not a test to be used for the diagnosis of OSA.

Drug induced sleep endoscopy (DISE) is performed following an instrumental diagnosis of OSA and in selected cases to reveal the site and type of respiratory obstruction [12].

Night-time pulse oximetry with memory is carried out at home and is therefore inexpensive and easy to perform. Its positive predictive value is 97% in cases of severe OSA [2,13]. The limitation of this method is the impossibility of diagnosing non-desaturating apneas or hypopneas, which justifies the low sensitivity of the examination. In order to limit this problem as much as possible, it is advisable to use pulse oximeters with a short averaging time (approx. 3 s), which also allow short duration desaturations to be captured. Pulse oximetry can also be affected by motion artefacts, especially in children. To overcome this technological limitation, it is advisable to use pulse oximeters equipped with artefact filtering systems and which allow visualisation of the plethysmographic wave in order to carry out a visual analysis of the events recorded. With these precautions we try to avoid falsely high values of the number of desaturations per hour (oxygen desaturation index (ODI)) which could generate false positives. It should be noted that, despite the limit of low sensitivity, a positive nocturnal saturation test avoids the need for other more complex and costly instrumental examinations, at least in OSA due to adenotonsillar hypertrophy [14]. If pulse oximetry is negative and there are symptoms suggestive of OSA, it is advisable to repeat pulse oximetry in the first instance. If the clinical-instrumental discrepancy persists, other, more complex diagnostic tests are recommended.

Cardiorespiratory monitoring in sleep identifies cardiorespiratory events (central or mixed obstructive apnoeas, hypopnoeas, respiratory periodicity, desaturations, electrocardiographic alterations, paradoxical breathing) occurring in sleep through a polygraphic recording [11]. It does not allow for the assessment of sleep architecture but has the advantage that it can also be performed at the patient's home. The recording can be made more accurate by adding CO_2 measurement (end-tidal or transcutaneous) and video recording to the polygraphic system.

Even in paediatrics, PSG represents the diagnostic gold standard for OSA as it allows recording of respiratory events occurring in relation to sleep phases and includes also electroencephalographic, electrooculographic and electromyographic derivations. As with cardiorespiratory monitoring, PSG may include a channel for CO_2 measurement and video recording [15]. PSG allows the detection of so-called Respiratory Events Related Arousals (RERA), which may be the only manifestation in milder forms of OSA. It is particularly indicated in patients with neuromuscular diseases, craniofacial abnormalities, obesity and in patients who have already undergone adenotonsillectomy in which OSA persists [15]. Furthermore, it should be used before and after the application of the maxillary expander and the application of continuous positive pressure (CPAP) or two-level positive pressure (BiPAP) [15]. The complexity of the method makes standard PSG an examination to be performed only in specialist centres and in selected cases.

DISE provides additional information on the site(s) and patterns of upper airway narrowing and obstruction in OSA [16]. It is performed in selected patients in whom this additional information regarding upper airway dynamics (VADS) is deemed useful. This method is an added value for the surgical or conservative therapeutic outcome [continuous positive pressure (C-PAP) or oral appliance (OA)] in strictly selected patients. The directions to DISE [16] are:

- Residual OSA after initial surgical treatment;
- OSA associated with syndromic disorders;
- Severe OSA and inconsistent ENT findings (adenoidal grade G1 or G2 and tonsillar grade G1 or G2).

All subjects who are candidates for DISE must first undergo an instrumental examination that is diagnostic for OSA. DISE is carried out at 2nd level centres in Phase 3 as an in-patient procedure.

5. Classification of OSA

The classification of OSA varies depending on the method used. The McGill oximetry score is frequently used to classify OSA based on pulse oximetry and consists of assessing the number of desaturations below 90%, 85%, and 80% of peripheral oxygen saturation (SpO2) and their organisation into clusters over a sleep duration of between 10 and 30 min. By definition, the cluster must include at least five desaturation events of at least 4% above the mean SpO2 [17].

A positive saturation test for OSA must include at least three clusters of desaturations. Mild forms of OSA are those with at least three desaturations below 90%, medium forms are those that also include three or more desaturations below 85% and severe forms are those that also include three or more desaturations below 80% [17]. Alterations in saturation that do not fall within these parameters are considered inconclusive and therefore require a PSG as the risk of false negatives is high [17].

According to the statements provided by the European Respiratory Society, the polysomnographic classification of OSA is based on the number of apneas-hypopneas per hour (Apnea Hypopnea Index, AHI) [2,3]. In children, mild forms are considered those in which the AHI is between 1 and 5 events/h with SpO2 nadir between 86% and 91%, medium forms are those with AHI between 5 and 10 events/h and nadir between 76% and 85%, and severe forms are those with AHI greater than 10 events/h and nadir less than 75% [2,3].

6. Therapy

Paediatric OSA is a multifactorial disease and therefore requires a multidisciplinary approach involving the family paediatrician, the paediatrician-pneumologist specialising in sleep-disordered breathing, the otorhinolaryngologist, the child neuropsychiatrist/neurologist (hospital and/or territorial), the orthodontist and other specialists (maxillofacial surgeon, cardiologist, anaesthetist) depending on the patient's clinical picture. Although watchful waiting as therapeutic approach can be useful in the majority of the cases, early treatment is essential in selected patients to improve the child's long-term outcome, especially when cognitive and/or behavioural problems coexist. Treatment of OSA has been shown to be associated with improvements in behaviour, attention and social relationships.

6.1. Medical Therapy

It is based on the use of topical nasal corticosteroids [17,18] in combination or not with oral antileukotrienes [19], which can be used in the treatment of mild forms, in residual forms after adenotonsillectomy (AT) surgery or as standby therapy before AT, orthodontic surgery or CPAP application.

6.2. Surgical Therapy

In the treatment of OSA due to adenotonsillar hypertrophy, AT is the first choice [20]. The surgical technique can be that of extracapsular, or classical AT or intracapsular AT/tonsillotomy with various technologies (i.e., debrider, plasma scalpel). Both techniques are considered effective in the literature with weak evidence of less bleeding and postoperative pain for intracapsular techniques [20].

The estimated effectiveness of surgical treatment ranges from 70% to 100% of cases [20]. The age of less than three years, obesity, the presence of structural or functional alterations of the upper airways (craniofacial syndromes, neuromuscular pathologies), the presence of cardiac co-morbidities, concomitant infections of the upper airways, are all conditions that must lead to careful post-operative in-patient monitoring [14]. The minimum age for performing adenoidectomy is three months, and for AT six months. Other organic alterations have the possibility of surgical correction, for example coanal atresia, laryngomalacia, labiopalatoschisis and craniofacial malformations (mid-facial hypoplasia and mandibular-retrognathic hypoplasia) [20]. In these cases, laser supraglottoplasty, endoscopic correction of coanal atresia, tracheostomy, or maxillofacial surgery are effective and must be planned and performed in a multidisciplinary and dedicated care setting. In patients with craniofacial malformations, surgery is aimed at widening the space of the upper airway. In this group of patients, the surgical indication and timing of surgery should be discussed in a multidisciplinary setting.

6.3. Orthodontic Therapy

It is aimed at widening the hard palate through the application of a fixed orthodontic appliance with an active phase of approximately two to four months followed by a stabilisation phase of at least 6 months to reduce the risk of recurrence. It is indicated in children with transverse contraction of the upper jaw and dental malocclusion [21].

Orthodontic therapy can be combined with both medical and surgical therapy. The orthodontic approach plays a significant role in cases with a narrow palate, mandibular hypoplasia or retruded mandible. The aim of the treatment is to enlarge the volume of the hard palate by means of a fixed orthodontic appliance called a rapid palatal expander, which acts actively by dislocating the median palatine suture for 3–4 months [21]. This can also be subsequently combined with an intraoral thruster, which allows the advancement of the mandible when it is not correctly positioned due to dental malocclusions.

6.4. Myofunctional Therapy

This is a rehabilitation intervention recommended in cases not completely resolved after AT or orthodontic treatment [22].

6.5. Ventilation Therapy with Positive Pressure Devices

The indication that surgery has failed occurs when it is contraindicated and when consent to surgery is refused [23–25]. It is a non-invasive technique that allows CPAP or BiPAP to be delivered through a mask into the airways. The aim of this treatment is to ensure patency of the upper airway during sleep. Titration of the device should be carried out by performing cardio-respiratory monitoring or PSG and, if possible, also continuous CO_2 measurement. Therapy with positive pressure devices is usually well tolerated (about 80% of cases). In the case of uncooperative children, several attempts should be made to improve compliance before moving on to an alternative treatment option. In general, multidisciplinary post-treatment follow-up involving the paediatrician of choice and the hospital centre team is essential to assess symptom improvement or to identify residual and/or persistent disorders.

7. Pathways Developed in Emilia-Romagna Region, Italy

In agreement with the European Respiratory Society recommendations) [2,3], the multidisciplinary group of experts on paediatric OSA from Emilia-Romagna Region, Italy, distinguished two different 'step' pathways, depending on the age group considered: under or over two years.

7.1. Stepwise Management of OSA in Patients ≤ 23 Months Old

STEP 1

Identification of individuals at risk of OSA

Presence of one or more of the following:

1. Symptoms of upper respiratory tract obstruction (i.e., snoring, apnoea, insomnia), history of apparent life threatening events (ALTE), with possible secondary role of gastroesophageal reflux and prematurity;
2. Growth retardation;
3. Feedback on adenoid or, less frequently, tonsillar hypertrophy; nasal obstruction; laryngeal obstruction; syndromic craniosynostosis and/or hypoplasia of the middle third of the face; cleft palate; mandibular hypoplasia (e.g., Pierre Robin sequence); neuromuscular disorders; complex pathologies (e.g., achondroplasia, Down syndrome);
4. Endoscopic finding of upper airway abnormalities.

STEP 2

Searching for comorbidities (pulmonary hypertension and pulmonary heart, growth retardation of the body, behavioural problems) and coexisting conditions (i.e., eating disorders).

STEP 3

Diagnosis and assessment of the severity of OSA with the involvement of a multidisciplinary team (involving paediatric pulmonologist, otorhinolaryngologist, paediatric neuropsychiatrist/neurologist, orthodontist, maxillofacial surgeon and others, depending on the individual case).

Use of diagnostic tools: PSG (gold standard) or, alternatively, polygraphy or nocturnal pulse oximetry

Cut-off values for defining OSA and its severity:

1. AHI < 1 and desaturations > 3% max 2.2/h in healthy children
2. Mild OSA: AHI 1–5 episodes/h
3. Moderate OSA: AHI > 5–10 episodes/h
4. Severe OSA: AHI > 10 episodes/h

STEP 4

Treatment of OSA if pathological PSG, polygraphy or pulse oximetry are associated with snoring, oral breathing or tachypnoea, ALTE, growth restriction, tonsillar hypertrophy, laryngeal or nasal obstruction, cleft lip and palate, syndromes with craniostenosis or facial hypoplasia, neuromuscular disorders. Treatment becomes a priority in cases of achondroplasia, Beck-Wiedemann syndrome, Chiari malformation, down syndrome, mucopolysaccharidosis or Prader-Willi syndrome.

STEP 5

Individualised approach of the multidisciplinary team based on aetiology, severity of OSA and comorbidities.

Examples: use of non-invasive ventilation such as CPAP, surgical correction in case of coanal atresia, severe laryngomalacia, mandibular hypoplasia, craniostenosis, need for tracheostomy in case of severe obstruction.

STEP 6

Follow-up and management of persistent OSA

After AT, OSA may recur usually after four to six months, sometimes requiring further adenoidectomy. In CPAP patients it is necessary to re-evaluate nocturnal saturation every 2–4 months during the first 12 months of treatment and then every 6 months (interface size, pressures, confirmation of necessity). In patients with BiPAP (e.g., neuromuscular patients) it is necessary to re-evaluate saturation every year. After supraglottoplasty it is necessary to re-evaluate the patient after one to six months. In children with Pierre Robin sequence hormone, therapy and ventilation with positive pressure devices are recommended and in patients operated with mandibular distraction it is necessary a frequent follow-up to evaluate the effectiveness of the intervention [26].

Suggested cases in selected clinical conditions

- Pierre Robin syndrome: if the McGill score is <2 and there is no pharyngeal collapse at endoscopy, try the prone position; if the score is >2 or AHI > 10 consider orthodontic appliance, nCPAP (if AHI > 10/h) or glossopexy. Collaboration with the maxillofacial surgeon is essential for an integrated approach to the pathology and to decide on the best timing for mandibular distraction and tracheostomy placement.
- Achondroplasia: AT, nCPAP.
- Down syndrome: AT, nCPAP, supraglottoplasty.
- Prader Willi syndrome: AT, oxygen therapy if central apnoea.
- Mucopolysaccharidosis: AT, nCPAP, replacement therapy.

7.2. Stepwise Management of OSA in Patients 2–18 Years Old

STEP 1

Identification of individuals at risk of respiratory or sleep-related disorders

Presence of one or more of the following:

- Symptoms of upper respiratory tract obstruction (e.g., snoring, apnoea, insomnia, shortness of breath);
- Feedback from:
  - ✓ Tonsillar hypertrophy;
  - ✓ Obesity
  - ✓ Hypoplasia of the middle third of the face
  - ✓ Mandibular hypoplasia (e.g., Pierre Robin sequence);
  - ✓ Neuromuscular disorders;
  - ✓ Complex pathologies (i.e., Prader Willi syndrome, Down syndrome); prematurity
  - ✓ Familiarity with obstructive sleep disorder;
- Endoscopic finding of upper airway abnormalities.

STEP 2

Search for comorbidities (e.g., pulmonary hypertension and pulmonary heart, day-time sleepiness, hyperactivity, learning disabilities, behavioural problems) and coexisting conditions (e.g., enuresis, poor growth, feeding disorders, recurrent wheezing, metabolic syndrome).

STEP 3

Identification of predictors of persistent obstructive sleep disorder:

- Obesity;
- Male;
- AHI > 5/h;
- African ethnic group;
- Untreated tonsillar hypertrophy.

STEP 4

Objective diagnosis of obstructive sleep breathing disorders and their severity

PSG (gold standard) or, alternatively, polygraphy. Nocturnal pulse oximetry when the above are not available, although taking into account the limitations mentioned above.

Definition of OSA 1: symptoms of obstructive sleep disorder in combination with AHI > 2/h or obstructive apnoea index > 1/h and/or adenotonsillar hypertrophy.

Definition of OSA 2: symptoms of obstructive sleep disorder in combination with AHI > 1/h and/or adenotonsillar hypertrophy.

Cut-off values for defining obstructive sleep breathing disorders and their severity:

- In children without sleep disturbance, it can be valued inside normal range AHI up to 2.5/h between 2 and 6 years and up to 2.1 from 6 to 18 years
- Mild OSA: AHI 1–5 episodes/h
- Moderate OSA: AHI > 5–10 episodes/h
- Severe OSA: AHI > 10 episodes/h

STEP 5

Possible indications for the treatment of obstructive sleep disorder

Cut-off values for the treatment of paediatric patients with OSA are reported below:

- AHI > 5/h
- AHI 1–5/h if cardiovascular or nervous morbidities, enuresis, growth retardation, risk factors for persistent obstructive sleep disorders
- Positive pulse oximetry + positive questionnaires

Treatment becomes a priority in cases of achondroplasia, Chiari malformation, Down's syndrome, mucopolysaccharidosis, Prader Willi syndrome.

STEP 6

Individualised approach of the multidisciplinary team based on aetiology, severity of OSA and comorbidities

Possible approached based on comorbidities are reported below:

- Weight loss if obese;
- Nasal steroids and/or montelukast;
- AT;
- Maxillary expander or orthodontic appliances;
- CPAP or BiPAP if hypoventilation;
- Maxillofacial surgery;
- Tracheostomy.

STEP 7

Follow-up and management of persistent OSA

In the follow-up of paediatric patients with OSA, clinical and polysomnographic control will always be anticipated in case of recurrence of nocturnal symptoms (in particular snoring and apnoeas reported by parents) and daytime symptoms. The procedures to be followed are outlined below:

- Six weeks after AT, repeat PSG;
- After 12 weeks of treatment with nasal steroids and/or Montelukast repeat PSG;
- After 12 months of maxillary expansion and after 6 months of orthodontic appliances, repeat PSG;

- In case of CPAP or BiPAP, repeat PSG annually.

The differences and similarities in the diagnosis and management of sleep-related obstructive respiratory disorders in young children (0–23 months) and older patients (2–18 years) are shown in Table 2.

Table 2. Differences and similarities in the diagnosis and management of sleep-related obstructive respiratory disorders in young children (0–23 months) and older patients (2–18 years).

Diagnosis	Patients 0–23 Months	Patients 2–18 Years
Symptoms of upper airway obstruction present in both wakefulness and sleep	Yes	No
Adenotonsillar hypertrophy and obesity as a cause of sleep-related obstructive respiratory disorders	Yes, but uncommon	Yes
Syndromes, congenital anomalies as a cause of sleep-related obstructive respiratory disorders	Yes	Yes
Feeding difficulties and poor growth can coexist with OSA	Yes	No
Pulmonary hypertension can complicate OSA	Yes	Yes
Polysomnography as the gold standard for OSA	Yes	Yes
Endoscopy useful for assessing upper airway collapse	Yes	No
Management	Yes	Yes
Adenotonsillectomy is the most useful treatment	Yes	Yes
Non-invasive ventilation is often used as a first treatment for dynamic airway collapse	Yes	No
Effective orthodontic appliances in cases of OSA with retrognathia and Malocclusion	No	Yes
Patients with complex conditions to be treated as a priority	Yes	Yes
Follow-up after surgery should detect persistent OSA	Yes	Yes
Patients on non-invasive ventilation undergo annual nocturnal saturation monitoring	Yes	Yes

8. Conclusions

The flow-chart of the diagnostic-therapeutic pathway for the management of OSA in paediatrics, regardless of the patient's age, can be divided into the three phases (Table 3).

In most cases, this pathway can be carried out by the primary care paediatrician, who represents the first filter for approaching the problem and who must be informed of the various phases of the diagnostic-therapeutic pathway. For this reason, it is essential that the primary care paediatrician receives adequate training on how to formulate the diagnostic suspicion of OSA and on what criteria to use to select patients to be sent to the hospital centre. The relationship between the paediatrician of the patient and her/his parents must see a synergy of behaviour between the various players in order to avoid uncertainty about the diagnostic and therapeutic decisions to be taken. The patient and her/his parents must see a synergy of behaviour between the various players in order to avoid uncertainty about the diagnostic and therapeutic decisions to be taken. It is particularly important to emphasise the follow-up phase following any treatment carried out, in which scheduled checks by hospital centres are necessary and in which the primary care paediatrician must remain the main reference point for families. The definition and evaluation of the organizational, process and outcome indicators of the developed flow chart and the impact of its implementation will remain fundamental.

Table 3. Flow chart for the classification of OSA in paediatric age.

Phase	Actors	Actions	Tools
Phase 1	Primary care Paediatricians clinic	Medical history/ objective examination	Pediatric sleep questionnaire Teenager STOP BANG
Phase 2	First level OSA outpatient clinic	ENT examination Rhinofibroscopy Pulse oximetry/polysomnography Specialist surgical examination Medical therapy Orthodontic therapy Myofunctional therapy Diet	
Phase 3	2nd level OSA outpatient clinic	Diagnosis (pulse oximetry, polygraphy, polysomnography, DISE) Therapy (medical and ventilatory) follow-up	Surgical therapy (regionally licensed hospital for paediatric surgery in children 2 years of age/presence of resuscitation paediatric)

This protocol has been developed by the multidisciplinary contribution of experts belonging to different specializations and represents, in our opinion, the most complete and up-to-date collection of recommendations regarding OSA management in paediatric age. The application of uniform and shared protocols aims to improve clinical practice, through the standardization of diagnostic procedures and therapeutic approaches. In order to overcome barriers or hurdles, a strong educational activity associated with tools such as audit and feedback as a moment of "self-analysis" of a health organization, focus groups that give space for discussion, and the support of the political decision-maker are key elements for the success of the implementations of these recommendations.

Supplementary Materials: The following are available online at https://www.mdpi.com/article/10.3390/life12050739/s1, Supplementary material S1: Pediatric Sleep Questionnaire. Supplementary material S2: Teenager STOP-BANG.

Author Contributions: S.E., G.R., and G.P. wrote the first draft of the manuscript and the revised version; R.G., C.V., F.C., S.P., and A.F. gave a substantial scientific contribution; S.F. coordinated and supervised the project. All authors have read and agreed to the published version of the manuscript.

Funding: This research received no external funding. It was endorsed by Direzione Generale Cura della Persona. Salute e Welfare of Emilia-Romagna Region.

Institutional Review Board Statement: Not applicable.

Informed Consent Statement: Not applicable.

Data Availability Statement: All of the data are included in the manuscript.

Conflicts of Interest: The authors declare no conflict of interest.

References

1. Alexander, N.S.; Schroeder, J.W. Pediatric obstructive sleep apnea syndrome. *Pediatric Clin. N. Am.* **2013**, *60*, 827–840. [CrossRef] [PubMed]
2. Kaditis, A.G.; Alonso Alvarez, M.L.; Boudewyns, A.; Alexopoulos, E.I.; Ersu, R.; Joosten, K.; Larramona, H.; Miano, S.; Narang, I.; Trang, H.; et al. Obstructive sleep disordered breathing in 2- to 18- year-old children: Diagnosis and management. *Eur. Respir. J.* **2016**, *47*, 69–94. [CrossRef] [PubMed]

3. Kaditis, A.G.; Alonso Alvarez, M.L.; Boudewyns, A.; Abel, F.; Alexopoulos, E.I.; Ersu, R.; Joosten, K.; Larramona, H.; Miano, S.; Narang, I.; et al. ERS statement on obstructive sleep disordered breathing in 1- to 23-month-old children. *Eur. Respir. J.* **2017**, *7*, 50. [CrossRef] [PubMed]
4. Ryan, S.; Taylor, C.T.; McNicholas, W.T. Selective activation of inflammatory pathways by intermittent hypoxia in obstructive sleep apnea syndrome. *Circulation* **2005**, *112*, 2660–2667. [CrossRef]
5. Ryan, S.; McNicholas, W.T. Intermittent hypoxia and activation of inflammatory molecular pathways in OSAS. *Arch. Physiol. Biochem.* **2008**, *114*, 261–266. [CrossRef]
6. Gozal, D. Sleep, sleep disorders and inflammation in children. *Sleep Med.* **2009**, *10*, 12–16. [CrossRef]
7. Marcus, C.L.; Brooks, L.J.; Draper, K.A.; Gozal, D.; Halbower, A.C.; Jones, J.; Schechter, M.S.; Ward, S.D.; Sheldon, S.H.; Shiffman, R.N.; et al. American Academy of Pediatrics. Diagnosis and management of childhood obstructive sleep apnea syndrome. *Pediatrics* **2012**, *130*, 714–755. [CrossRef]
8. Arens, R.; Muzumdar, H. Childhood obesity and obstructive sleep apnea syndrome. *J. Appl. Physiol.* **2010**, *108*, 436–444. [CrossRef]
9. Hull, J.; Aniapravan, R.; Chan, E.; Chatwin, M.; Forton, J.; Gallagher, J.; Gibson, N.; Gordon, J.; Hughes, I.; McCulloch, R.; et al. BTS guideline for respiratory management of children with neuromuscular weakness. *Thorax* **2012**, *67*, 1–40. [CrossRef]
10. Chervin, R.D.; Hedger, K.; Dillon, J.E.; Pituch, K.J. Pediatric sleep questionnaire (PSQ): Validity and reliability of scales for sleep-disordered breathing, snoring, sleepiness, and behavioral problems. *Sleep Med.* **2000**, *1*, 21–32. [CrossRef]
11. Combs, D.; Goodwin, J.L.; Quan, S.F.; Morgan, W.J.; Parthasarathy, S. Modified STOP-Bang tool for stratifying obstructive sleep apnea risk in adolescent children. *PLoS ONE* **2015**, *10*, e0142242. [CrossRef]
12. American Thoracic Society. Standards and indications for cardiopulmonary sleep studies in children. *Am. J. Respir. Crit. Care Med.* **1996**, *153*, 866–878. [CrossRef]
13. Kaditis, A.; Kheirandish-Gozal, L.; Gozal, D. Pediatric OSAS: Oximetry can provide answers when polysomnography is not available. *Sleep Med. Rev.* **2016**, *27*, 96–105. [CrossRef]
14. Nixon, G.M.; Kermack, A.S.; Davis, G.M.; Manoukian, J.J.; Brown, K.A.; Brouillette, R.T. Planning adenotonsillectomy in children with obstructive sleep apnea: The role of overnight oximetry. *Pediatrics* **2004**, *113*, 19–25. [CrossRef]
15. Tan, H.L.; Gozal, D.; Ramirez, H.M.; Bandla, H.P.R.; Kheirandish-Gozal, L. Overnight polysomnography versus respiratory polygraphy in the diagnosis of pediatric obstructive sleep apnea. *Sleep* **2014**, *37*, 255–260. [CrossRef]
16. De Vito, A.; Carrasco Llatas, M.; Vanni, A.; Bosi, M.; Braghiroli, A.; Campanini, A.; De Vries, N.; Hamans, E.; Hohenhorst, W.; Kotecha, B.T.; et al. European position paper on drug-induced sedation endoscopy (DISE). *Sleep Breath.* **2014**, *18*, 453–465. [CrossRef]
17. Brouillette, R.T.; Manoukian, J.J.; Ducharme, F.M.; Oudjhane, K.; Earle, L.G.; Ladan, S.; Morielli, A. Efficacy of fluticasone nasal spray for pediatric obstructive sleep apnea. *J. Pediatrics* **2001**, *138*, 838–844. [CrossRef]
18. Kheirandish-Gozal, L.; Gozal, D. Intranasal budesonide treatment for children with mild obstructive sleep apnea syndrome. *Pediatrics* **2008**, *122*, 149–155. [CrossRef]
19. Bao, J.L.; Xinyuan-Gao Han, Y.B.; Zhang, K.; Liu, L. Efficacy and safety of montelukast for pediatric obstructive sleep apnea syndrome: A protocol for systematic review and meta-analysis. *Medicine* **2021**, *100*, e23958. [CrossRef]
20. Bhattacharjee, R.; Kheirandish-Gozal, L.; Spruyt, K.; Mitchell, R.B.; Promchiarak, J.; Simakajornboon, N.; Kaditis, A.G.; Splaingard, D.; Splaingard, M.; Brooks, L.J.; et al. Adenotonsillectomy outcomes in treatment of obstructive sleep apnea in children: A multicenter retrospective study. *Am. J. Respir. Crit. Care Med.* **2010**, *182*, 676–683. [CrossRef]
21. Carvalho, F.R.; Lentini-Oliveira, D.A.; Prado, L.B.; Prado, G.F.; Carvalho, L.B. Oral appliances and functional orthopaedic appliances for obstructive sleep apnoea in children. *Cochrane Database Syst. Rev.* **2016**, *10*, CD005520. [CrossRef]
22. Zhang, F.; Tian, Z.; Shu, Y.; Zou, B.; Yao, H.; Li, S.; Li, Q. The Efficiency of Orofacial Myofunctional Therapy in Treating Obstructive Sleep Apnea: A Meta-Analysis of Observational Studies. *J. Oral Rehabil.* **2022**. [CrossRef]
23. Downey, R.; Perkin, R.M.; MacQuarrie, J. Nasal continuous positive airway pressure use in children with obstructive sleep apnea younger than 2 years of age. *Chest* **2000**, *117*, 1608–1612. [CrossRef]
24. Massa, F.; Gonsalez, S.; Laverty, A.; Wallis, C.; Lane, R. The use of nasal continuous positive airway pressure to treat obstructive sleep apnoea. *Arch. Dis. Child.* **2002**, *87*, 438–443. [CrossRef]
25. Lynch, M.K.; Elliott, L.C.; Avis, K.T.; Schwebel, D.C.; Goodin, B.R. Quality of life in youth with obstructive sleep apnea syndrome (OSAS) treated with continuous positive airway pressure (CPAP) therapy. *Behav. Sleep Med.* **2019**, *17*, 238–245. [CrossRef]
26. Salvatoni, A.; Nosetti, L.; Salvatore, S.; Agosti, M. Benefits of multidisciplinary care in Prader-Willi syndrome. *Expert Rev. Endocrinol. Metab.* **2021**, *16*, 63–71. [CrossRef]

life

MDPI

Article

Nasal Symptoms in Patients with Obstructive Sleep Apnoea and Their Association with Continuous Positive Airway Pressure Usage

Konstantinos Chaidas [1,*], Kallirroi Lamprou [2], Amberley Munnings [1], John R. Stradling [2] and Annabel H. Nickol [2]

1 Ear, Nose and Throat Department, Oxford University Hospitals NHS Foundation Trust, Oxford OX3 9DU, UK; amberley.munnings@nhs.net

2 Oxford Centre for Respiratory Medicine, Oxford University Hospitals NHS Foundation Trust, Oxford OX3 7LE, UK; lamproukallirroi@gmail.com (K.L.); john.stradling@ouh.nhs.uk (J.R.S.); annabel.nickol@ouh.nhs.uk (A.H.N.)

* Correspondence: konchaidas@gmail.com

Abstract: The role of nasal symptoms in continuous positive airway pressure (CPAP) tolerance is not completely clear. This study aimed to investigate the association between CPAP usage and nasal symptoms, either prior to, or developing during, CPAP use in patients with obstructive sleep apnoea (OSA). Two hundred thirty patients were studied and divided into high-, low-, and non-CPAP users. Nasal symptoms and related quality of life parameters were evaluated prior to CPAP initiation and after three months. We also investigated predictive factors for CPAP usage. Non-CPAP users had significantly worse baseline scores for runny nose compared with high and low users (1.34 vs. 0.68 and 0.75, respectively, $p = 0.006$). There were no other significant differences between the groups. Runny nose was an independent predictive factor for lower CPAP usage ($p = 0.036$). An evaluation after three months showed worsening in runny nose score in high-CPAP users ($p = 0.025$) but not in low- and non-users. There were no significant changes in other nasal symptoms. Our study demonstrates that nasal symptoms were very common in this population but rhinorrhoea was the only symptom associated with poorer CPAP adherence. Moreover, rhinorrhoea worsened after a three-month trial of high-CPAP usage.

Keywords: nasal complaints; rhinorrhoea; nasal obstruction; allergic rhinitis; CPAP adherence; CPAP predictors

Citation: Chaidas, K.; Lamprou, K.; Munnings, A.; Stradling, J.R.; Nickol, A.H. Nasal Symptoms in Patients with Obstructive Sleep Apnoea and Their Association with Continuous Positive Airway Pressure Usage. *Life* **2022**, *12*, 305. https://doi.org/10.3390/life12020305

Academic Editors: David Slouka and Milan Štengl

Received: 18 January 2022
Accepted: 16 February 2022
Published: 17 February 2022

1. Introduction

Obstructive sleep apnoea (OSA) is a common respiratory disorder affecting 9–38% of the adult population [1] and is associated with daytime sleepiness, impaired cognitive ability, and serious sequelae such as road traffic accidents, cardiovascular morbidity, and all-cause mortality [2]. Continuous positive airway pressure (CPAP) is considered as the gold-standard treatment, especially in moderate and severe cases. CPAP is highly efficacious for the majority of patients, with good long-term adherence up to 80% and 71% at 5 and 10 years, respectively [3]. However, this still leaves a substantial minority of patients poorly tolerant of CPAP.

Many factors, such as OSA severity, have been identified as influencing CPAP usage but much of the variance is left unexplained [3]. Side effects of CPAP are present in 15% to 45% of patients and complaints such as skin irritation, dry mouth, air leak, mask discomfort, claustrophobia, and nasal symptoms have been reported as reasons for poor adherence to CPAP therapy [4]. Nasal symptoms are common in patients with OSA, but their role in OSA and CPAP tolerance is not completely clear [5]. Moreover, the actual effect of CPAP on the nasal cavity is not fully understood. There are conflicting data on the relevance of nasal symptoms, either prior to, or developing during, CPAP usage. This study looks at both these issues by studying a cohort of patients diagnosed with symptomatic OSA and going on to CPAP. We hypothesised that a higher burden of nasal symptoms would be associated with poorer adherence to CPAP, particularly if nasal symptoms worsened following initiation on CPAP.

2. Materials and Methods

2.1. Study Protocol

Patients aged over 18 years old with a new diagnosis of OSA based on their history and a sleep study, in whom a trial of CPAP was indicated, were recruited consecutively and verbal consent was obtained. Patients were excluded from the study if they were on treatment with systemic or topical medications that might affect nasal symptoms such as oral or nasal corticosteroids, antihistamines, and nasal decongestants.

Prior to CPAP initiation, a full medical history was obtained, and the following parameters were evaluated: age, sex, smoking history, body mass index (BMI), neck circumference, Mallampati grade, co-morbidities, and current medical treatment. The severity of OSA was evaluated based on the oxygen desaturation index (ODI) during an overnight sleep study. Patients were set-up on CPAP, with heated humidification and oronasal mask. Three months after initiating CPAP therapy, CPAP pressure and adherence were noted, and these data were available for all patients through remote review of data from their machine with patient consent.

Patients completed questionnaires at baseline and three months after initiating CPAP therapy where feasible, including the following:

(a) Daytime sleepiness was assessed using the Epworth Sleepiness Scale (ESS) [6].

(b) Nasal breathing was evaluated with a visual analogue scale (VAS), where a score of 0 indicates the absence of nasal obstruction and a score of 10 indicates complete nasal obstruction. A score of 2 or less defines asymptomatic individuals, a score between 2 and 5 mild symptoms, and a score greater than 5 moderate-severe nasal obstruction.

(c) Assessment for the presence of allergic rhinitis was performed via the 'score for allergic rhinitis' questionnaire (SFAR) [7]. SFAR is a validated questionnaire aiming to identify patients with allergic rhinitis (AR), which consists of eight questions with a total score ranging between 0 and 16. Individuals with a score of 7 or higher are considered as positive for AR.

(d) Nasal side effects and related quality of life were assessed by using the validated Mini Rhinoconjunctivitis Quality of Life Questionnaire (Mini RQLQ) [8]. The Mini RQLQ consists of 14 items in 5 domains (activity limitations, practical problems, nose symptoms, eye symptoms, and other problems) and each question can be answered on a 7-point scale (0 = not troubled, 6 = extremely troubled).

The whole group was divided into 'high', 'low', and 'non' users based on CPAP adherence and comparisons of various parameters were made. High usage was arbitrarily defined as CPAP use for at least 4 h per night for over 70% of the nights [9], whereas patients using CPAP for less than 4 h per night were considered as low users. Those who discontinued CPAP therapy or declined its use were considered as non-users. We also investigated the presence of predictive factors for CPAP usage. We evaluated changes prior to and after the CPAP trial in each group over a three-month period. Changes between baseline and follow-up in CPAP users with or without allergic rhinitis were also assessed.

2.2. Statistical Analysis

2.2.1. Prevalence of Nasal Symptoms and Association with CPAP Adherence

Data are presented as mean ± standard deviation or as a percentage. Continuous and categorical variables were compared using one-way analysis of variance or chi-squared test, as appropriate. One-way analysis of variance was followed by post hoc tests for pair comparisons between groups. Independent predictive factors of CPAP usage were identified by performing multiple linear regression analysis starting with a model of all variables with a $p < 0.10$ associated with their coefficients in the unadjusted univariate analysis. The selection of variables included in the final model was based on the statistical significance of their coefficients.

2.2.2. Changes in Nasal Symptoms after a Three-Month Course of CPAP

Changes in outcome variables before and after CPAP were compared by Wilcoxon signed-rank test. Categorical variables were compared using a chi-squared test. Analysis of changes was performed on high-, low-, and non-CPAP users. The multiple linear regression model was performed again, as above, but now including change in nasal symptoms as a further independent predictor.

p values < 0.05 were considered statistically significant. No adjustment was made for multiple comparisons in this exploratory study. All data were statistically analysed using SPSS software for Windows version 19.0 (SPSS Inc., Chicago, IL, USA).

3. Results

3.1. Prevalence of Nasal Symptoms and Association with CPAP Adherence

Two hundred thirty patients were recruited. Table 1 shows the baseline characteristics of all study participants also divided into high-, low-, and non-CPAP users. One hundred and sixty five of 230 patients (71.7%) continued using CPAP after three months and 109 patients (47.4%) had high CPAP adherence. The percentage of study participants with nasal obstruction prior to CPAP initiation was 57.9%, of whom it was mild in 37% and moderate-severe in 20.9%. Fifty-four patients (23.5%) had an SFAR score indicative of allergic rhinitis. There was no statistically significant difference in most nasal symptoms between the groups, with the exception of runny nose ($p = 0.006$). Specifically, non-CPAP users had significantly worse baseline scores for runny nose compared with high (1.34 vs. 0.68, $p = 0.006$) and low users (1.34 vs. 0.75, $p = 0.047$). There were no other statistically significant differences between the groups apart from CPAP pressure delivered, but this difference is not clinically significant. In our centre, CPAP pressure is determined in part by OSA severity, and the high-CPAP users had numerically more severe OSA, though this was not statistically significant.

Table 1. Characteristics of study population at baseline and comparison between high-, low-, and non-CPAP users.

Variables	Study Participants (n = 230)	High CPAP Users (n = 109)	Low CPAP Users (n = 56)	Non-CPAP Users (n = 65)	p-Value
Age, years	51.4 ± 13.1	52.9 ± 11.8	49.2 ± 13.1	50.7 ± 15.1	0.20
Male sex	171 (74.3)	82 (75.2)	43 (76.8)	46 (70.8)	0.72
Smoking					0.58
Non-smokers	194 (84.3)	95 (87.2)	44 (78.6)	55 (84.6)	
Ex-smokers	9 (3.9)	3 (2.8)	4 (7.1)	2 (3.1)	
Smokers	27 (11.7)	11 (10.1)	8 (14.3)	8 (12.3)	
Previous nasal surgery	16	7	5	4	0.80
BMI, kg/m^2	36.6 ± 8.9	35.9 ± 8.8	36.7 ± 8.4	37.9 ± 9.3	0.34
Neck circumference, inches	17.1 ± 1.7	17.1 ± 1.7	17.3 ± 1.6	17.1 ± 1.9	0.76
Mallampati scale					0.72
Grade 1	33 (14.3)	13 (11.9)	11 (19.6)	9 (13.8)	
Grade 2	104 (45.2)	49 (45.0)	25 (44.6)	30 (46.2)	
Grade 3	87 (37.8)	43 (39.4)	20 (35.7)	24 (36.9)	
Grade 4	6 (2.6)	4 (3.7)	0 (0)	2 (3.1)	
ODI, episodes/hour	40.9 ± 29.2	43.7 ± 32.9	34.9 ± 23.1	41.3 ±27.1	0.19

Table 1. *Cont.*

Variables	Study Participants (n = 230)	High CPAP Users (n = 109)	Low CPAP Users (n = 56)	Non-CPAP Users (n = 65)	p-Value
OSA severity					
Mild	29 (12.6)	13 (11.9)	8 (14.3)	8 (12.3)	0.72
Moderate	74 (32.2)	31 (28.4)	21 (37.5)	22 (33.8)	
Severe	127 (55.2)	65 (59.6)	27 (48.2)	35 (53.8)	
CPAP use per day, hours	3.7 ± 2.8	6.2 ± 1.2	2.8 ± 1.1	0.2 ± 0.2	
CPAP pressure, cm H_2O	10.5 ± 1.7	10.9 ± 1.9	10.2 ± 1.2	10.1 ± 1.6	0.005 *
Patient questionnaires					
ESS score/24	11.3 ± 5.2	11.3 ± 5.3	11.1 ± 4.4	11.6 ± 5.6	0.88
VAS score/10	3.06 ± 2.93	3.01 ± 2.97	2.84 ± 2.45	3.34 ± 3.27	0.63
Nasal obstruction category					
Asymptomatic (VAS ≤ 2)	97 (42.2)	49 (45.0)	20 (35.7)	28 (43.1)	0.29
Mild (2 < VAS ≤ 5)	85 (37.0)	38 (34.9)	27 (48.2)	20 (30.8)	
Moderate-severe (VAS > 5)	48 (20.9)	22 (20.2)	9 (16.1)	17 (26.2)	
Allergic rhinitis (SFAR score ≥ 7/16)	54 (23.5)	25 (22.9)	13 (23.2)	16 (24.6)	0.97
Total Mini RQLQ score (0–84)	21.53 ± 16.28	20.88 ± 15.87	20.21 ± 14.46	23.74 ± 18.33	0.42
Activity limitations (0–18)	5.15 ± 4.54	4.86 ± 4.55	4.71 ± 4.12	6.00 ± 4.80	0.20
Regular activities at home (0–6)	1.18 ± 1.59	1.02 ± 1.60	1.05 ± 1.38	1.55 ± 1.69	0.08
Recreational activities (0–6)	1.20 ± 1.63	1.15 ± 1.69	0.96 ± 1.35	1.51 ± 1.72	0.17
Sleep (0–6)	2.77 ± 2.15	2.70 ± 2.16	2.70 ± 2.07	2.94 ± 2.24	0.75
Practical problems (0–12)	2.46 ± 2.88	2.49 ± 2.86	2.43 ± 2.79	2.43 ± 3.03	0.99
Need to rub nose/eyes (0–6)	1.31 ± 1.57	1.44 ± 1.62	1.29 ± 1.53	1.11 ± 1.53	0.40
Need to blow nose repeatedly (0–6)	1.15 ± 1.57	1.05 ± 1.48	1.14 ± 1.57	1.32 ± 1.71	0.53
Nose symptoms (0–18)	3.64 ± 3.92	3.27 ± 3.57	3.43 ± 3.36	4.45 ± 4.80	0.14
Sneezing (0–6)	1.13 ± 1.43	1.04 ± 1.37	1.05 ± 1.35	1.37 ± 1.59	0.30
Stuffy/blocked nose (0–6)	1.62 ± 1.82	1.55 ± 1.82	1.63 ± 1.66	1.74 ± 1.96	0.81
Runny nose (0–6)	0.88 ± 1.38	0.68 ± 1.15	0.75 ± 1.10	1.34 ± 1.80	0.006 *
Eye symptoms (0–18)	3.31 ± 4.08	3.42 ± 4.39	3.09 ± 3.51	3.31 ± 4.05	0.89
Itchy eyes (0–6)	1.00 ± 1.43	1.05 ± 1.62	0.96 ± 1.18	0.94 ± 1.32	0.88
Sore eyes (0–6)	1.07 ± 1.53	1.14 ± 1.68	0.96 ± 1.29	1.05 ± 1.46	0.78
Watery eyes (0–6)	1.24 ± 1.59	1.24 ± 1.65	1.16 ± 1.44	1.32 ± 1.64	0.86
Other symptoms (0–18)	6.96 ± 4.80	6.81 ± 4.78	6.57 ± 4.39	7.55 ± 5.17	0.48
Tiredness/fatigue (0–6)	3.11 ± 1.97	3.15 ± 1.99	3.00 ± 1.87	3.14 ± 2.07	0.89
Thirst (0–6)	1.78 ± 1.76	1.66 ± 1.65	1.68 ± 1.76	2.08 ± 1.92	0.28
Feeling irritable (0–6)	2.09 ± 1.84	2.04 ± 1.86	1.89 ± 1.64	2.34 ± 1.98	0.39

Values are given as mean ± SD or number (%). *: $p < 0.05$; CPAP: continuous positive airway pressure, BMI: body mass index, ODI: oxygen desaturation index, OSA: obstructive sleep apnoea, ESS: Epworth Sleepiness Scale, VAS: visual analogue scale, SFAR: score for allergic rhinitis, RQLQ: Rhinoconjunctivitis Quality of Life Questionnaire.

Multiple regression analysis confirmed that runny nose was the only independent predictive factor for CPAP usage, with a high baseline score having a negative impact on use ($p = 0.036$; Table 2).

Table 2. Association between baseline characteristics and CPAP usage after three months.

Predictive Factors	Coefficients (Unstandardised)		CI 95%		p Value
	B	SE	Lower Bound	Upper Bound	
Runny nose	−0.294	0.139	−0.568	−0.020	0.036 *

*: $p < 0.05$; CPAP: continuous positive airway pressure.

3.2. Changes in Nasal Symptoms after a Three-Month Course of CPAP

Of 150 consecutive patients initially recruited, 103 individuals completed the follow-up assessment of nasal symptoms after three months. There was no difference in baseline characteristics in this sub-group compared to the group as a whole. Table 3 shows the changes for high-, low-, and non-users. As expected, symptoms associated with OSA (ESS, sleep, tiredness, and feeling of irritability) improved with a high degree of statistical significance in high-CPAP users. Interestingly, there was an improvement in ESS score in low- and non-CPAP users, as well, although to a less degree. In general, there were no significant changes in nasal symptoms except for an increase in runny nose score in high-CPAP users ($p = 0.025$).

Table 3. Changes between baseline and follow-up assessment after 3 months.

Clinical Characteristics	High-CPAP Users (n = 53)		Low-CPAP Users (n = 23)		Non-CPAP Users (n = 27)	
	Paired Difference (FU—Baseline)	*p* Value	Paired Difference (FU—Baseline)	*p* Value	Paired Difference (FU—Baseline)	*p* Value
ESS score	−4.55 ± 4.82	0.000 **	−3.74 ± 5.46	0.006 *	−2.63 ± 5.35	0.017 *
VAS score	−0.08 ± 3.19	0.90	0.43 ± 2.66	0.48	0.81 ± 2.76	0.21
Total Mini RQLQ score	−1.04 ± 17.68	0.96	0.35 ± 11.19	0.69	1.11 ± 12.46	0.83
Activity limitations	−1.08 ± 4.58	0.12	−1.26 ± 4.20	0.28	−0.96 ± 4.38	0.22
Regular activities at home	−0.02 ± 1.66	0.95	−0.52 ± 1.41	0.10	−0.22 ± 1.48	0.47
Recreational activities	0.13 ± 1.80	0.60	−0.04 ± 1.43	0.96	−0.33 ± 1.82	0.31
Sleep	−1.19 ± 2.48	0.001 **	−0.70 ± 2.08	0.11	−0.41 ± 2.31	0.25
Practical problems	0.89 ± 3.74	0.11	0.74 ± 2.53	0.12	0.30 ± 3.00	0.56
Need to rub nose/eyes	0.57 ± 2.14	0.07	0.35 ± 2.14	0.29	0.63 ± 1.62	0.06
Need to blow nose repeatedly	0.32 ± 1.92	0.28	0.39 ± 1.37	0.22	−0.33 ± 2.22	0.42
Nose symptoms	0.60 ± 4.29	0.26	1.22 ± 2.39	0.19	0.52 ± 3.12	0.34
Sneezing	−0.02 ± 1.78	0.66	0.13± 1.01	0.49	0.48 ± 1.25	0.07
Stuffy/blocked nose	0.11 ± 1.90	0.58	0.61 ± 1.85	0.15	0.07 ± 1.73	0.93
Runny nose	0.49 ± 1.59	0.025 *	0.48 ± 1.08	0.056	−0.04 ± 1.60	0.73
Eye symptoms	0.06 ± 5.33	0.86	0.00 ± 3.95	0.94	0.41 ± 4.77	0.84
Itchy eyes	0.17 ± 1.93	0.53	−0.04 ± 1.30	0.86	0.26 ± 1.75	0.69
Sore eyes	0.02 ± 2.13	0.97	0.04 ± 1.69	0.82	−0.15 ± 1.92	0.63
Watery eyes	−0.13 ± 1.93	0.66	0.00 ± 1.48	0.87	0.30 ± 2.03	0.68
Other symptoms	−1.42 ± 5.42	0.042 *	−0.39 ± 5.20	0.30	0.78 ± 3.71	0.46
Tiredness/fatigue	−1.06 ± 2.65	0.005 *	−0.61 ± 2.15	0.16	0.41 ± 2.32	0.32
Thirst	0.15 ± 1.92	0.75	0.78 ± 1.93	0.08	0.48 ± 1.16	0.04 *
Feeling irritable	−0.58 ± 1.85	0.038 *	−0.57 ± 2.11	0.07	−0.04 ± 1.65	0.83

Values are given as mean ± SD. *: $p < 0.05$, **: $p < 0.005$; CPAP: continuous positive airway pressure, ESS: Epworth Sleepiness Scale, VAS: visual analogue scale, RQLQ: Rhinoconjunctivitis Quality of Life Questionnaire.

An evaluation of changes after CPAP in high users with and without allergic rhinitis (AR) revealed that non-AR patients (n = 40) had a significant worsening in rhinorrhoea after 3 months (paired difference FU-baseline: 0.55 ± 1.34, $p = 0.048$). Patients with allergic rhinitis (n = 13) also had worse scores for runny nose at follow-up (paired difference FU-baseline: 0.31 ± 2.25, $p = 0.57$), but this did not reach statistical significance, probably due to the low number of patients in the AR group. The VAS scores for nasal obstruction were relatively unchanged in both groups with a paired difference (FU-Baseline) of −0.15 ± 4.14

($p = 0.94$) in AR patients and -0.05 ± 2.87 ($p = 0.64$) in non-AR patients. When changes in nasal symptoms were included in the linear regression model of predictors of CPAP usage, there were no significant new predictors, apart from a change in tiredness, as would be expected (Table 4).

Table 4. Association between change in nasal symptoms and CPAP usage after 3 months.

Predictive Factors	Coefficients (Unstandardised)		CI 95%		p Value
	B	SE	Lower Bound	Upper Bound	
Tiredness	−0.262	0.106	−0.473	−0.052	0.015 *

*: $p < 0.05$; CPAP: continuous positive airway pressure.

4. Discussion

We have demonstrated that the prevalence of nasal symptoms is high in patients initiating treatment with CPAP for OSA, even once those with pre-diagnosed nasal conditions have been excluded. Almost 58% of patients reported nasal obstruction and 23.5% reported symptoms in keeping with allergic rhinitis. Our hypothesis that nasal symptoms would be a barrier to CPAP use was not borne out by our data for all factors apart from a runny nose, with this being the only symptom where higher baseline symptom burden was associated with subsequent lower hours of CPAP use. We speculate that patients persevere with CPAP because of the benefit they derive in sleepiness and general well-being.

Nasal obstruction secondary to allergic rhinitis or other underlying pathology has been identified as a risk factor for sleep apnoea [10]. Allergic rhinitis is an inflammatory condition of the nasal mucosa affecting 10–30% of the adult population, and is characterised by one or more symptoms, including sneezing, itching, nasal congestion, and rhinorrhoea [11]. Almost one in four (23.5%) study participants suffered from AR even though those with a pre-existing diagnosis of rhinitis had been excluded from this observational study, suggesting that the actual prevalence in the local OSA population is even higher. Shadan et al. found that 37% of OSA patients were affected by allergic rhinitis [12]. There was no difference in AR incidence at baseline between high-, low-, and non-CPAP users in our study.

The use of CPAP affects nasal symptoms, but its effect remains controversial as previous studies have shown mixed results. The presence of positive airway pressure can lead to nasal complaints, such as nasal obstruction, rhinorrhoea, nasal dryness, and sneezing in up to 44–65% of CPAP users [4,13]. Balsalobre et al. showed that CPAP use by awake healthy individuals resulted in worsening of nasal obstruction, which was more evident in those with a known history of AR [14]. Yang et al. showed exacerbation of rhinitis-related symptoms after CPAP within the first year in patients without pre-existing rhinitis and within the second year in known rhinitic patients, suggesting that CPAP therapy may increase the incidence of AR [15].

Interestingly, some studies have revealed improvement of nasal symptoms post-CPAP. Willing et al. demonstrated a decrease in nasal resistance during CPAP therapy in healthy individuals [16]. Cisternas et al. showed a worsening of nasal dryness in non-rhinitic patients, but an improvement of nasal complaints in AR patients [17]. Pitts et al. revealed an improvement in nasal patency in OSA patients using long-term CPAP and even in those with limited CPAP use, with greater improvement in patients with narrower nasal cavities [18].

Our study revealed a significant increase in the score for runny nose after high-CPAP usage, though we acknowledge that there is the possibility that this may have occurred due to chance in view of the large number of observations made. Worsening rhinorrhoea was also noted in low-CPAP users, although this was not statistically significant. In contrast, the score for runny nose remained unchanged in non-users, suggesting that positive airway pressure contributes to rhinorrhoea. Runny nose worsened after CPAP in both AR and non-AR groups, although it did not reach statistical significance in the AR group, probably due to the relatively low number of subjects in this group. On the other hand, there was no significant change in other nasal symptoms including nasal obstruction. Previous studies have shown that CPAP treatment can lead to nasal inflammation which, however, does not necessarily have any clinical implications [17,19]. The addition of humidification may be associated with improved neutrophilic infiltration [20] and a decrease in nasal side effects [21]. It is, thus, likely that the use of CPAP with heated humidification by our study participants has played a role in the prevention of nasal dryness and oedema, although it did not have a favourable effect on rhinorrhoea.

CPAP adherence is variable and has been identified as a limiting factor of this therapeutic modality for patients with OSA. Our study showed that high-CPAP adherence was 47.4%, whereas 71.7% of patients continued using CPAP after a three-month period. This is comparable with previous studies showing a CPAP compliance rate between 30% and 60% [9].

Several factors have been identified as affecting adherence to CPAP including claustrophobia or discomfort caused by the mask, excessive air leak, mouth dryness, and choking sensation [22,23]. Nasal complaints present either prior or secondary to CPAP have also been identified as predictors for CPAP adherence. Others have shown that nasal obstruction is an important factor for CPAP intolerance and increased nasal resistance prior to CPAP initiation is associated with early CPAP discontinuation [24,25]. Li et al. found that CPAP use was significantly lower in patients with smaller nasal passages [26].

Interestingly, our study did not reveal an association between nasal obstruction and CPAP usage. Our findings concur with those published by Skoczynski et al. who observed that acceptance of CPAP therapy was not correlated with nasal patency and Pitts et al. who found that CPAP adherence had no correlation with any measure of nasal patency [18,19]. There was a significant difference in the baseline score for runny nose between CPAP users and non-users and our model shows that rhinorrhoea is an independent predictive factor for low-CPAP usage. There was also significant worsening of rhinorrhoea after high-CPAP usage for three months, but there was no association between changes in nasal symptoms and CPAP adherence. Change in tiredness was the only additional predictor, as would be expected, as it is well demonstrated that CPAP therapy significantly improves daytime sleepiness and tiredness [27].

In most previous studies, a nasal mask was the primary mask utilised, as nasal CPAP is associated with better adherence, lower residual AHI, and higher therapeutic levels compared to oronasal CPAP [28]. However, nasal obstruction and oral breathing are common among patients with OSA and can lead to oral air leak due to mouth opening as a potential adverse effect of CPAP [29]. It has been suggested that patients with oral breathing may be less adherent to nasal CPAP [22]. Patients with nasal pathology and/or nasal symptoms may find it easier to use a full-face mask instead of a nasal mask as the positive air pressure can bypass reduced nasal patency and minimise CPAP-related nasal side effects. All study participants were given an oronasal mask and the delivery route may have influenced our findings and, specifically, the absence of increased nasal complaints after CPAP apart from rhinorrhoea and the lack of association between nasal obstruction and CPAP usage.

The initial adherence to CPAP has been the best predictor for long-term compliance and CPAP usage gradually decreases in patients with nasal complaints [30]. Therefore, it is important to evaluate OSA patients prior to CPAP initiation aiming to treat nasal symptoms and CPAP-related adverse effects such as rhinorrhoea before and during CPAP therapy. Despite not currently being part of our Unit's standard practice, a clinical assessment by an otorhinolaryngologist prior to CPAP initiation may be beneficial in these patients in order to identify and manage underlying nasal pathology potentially affecting CPAP usage. Humidification, nasal douching, nasal steroids, and nasal surgery are considered the main pillars of the management of nasal symptoms. A systematic review and meta-analysis by Camacho et al. [31] showed that nasal surgery results in a significant overall reduction in CPAP pressures and increase in CPAP adherence with the most effective surgery type being the combination of septoplasty with turbinoplasty. In addition to nasal examination, a full upper airway assessment is necessary in order to exclude other pathologies potentially contributing to upper airway obstruction, breathing difficulties, high CPAP pressure requirements and low-CPAP usage such as tonsillar hypertrophy, tongue base prominence, and/or epiglottic collapse [32].

This study has certain limitations. First, nasal examination was not performed; however, the assessment of nasal symptoms was carried out via validated questionnaires. The SFAR questionnaire was used for the diagnosis of allergic rhinitis without performing radioallergosorbent (RAST) or skin-prick tests. The SFAR score has 74% sensitivity and 83% specificity but is a quick and relatively reliable tool [7]. In order to better evaluate the effect of CPAP on the nasal cavity, certain exclusion criteria were used and for that reason, study participants may differ from the general population. Namely, we excluded patients with known nasal pathology or current medication affecting nasal mucosa and, therefore, the prevalence of nasal symptoms and AR in the general OSA population may be higher than in our study. In contrast with most of the similar studies, we used ODI instead of the apnoea-hypopnoea index (AHI) to make the diagnosis of OSA and classify its severity. The symptoms of OSA are closely related to oxygen desaturations and although AHI is widely used, ODI is as valuable as AHI in diagnosing and grading OSA [33] with a strong correlation of 0.97 with AHI [34]. The follow-up assessment was performed 3 months after initiating CPAP therapy and although this could be considered as a potential weakness, the vast majority of patients demonstrate intolerance to CPAP within the first month of use with early CPAP adherence being reported as the greatest predictor for long-term CPAP adherence [35]. We therefore presume that our results are also related to long-term effects. Likewise, nasal side effects, if any, are expected to be present after three months of CPAP use.

5. Conclusions

Despite the high prevalence of nasal symptoms in OSA patients, our study did not demonstrate an association between most nasal symptoms and CPAP usage, apart from rhinorrhoea. Runny nose seems to be an independent predictive factor for poorer CPAP adherence. Moreover, rhinorrhoea worsens after a three-month trial of CPAP, especially in high users. However, given the high frequency of nasal symptoms, it may still be worth sleep practitioners enquiring about their presence and considering treatment. The pathophysiological mechanism generating increased rhinorrhoea in CPAP therapy is not fully understood and further studies to evaluate the effect of humidification and oronasal CPAP masks on the nasal mucosa will be required.

Author Contributions: Conceptualisation, K.C.; methodology, K.C., J.R.S. and A.H.N.; formal analysis, K.C.; investigation, K.C., K.L. and A.M.; resources, K.C., K.L. and A.H.N.; data curation, K.C.; writing—original draft preparation, K.C.; writing—review and editing, K.L., A.M., J.R.S. and A.H.N.; visualisation, K.C.; supervision, K.C., J.R.S. and A.H.N.; project administration, K.C., K.L. and A.H.N. All authors have read and agreed to the published version of the manuscript.

Funding: This research received no external funding.

Institutional Review Board Statement: This is an observational service evaluation study which was reviewed and deemed exempt by the Oxford Review Board (Caldicott Guardian/Data protection office). The study was conducted according to the guidelines of the Declaration of Helsinki.

Informed Consent Statement: Informed consent was obtained from all subjects involved in the study.

Data Availability Statement: The data presented in this study are available on request from the corresponding author. The data are not publicly available due to privacy-related restrictions.

Conflicts of Interest: The authors declare no conflict of interest.

References

1. Senaratna, C.V.; Perret, J.L.; Lodge, C.J.; Lowe, A.J.; Campbell, B.E.; Matheson, M.C.; Hamilton, G.S.; Dharmage, S.C. Prevalence of obstructive sleep apnea in the general population: A systematic review. *Sleep Med. Rev.* **2017**, *34*, 70–81. [CrossRef] [PubMed]
2. Marshall, N.S.; Wong, K.K.; Liu, P.Y.; Cullen, S.R.; Knuiman, M.W.; Grunstein, R.R. Sleep apnea as an independent risk factor for all cause mortality: The Busselton Health Study. *Sleep* **2008**, *31*, 1079–1085. [PubMed]
3. Kohler, M.; Smith, D.; Tippett, V.; Stradling, J.R. Predictors of long-term compliance with continuous positive airway pressure. *Thorax* **2010**, *65*, 829–832. [CrossRef] [PubMed]
4. Pepin, J.L.; Leger, P.; Veale, D.; Langevin, B.; Robert, D.; Levy, P. Side effects of nasal continuous positive airway pressure in sleep apnea syndrome. Study of 193 patients in two French sleep centers. *Chest* **1995**, *107*, 375–381. [CrossRef]
5. Brimioulle, M.; Chaidas, K. Nasal function and CPAP use in patients with obstructive sleep apnoea: A systematic review. *Sleep Breath* **2021**. [CrossRef]
6. Johns, M.W. A new method for measuring daytime sleepiness: The Epworth sleepiness scale. *Sleep* **1991**, *14*, 540–545. [CrossRef]
7. Annesi-Maesano, I.; Didier, A.; Klossek, M.; Chanal, I.; Moreau, D.; Bousquet, J. The score for allergic rhinitis (SFAR): A simple and valid assessment method in population studies. *Allergy* **2002**, *57*, 107–114. [CrossRef]
8. Juniper, E.F.; Thompson, A.K.; Ferrie, P.J.; Roberts, J.N. Development and validation of the mini Rhinoconjunctivitis Quality of Life Questionnaire. *Clin. Exp. Allergy* **2000**, *30*, 132–140. [CrossRef]
9. Weaver, T.E.; Sawyer, A.M. Adherence to continuous positive airway pressure treatment for obstructive sleep apnoea: Implications for future interventions. *Indian J. Med. Res.* **2010**, *131*, 245–258.
10. Young, T.; Finn, L.; Kim, H. Nasal obstruction as a risk factor for sleep-disordered breathing. The University of Wisconsin Sleep and Respiratory Research Group. *J. Allergy Clin. Immunol.* **1997**, *99*, 757–762. [CrossRef]
11. Mims, J.W. Epidemiology of allergic rhinitis. *Int. Forum Allergy Rhinol.* **2014**, *4* (Suppl. S2), 18–20. [CrossRef]
12. Shadan, F.F.; Jalowayski, A.A.; Fahrenholz, J.; Kline, L.E.; Dawson, A. Nasal cytology: A marker of clinically silent inflammation in patients with obstructive sleep apnea and a predictor of noncompliance with nasal CPAP therapy. *J. Clin. Sleep Med.* **2005**, *1*, 266–270. [CrossRef]
13. Hoffstein, V.; Viner, S.; Mateika, S.; Conway, J. Treatment of obstructive sleep apnea with nasal continuous positive airway pressure. Patient compliance, perception of benefits, and side effects. *Am. Rev. Respir. Dis.* **1992**, *145*, 841–845. [CrossRef]
14. Balsalobre, L.; Pezato, R.; Gasparini, H.; Haddad, F.; Gregorio, L.C.; Fujita, R.R. Acute impact of continuous positive airway pressure on nasal patency. *Int. Forum Allergy Rhinol.* **2017**, *7*, 712–717. [CrossRef]
15. Yang, Q.; Li, H.; Wu, W.; Huang, X.; Tu, B.; Ma, Y. Effect of continuous positive airway pressure on allergic rhinitis in patients with obstructive sleep apnea-hypopnea syndrome. *Ther. Clin. Risk Manag.* **2018**, *14*, 1507–1513. [CrossRef]
16. Willing, S.; San Pedro, M.; Driver, H.S.; Munt, P.; Fitzpatrick, M.F. The acute impact of continuous positive airway pressure on nasal resistance: A randomized controlled comparison. *J. Appl. Physiol.* **2007**, *102*, 1214–1219. [CrossRef]
17. Cisternas, A.; Aguilar, F.; Montserrat, J.M.; Avila, M.; Torres, M.; Iranzo, A.; Berenguer, J.; Vilaseca, I. Effects of CPAP in patients with obstructive apnoea: Is the presence of allergic rhinitis relevant? *Sleep Breath* **2017**, *21*, 893–900. [CrossRef]
18. Pitts, K.D.; Arteaga, A.A.; Hardy, E.T.; Stevens, B.P.; Spankovich, C.S.; Lewis, A.F. The effect of continuous positive airway pressure therapy on nasal patency. *Int. Forum Allergy Rhinol.* **2018**, *8*, 1136–1144. [CrossRef]
19. Skoczynski, S.; Ograbek-Krol, M.; Tazbirek, M.; Semik-Orzech, A.; Pierzchala, W. Short-term CPAP treatment induces a mild increase in inflammatory cells in patients with sleep apnoea syndrome. *Rhinology* **2008**, *46*, 144–150.
20. Gelardi, M.; Carbonara, G.; Maffezzoni, E.; Marvisi, M.; Quaranta, N.; Ferri, R. Regular CPAP utilization reduces nasal inflammation assessed by nasal cytology in obstructive sleep apnea syndrome. *Sleep Med.* **2012**, *13*, 859–863. [CrossRef]
21. Ryan, S.; Doherty, L.S.; Nolan, G.M.; McNicholas, W.T. Effects of heated humidification and topical steroids on compliance, nasal symptoms, and quality of life in patients with obstructive sleep apnea syndrome using nasal continuous positive airway pressure. *J. Clin. Sleep Med.* **2009**, *5*, 422–427. [CrossRef]
22. Bachour, A.; Maasilta, P. Mouth breathing compromises adherence to nasal continuous positive airway pressure therapy. *Chest* **2004**, *126*, 1248–1254. [CrossRef] [PubMed]
23. Borel, J.C.; Tamisier, R.; Dias-Domingos, S.; Sapene, M.; Martin, F.; Stach, B.; Grillet, Y.; Muir, J.F.; Levy, P.; Series, F.; et al. Type of mask may impact on continuous positive airway pressure adherence in apneic patients. *PLoS ONE* **2013**, *8*, e64382. [CrossRef] [PubMed]
24. Sugiura, T.; Noda, A.; Nakata, S.; Yasuda, Y.; Soga, T.; Miyata, S.; Nakai, S.; Koike, Y. Influence of nasal resistance on initial acceptance of continuous positive airway pressure in treatment for obstructive sleep apnea syndrome. *Respiration* **2007**, *74*, 56–60. [CrossRef]
25. Morris, L.G.; Setlur, J.; Burschtin, O.E.; Steward, D.L.; Jacobs, J.B.; Lee, K.C. Acoustic rhinometry predicts tolerance of nasal continuous positive airway pressure: A pilot study. *Am. J. Rhinol.* **2006**, *20*, 133–137. [CrossRef] [PubMed]
26. Li, H.Y.; Engleman, H.; Hsu, C.Y.; Izci, B.; Vennelle, M.; Cross, M.; Douglas, N.J. Acoustic reflection for nasal airway measurement in patients with obstructive sleep apnea-hypopnea syndrome. *Sleep* **2005**, *28*, 1554–1559. [CrossRef] [PubMed]
27. Engleman, H.M.; Asgari-Jirhandeh, N.; McLeod, A.L.; Ramsay, C.F.; Deary, I.J.; Douglas, N.J. Self-reported use of CPAP and benefits of CPAP therapy: A patient survey. *Chest* **1996**, *109*, 1470–1476. [CrossRef] [PubMed]

28. Genta, P.R.; Kaminska, M.; Edwards, B.A.; Ebben, M.R.; Krieger, A.C.; Tamisier, R.; Ye, L.; Weaver, T.E.; Vanderveken, O.M.; Lorenzi-Filho, G.; et al. The Importance of Mask Selection on Continuous Positive Airway Pressure Outcomes for Obstructive Sleep Apnea. An Official American Thoracic Society Workshop Report. *Ann. Am. Thorac. Soc.* **2020**, *17*, 1177–1185. [CrossRef]
29. Lebret, M.; Arnol, N.; Contal, O.; Martinot, J.B.; Tamisier, R.; Pepin, J.-L.; Borel, J.-C. Nasal obstruction and male gender contribute to the persistence of mouth opening during sleep in CPAP-treated obstructive sleep apnoea. *Respirology* **2015**, *20*, 1123–1130. [CrossRef]
30. Inoue, A.; Chiba, S.; Matsuura, K.; Osafune, H.; Capasso, R.; Wada, K. Nasal function and CPAP compliance. *Auris Nasus Larynx* **2019**, *46*, 548–558. [CrossRef]
31. Camacho, M.; Riaz, M.; Capasso, R.; Ruoff, C.M.; Guilleminault, C.; Kushida, C.A.; Certal, V. The effect of nasal surgery on continuous positive airway pressure device use and therapeutic treatment pressures: A systematic review and meta-analysis. *Sleep* **2015**, *38*, 279–286. [CrossRef]
32. Virk, J.S.; Kotecha, B. When continuous positive airway pressure (CPAP) fails. *J. Thorac. Dis.* **2016**, *8*, E1112–E1121. [CrossRef]
33. Temirbekov, D.; Gunes, S.; Yazici, Z.M.; Sayin, I. The Ignored Parameter in the Diagnosis of Obstructive Sleep Apnea Syndrome: The Oxygen Desaturation Index. *Turk. Arch. Otorhinolaryngol.* **2018**, *56*, 1–6. [CrossRef]
34. Levendowski, D.J.; Hamilton, G.S.; St Louis, E.K.; Penzel, T.; Dawson, D.; Westbrook, P.R. A comparison between auto-scored apnea-hypopnea index and oxygen desaturation index in the characterization of positional obstructive sleep apnea. *Nat. Sci. Sleep* **2019**, *11*, 69–78. [CrossRef]
35. Van Ryswyk, E.; Anderson, C.S.; Antic, N.A.; Barbe, F.; Bittencourt, L.; Freed, R.; Heeley, E.; Liu, Z.; Loffler, K.A.; Lorenzi-Filho, G.; et al. Predictors of long-term adherence to contimuous positive airway pressure in patients with obstructive sleep apnea and cardiovascular disease. *Sleep* **2019**, *42*, 1929–1937. [CrossRef]

MDPI AG
Grosspeteranlage 5
4052 Basel
Switzerland
Tel.: +41 61 683 77 34

Life Editorial Office
E-mail: life@mdpi.com
www.mdpi.com/journal/life

www.ingramcontent.com/pod-product-compliance
Lightning Source LLC
Chambersburg PA
CBHW040823250926
51850CB00007BA/1244
9783725820641